The Constellations

The Constellations

An Enthusiast's Guide to the Night Sky

LLOYD MOTZ and
CAROL NATHANSON

AURUM PRESS

to
Minne, Robin, and Julie
—L.M.

in Memory of
Prof. Jonas B. Nathanson
—C.N.

Acknowledgments

The Constellations, the product of a lifetime in the field of astronomy, has been under contract with Doubleday since 1980. There have been many in-house shepherds, each one competent, dedicated, and appreciated. They are: Peter Kruzan, Patrick LoBrutto, James Menick, Jennifer O'Grady, Terrence Rafferty, Al Sarrantonio, and Clifford Thompson.

—L.M.
C.N.

First published in Great Britain 1991 by Aurum Press Ltd,
10 Museum Street, London WC1A 1JS
Copyright © 1988 by Lloyd Motz and Carol Nathanson

First published in the USA by Doubleday, New York

Drawings, diagrams and starcharts designed by Carol Nathanson, co-author

British Library Cataloguing in Publication Data

Motz, Lloyd *1910–*
    The constellations.
    1. Constellations
    I. Title   II. Nathanson
    523.8

ISBN 1 85410 088 2

Printed in Great Britain by
Butler & Tanner Ltd, Frome and London

Contents

III *Spring*

IV *Summer*

VI *The Southern Skies*

LOOKING AT THE STARS

In ancient times, the Earth was considered to be all the Universe to speak of. There was the sky, of course, but that was just a lid that fit over the Earth. There were the Sun and the Moon, but those were just lamps that gave us light by day and night. Finally, there were the stars, but they were just decoration, little dots of light fixed to the firmament.

You couldn't look at the Sun for long, but the night sky was beautiful, and in the absence of television, movies, radio, or even books, there was nothing much to do at night (if you were alone and had a spare hour) except look at the sky.

And if you did, you couldn't help notice that the Moon changed its shape from night to night and did so in a very regular way. And if you watched it further you noticed that it changed its position against the stars, making a complete circle of the sky in 29½ days—which meant it wasn't fixed to the firmament but moved under it.

If that roused your curiosity, you would watch still more carefully and notice that the stars just after sunset changed from night to night as though the entire firmament were slowly shifting, making a complete circle in a little over 365 days.

And then, just to put the cherry on top, you would notice that five of the brighter stars (the ones we call Mercury, Venus, Mars, Jupiter, and Saturn) also shift position among those stars that *are* fixed to the firmament (we still call the thousands of unmoving little dots of light the "fixed stars" sometimes) and what's more do so in a more complicated fashion than the Sun and Moon do. The starlike objects backtrack every once in a while. Saturn, which circles the sky very slowly, backtracks twenty-nine times while doing so.

The ancient Sumerians (who were very clever—they invented writing, for instance) were probably the first to notice all this and to wonder what it could possibly mean. We don't know exactly how they thought it out, but my own feeling is that they felt that the gods wouldn't set up the sky in that fashion without a reason. The most logical reason would be that it was a cryptogram. The changing position of the seven objects that wandered among the stars (they were called planets by later peoples, from the Greek word for "wanderers") must represent a complicated message that told people what the future would hold, what they ought to do, and so on. It was just necessary to figure it out.

First, you would have to notice exactly the details of the changes—where the various planets are relative to the stars. To do that, you would have to mark out the stars, in order to recognize them when you see them and to be able to talk about them to each other. It would be simple to divide them up as squares, triangles, crosses, and other shapes, but that's not distinctive enough, and besides it doesn't satisfy human beings, who are, to a surprising degree, poets and artists at heart.

Therefore, they chose combinations of stars that looked (if you have a terrific imagination) like bears, or hunters, or eagles, or scorpions. These are now called "constellations" from Latin words meaning "with stars"; that is, they are figures marked out with stars.

At first, the sky-watchers just did this for the portion of the sky (a thick band around the middle) along which the seven planets moved. This band was divided into twelve constellations, so that every time the Moon circled the entire sky, the Sun moved one constellation. (This was considered very neat.) Since the constellations in this band were mostly taken to represent animals, it is called the "zodiac" from a Greek phrase meaning "circle of animals."

Eventually, of course, as the sky-watchers continued to observe the sky, it seemed to make sense to fill the rest of the sky with constellations, too. At first there were some stars that fell in between the constellations and weren't included in any, but modern astronomers would not have anything that slipshod. They have drawn boundaries that fit perfectly so that every object in the sky is in one constellation or another. They have even extended the system to that part of the sky that can only be seen in the southern hemisphere and that the ancients never saw.

Naturally, once you said (for instance) that Jupiter is in the sign of the Scorpion, this would fascinate the average person who would consequently assume there was a real scorpion in the sky and wonder how it got there—to say nothing of winged horses and lyres and swans and crabs and lions and rivers and human beings of one kind or another.

It seemed only natural to make up stories, or to adapt stories that already existed, to account for it. We are most acquainted with the Greek myths in this connection.

Now when I was young, the astronomy books meant for children talked mostly of the constellations and the old myths surrounding them with some mention of the brightest one or two stars in each. It was fascinating, but it didn't teach me much astronomy.

However, Lloyd Motz and Carol Nathanson know exactly what to do. This book is charmingly told and presupposes very little knowledge to begin with, but it is *not* for children, but for curious adults.

Motz and Nathanson go carefully through each constellation, not skipping one. They tell you the myths for those that have myths attached, and they tell you about every interesting object in the constellation that you can see with a home telescope.

What's more they tell you the actual astronomy of it. If they point out a red giant or a white dwarf, they tell you why the former is red and bright and the latter white and dim, how the truth about them came to be known, and what is likely to happen to them in the future. They talk about variable stars, about galaxies, about nebulas, about quasars, about everything astronomers talk about.

They bring in tales from history and tales of modern science, and when you are through you will find that while the sky is *still* a charming spectacle, it is also an encyclopedia of the universe. And we can be thankful that the sky exists with just enough in the way of peculiarity to have caught the eyes of the ancients and so begin the process that has led us to so much.

—Isaac Asimov

Introduction

HOW SKY PICTURES WERE BORN

We do not know when the constellations began to assume special significance for our ancestors, because the delineation of pictures formed by groupings of stars is an archaic concept, its origins lost in the mists of prehistory. We suspect, however, that stellar imaging, perhaps nearly as old as humanity itself, was well known to primitive peoples. In pursuit of this idea, we invoke the aid of a logical fantasy: instantly, transformed into investigative journalists, we take a long, magical step into the past, where we witness a primeval jungle scene. Our up-to-the-minute news report follows:

"It is a torrid summer evening, well before the time of recorded history. A young tribesman rests from the exertion of a recent hunting foray. Gazing languidly at darkening skies, he observes myriad points of light that gradually appear in the heavenly canopy. To his sharp and inquisitive eye, it seems as if they are grouped together into familiar shapes and patterns. Allowing his vivid and creative imagination free rein, the reclining hunter amuses himself by projecting onto the heavens the events of a day in prehistoric wilderness, his self-image, and his many occupations. Among the bright stars, he easily recognizes the visages of his tribal family, his hunting companions, his prey, and even the crudely wrought weapons and utensils he uses.

"After his group returns from its successful hunt and the tribal members all sit feasting around bubbling pots of meat, he points out to his family two conspicuous and well-shaped sky pictures: one seems to hang low in the northern heavens, reminding them of the simple, long-handled dippers with which they scoop up liquids; the second picture dominates the southern horizon and resembles the hunter himself, adorned with a glistening belt and sword. Other tribal families notice the same star formations; thus a tradition is born."

Although we dipped our journalistic pen into the realm of fantasy to get this scoop on the constellations' possible origin, the prehistoric scene we conjured up has a factual basis where the tribesman's activities are concerned. Hunting tools and human bones and teeth found near fossilized remains of animals all dating back thousands of years ago as well as prehistoric animals skillfully painted on the walls of ancient caves confirm the theory that early man's principal means of obtaining vital nourishment was by hunting the familiar beasts of his tropical world.

Primitive peoples, therefore, were thoroughly familiar with the life-sustaining wild creatures that shared their territory. Tribal societies gradually began to identify themselves with various natural objects: plants, minerals, and most often, animals that they believed had favorably affected their destinies. They regarded these creatures as incarnations of ancestral beings, helpful links to the unseen world of spirits. A system of belief developed in which all individuals, regardless of tribe, who had as their ancestral sign, or totem, the same object or animal, considered themselves related by blood.

The totem was sacred; this belief was part of a broader religious system called animism, the peopling of natural objects with spirits. When these early societies learned to raise and care for certain suitable animals, the domesticated beasts assumed an even greater importance in human consciousness than former wild prey. Representations of the bull, ram, horse, dog, and even the goose took on sacred significance.

In the slow process of our ancestors' early religious development, the heavens became the natural home of the gods, therefore a realm of immortality. Many of these pagan deities still bore the appearances of earlier sacred animals, or were half human and half animal in form. In the course of centuries, all these symbols, man and beast alike, came to be associated with the striking star patterns that rose and set with harmonious regularity in a clear and unpolluted "heavenly dome."

Some sociologists believe that the tribal totem animals found their way into the skies via the *zodiac* (a band of twelve constellations lying along the Sun's apparent seasonal path through the heavens), which at first, in the archaic Euphratean astronomy, may have consisted of only six symbols: bull, crab, maiden, scorpion, sea goat, and fishes. Other peoples of antiquity, in particular the early Egyptians, the Israelites of biblical times, and the Chinese, developed their own zodiacal figures. Native Egyptian stellar groups included "Hippopotamus" (Draco), "Thigh" (Ursa Major), and *Sahu,* which was identified with Orion as a holy constellation occupied by the soul of the god Osiris. The ten "lost tribes" of Israel are said to have been represented by various symbols, among these the familiar Judaic lion, the scorpion of Dan, and the bow of Ephraim and Manassah (symbolizing Sagittarius). The Chinese, in their zodiacal signs, or *kung,* which emerged between 2700 B.C. and 700 B.C., displayed a broad assortment of beasts: tiger, horse, dragon, serpent, hare, ram, ape, cock, dog, boar, rat, and ox. The Babylonian *Epic of Creation,* found on tablets dating from the reign of Assurbanipal (7th cent. B.C.) and perhaps originating in 2350 B.C., consists of twelve books corresponding to the twelve zodiacal signs, or *Mizrātā,* which probably evolved into the biblical *Mazzārōth,* and later, the Hebrew peoples' *Galgal ha-Mazzālōth,* or "Circle of Signs." In Arabia, the zodiac was *Al Minṭakah al Burūj,* "The Girdle of Signs"; Aristotle, in the fourth century B.C., referred to it as the "Circle of Little Animals," and centuries later this idea was reflected in the German *Tierkreis* and the English "Bestiary."

Thus an astrological tradition, that had its shadowy beginnings in primitive hunting society, manifested itself aeons later among the near-prehistoric Euphrateans, the early Chaldeans, Babylonians, and Egyptians, and ultimately in all national cultures. From about the fourth century B.C., the Greeks—and later on the Romans—began to associate the starry sky pictures with their own fabled heroes, heroines, and demigods. It is interesting that the brighter appearance of the nearby planets of our Solar System as compared to the distant stars had a curious and amusing result: a wholly unintentional subordination of the great divinities of antiquity to their human subjects, at least where the gods' living quarters were concerned! Owing to the incorrect assumptions of early astronomy, the ancients began to associate their most important gods and goddesses with the bright planets, or wanderers, as they were called (because they seemed to hold no fixed position in the skies). We now know, however, that the classical deities' planetary "abodes," insignificant and tiny compared to the stars, are in fact mere satellites of cold material, illuminated solely by our dwarf Sun and circling it in a remote corner of the Galaxy; and that brilliant Venus and mighty Jupiter are but microscopic specks compared to the flaming girth of Orion's faintest star!

Because peoples of antiquity accepted natural phenomena exactly as they appeared to the unaided eye, superhuman heroes, ordinary frail mortals, and all the celebrated animals of ancient tradition were assigned to an unimaginably vaster realm than were their divine rulers, the gods. As legendary mortals who were transported to the domain of naked-eye stars, these characters of folk tradition and poetic fantasy had conquered the limitless reaches of interstellar space! We still retain them as identifying figures in our star charts, and we enjoy reading the colorful stories of their life histories and improbable exploits.

Thus ancient mythology, which took its place long ago in the heavens, remains happily enthroned there in modern times.

1. *Indian zodiac* (reconstruction). In the ancient *Rig Veda,* the wheel of the zodiac of India shows the twelve traditional figures in ovals within the outer rim and nine additional figures in circles surrounding the hut. Pictured in the center is the ancient Hindu concept of the Earth as a tall mountain surrounded and supported by eternal seas that extend into the heavens, where the gods abide. (*The Bettmann Archive*)

CHARTING THE HEAVENS

In 366 B.C., when the Greek astronomer Eudoxos published his work *Phainomena*, describing the very ancient Egyptian constellations, such star patterns were confined to the naturalistic outlines of their identifying forms: stars lying outside these figures were called "unformed" or "scattered"; or as the Arabians said, *al h'arij min al surah*, "outside the image." And so they remained for many centuries. In 1930, however, the International Astronomical Union adopted standardized boundaries for all the recognized constellations (prepared by E. Delporte and following leading atlases); the formerly adjacent stars are included in these areas.

The Greek poet Aratos, in his astronomical poem *Phainomena* (270 B.C.), based upon Eudoxos' prose work, included forty-five constellations of archaic origin, which represented the heavens of 2200 to 2000 B.C. Thirty years later, Eratosthenes recorded forty-two. The beginning of the Christian Era was marked with the publication of the *Poeticon Astronomicon*, reputedly by Gaius Julius Hyginus (Historia). Hipparchus of Rhodes, inspired by the appearance of the nova of 134 B.C., assembled his famous catalogue, based on observations made between A.D. 127 and 151 and containing forty-nine constellations, with 1,080 stars. In his *Syntaxis* (the *Almagest*), Ptolemy of Alexandria revised Hipparchus' work, listing forty-eight constellations (known as "the ancient 48") including 1,028 stars; this was quite fortunate, because Hipparchus' catalogue was eventually lost, and no known copy exists.

The Alfonsine Tables, of Spain (A.D. 1252), adapted from Ptolemy by Arabian or Moorish astronomers, was the only notable compilation in a barren period lasting thirteen hundred years from his time. Ulug-Beg's tables, also based upon Ptolemy's *Syntaxis* and published at Samarkand in 1437, was the next great listing and is still referred to. In 1548–51 the celestial globes of Mercator made their appearance. Tycho Brahe's catalogue followed in 1602; and then the famed *Uranometria*, of Johann Bayer, enhanced with copies of constellation figures designed by the great German artist Albrecht Dürer for his own star charts. These drawings were subsequently used by many cataloguers. Until the time of Bayer, Arabic names were used to designate the stars; Bayer greatly simplified this system with the introduction of Greek and Roman letters. From the seventeenth century onward, Bayer and other astronomers also began to add new constellations, especially in the southern skies, and to revise some of the more unwieldy traditional figures. Other notable catalogues of that time were the *Planisphaerium Stellatum*, of Jakob Bartsch, in 1624; the *Catalogue of Southern Stars*, of Edmund Halley; and the *Prodromus Astronomiae*, of Hevelius (1690).

After the invention of the telescope and its resultant disclosures of an ever increasing number of stars, Bayer's Greek letters were soon exhausted for those stars in any given constellation; therefore in the Rev. Dr. John Flamsteed's *Historia Coelestis Britannica*, completed posthumously in 1725 (or 1729), a system of numbers running from west to east in each constellation was introduced. Flamsteed's numbers are still used. Lacaille, in 1752 and 1763, added modern groups in the south with names related to science and art; and Lalande's *Astronomie* (1776–92) contained the currently accepted total of eighty-eight constellations. In 1800, Johann Elert Bode's *Uranographia* appeared, and a year later he drew the first modern boundaries for the constellations.*

From Flamsteed's time until nearly the end of the nineteenth century, much work was done in the measurement of star positions, and the principal observatories published many catalogues. The most important of these was Argelander's Bonn Catalogue, which appeared somewhat after the mid-nineteenth century, giving positions of 324,289 stars down to the tenth magnitude. Argelander also defined useful boundaries for the northern constellations and published star charts that remain of very great value. The British Association Catalogue of 1845 is also important, and references are still made to its listings.

In 1841, the Royal Astronomical Society asked Sir John Herschel and Francis Baily to attempt a reform of the constellations. These had been subdivided into numerous new asterisms and many minor figures added by modern astronomers until, as a result, the existing sky maps were in a state of great confusion. Herschel's reforms, however, never found popular acceptance; the traditional figures refused to submit to drastic change and have retained their age-old identities, even within the officially recognized boundaries.

In the current century, numerous catalogues and charts have appeared, and there have been many listings of telescopic and spectroscopic objects for the professional astronomer's use. Recommended to the more advanced observing amateur is Arthur P. Norton's *Star Atlas,* of Sky Publishing Corporation, Cambridge, Massachusetts; useful monthly charts also appear in Sky Publishing's magazine, *Sky & Telescope.* Helpful guides for the beginner are the *Edmund Magnitude 5 Star Atlas* (which may also be ordered from Sky Publishing Corp.); *Putnam's Field Book of the Skies* (W. T. Olcott and S. W. Putnam), a popular older book that may be obtained in the public library if out of print; and the compact and handy little *Spotter's Guide to the Night Sky* (N. Henbest and L. Motz, Mayflower Books, 1979), currently available.

Happy watching!

---

* *Bode is known for a formula, called Bode's law, that demonstrates a predictable regularity in the mean distances of the planets from the Sun.*

Where has the Pleiad gone?
Where have all the missing stars found light and home?
Who bids the Stella Mira go and come?
Why sits the Pole-star lone?
And why, like banded sisters, through the air
Go in bright troops the constellations fair?
—Nathaniel Parker Willis (1806–67)

Stars for All Seasons

2. *The Dioscuri, or Castor and Pollux, the Guardians of Mariners*. Castor, helmeted, steadies the storm-tossed ship, while Pollux raises his arms in supplication; a rolling wave reveals the Twins' images carved on the prow. Engraving by B. Picart. (*The Bettman Archive*)

ONE

Two Bears and a Dragon

On thy unaltering blaze
The half wrecked mariner, his compass lost,
    Fixes his steady gaze,
And steers, undoubting, to the friendly coast . . .
            W. C. Bryant, *Hymn to the North Star*

LOST AT SEA

With terrifying screams, the wind lashes angry water into great explosions of froth. Giant waves threaten to crush the fragile wooden hull of the sailship as it pitches to and fro. Huddled below, in the forecastle, the frightened sailors have already hauled in the sails, but this storm's appearance was so sudden that it has washed away most of their provisions and personal possessions. Among these lost articles is an object of life-and-death value to the wretched men and their captain: his compass!

In this year, A.D. 1443, there is no other navigational instrument to guide them. Gripping the helm, the captain no longer knows whether his ship is headed in the direction of land or out toward open sea. He tries to determine his position by scanning the blackened skies for a familiar star, but his anguished gaze is met only by driving rain and salt spray. There is nothing else he can do at this crucial moment but bend his efforts toward riding out the storm and assuring immediate survival for himself and his crew.

Suddenly the wind's howling sinks to a moan, and the rainfall lightens noticeably; the raging squall is gone as quickly as it came. Hardly daring to believe that his damaged ship is still seaworthy, the captain glances upward toward a rift in the clouds and sees a flickering silver point of light which soon vanishes in the murky skies.

" 'Star of the Sea,' save me!" pleads the desperate captain as he focuses upon the spot where the point of light had danced. A tense moment passes, then the clouds part again and his heart leaps with hope as a gentle beacon of salvation is revealed: the faithful glow of his *Stella Maris*—the polestar, along with the identifying pattern of the Little Dipper, known to him as the Lesser Bear.

Now the captain knows without doubt the northerly direction and, from it, his proper course through the vast world of water. As the rainclouds disperse, the reassuring form of the Lesser Bear is joined by its companion the Greater Bear. Clearing skies also reveal the fainter stars of Draco, a fearsome dragon circling the Lesser Bear; but Draco's writhing coils are a comforting sight to the hard-pressed sailors and their captain. They know that these familiar figures in the northernmost part of the heavens will remain visible throughout the long night and guide their ship safely into port.

Another century will pass, however, before seamen learn that such steadfastness is not a quality inherent to their Stella Maris and the other circumpolar stars but, rather, a direct result of Earth's motion in the Solar System. As the science of astronomy progresses, in fact, one lesson will be repeated many times: the intrinsic must be distinguished from the merely apparent in all phenomena.

EARTH, A WORLD IN MOTION

Early Concepts and How Astronomy Evolved

The rotation of the Earth from west to east, which causes the stars, the Sun, and all other heavenly bodies to pass across our field of vision from east to west, led nonscientific people in ages past to the erroneous belief that the heavens are actually rotating about the Earth.

Before the time of the Polish astronomer Nicolaus Copernicus (1473–1543), astronomers also regarded the east-to-west movement of the night skies exactly as it appeared to them, and many thought that a very large material globe was rotating about the Earth, carrying with it all the stars: tiny points of light embedded in its crystalline substance. Aristotle (384–322 B.C.), one of the foremost Greek philosophers, believed that the entire universe, in the form of this encompassing divine sphere and fifty-five inner spheres driven by it, rotated about the Earth. Owing to the strength of his great influence upon his contemporaries as well as on thinkers of subsequent ages, this *geocentric*, or Earth-centered, view of the universe was widely held until relatively recent times, despite the conclusions of Aristarchus of Samos (291 B.C.), who correctly reasoned that the Sun is the center of the Solar System and that the Earth and other planets revolve around it.

The Egyptian astronomer Claudius Ptolemaeus, popularly known as Ptolemy of Alexandria (A.D. 121–61), postulated Aristotle's geocentric universe; Ptolemaic theory dominated astronomy for fourteen centuries.

In the sixteenth century, the first major challenge to the Aristotelian universe after Aristarchus was offered by Copernicus in his book *De Revolutionibus,* in which he presented his daring ideas of a heliocentric, or Sun-centered, system, wherein the Earth not only revolves around a stationary, central Sun but also rotates on its own axis like a spinning top. With his discovery of the laws of planetary motion, the German astronomer and mathematician Johannes Kepler (1571–1630) strengthened the Copernican theory; and the great Italian physicist and pioneer telescope builder Galileo Galilei (1564–1642) brought further credibility to it with his discovery that the planet Jupiter is the center of at least four satellites: a model, therefore, for the sun-centered Solar System. Galileo's observation of the crescent phases of Venus (duplicating those of the Moon) offered even more convincing evidence, because such phases can occur only if the planet revolves around the Sun, being illuminated by it from different angles.

Alarmed authorities suppressed Galileo's great work *Dialogue Concerning the Two Chief World Systems—Ptolemaic and Copernican* soon after its publication in 1632, and early the next year its daring author was summoned to Rome by the papal Inquisition, where he was forced to disavow his beliefs in a written renunciation. Under permanent house arrest, aging and in poor health, Galileo managed to complete a subsequent work, *Dialogue Concerning Two New Sciences,* which was smuggled abroad for eventual publication in Holland. News of Galileo's recantation had a shattering effect upon those who opposed geocentric belief and caused the great French mathematician René Descartes (1596–1650) to withdraw from publication his own treatise *Le Monde* (The World), in which he cites arguments strongly supportive of Copernican theory.

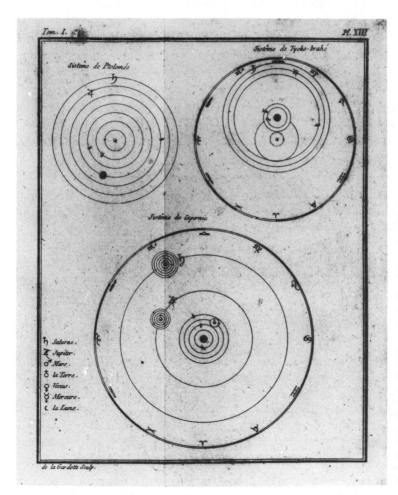

3. *Three systems portraying the universe.* The Ptolemaic, upper left, depicts the Moon, Sun, planets, and fixed stars revolving about a central Earth. Tycho Brahe's concept, upper right, shows the Moon, Sun, and stars orbiting the Earth, with the planets circling the Sun. The Copernican system (bottom) depicts the Sun at the center, orbited by all the planets, including the Earth, and by the stars. The Moon is correctly shown orbiting the Earth. *(Yerkes Observatory, University of Chicago)*

Thus, despite the correct methods of observation and reasoning upon which they are based, Copernicus' concepts of a revolving and rotating Earth were not generally accepted until after Galileo's time. Later in the seventeenth century, the renowned and brilliant English physicist Sir Isaac Newton (1642–1727) illuminated scientific understanding with his discovery of the laws of gravity and motion, set forth in his work *Mathematical Principles of Natural Philosophy,* or, simply, the *Principia.* From these principles, Newton showed how he could mathematically deduce Kepler's laws of planetary motion. With the heliocentric theory now fully enthroned in the scientific world, the lengthy reign of Aristotelian belief had come to its end.

The Great Celestial Coordinates

To understand how the revolution of the Earth, and its rotation on a very nearly fixed axis, respectively, give the Sun its apparent motion among the stars, and the sky its apparent rotation from east to west, we first note that the Earth revolves around the Sun, in a very nearly fixed plane, once a year. Called the *plane of the ecliptic,* it cuts the sky, or the celestial sphere, in a large circle known as the *ecliptic;* this great circle is the path along which the Sun appears to shift its position eastwardly by about one degree per day, an apparent motion corresponding exactly to the path of the Earth around the Sun. The axis about which the Earth rotates determines yet another plane: the *plane of the Earth's equator,* which, when extended, cuts the sky, or celestial sphere, in what astronomers call the *celestial equator.* The plane of the celestial equator and the plane of the ecliptic cut each other at an angle of 23½ degrees. This is commonly referred to as the obliquity of the ecliptic, or the angle of tilt of the Earth's axis with respect to the *pole of the ecliptic* (see fig. 4).

The ecliptic and the celestial equator intersect in two diametrically opposite points on the celestial sphere called the vernal and the autumnal equinoxes. In its apparent motion along the ecliptic, the Sun coincides with the vernal equinox when spring begins on March 21 and with the autumnal equinox when autumn begins on September 23. Thus the Sun is north of the equator from March 21 to September 23 in its apparent eastward journey and south of the equator from September 23 to March 21.

. . . Wilt thou be able to make thy way against the whirling poles
that their swift axis sweep thee not away? . . .

Ovid, *Metamorphoses*

All stars in the northern part of the sky appear to revolve in concentric circles around a point called the *north celestial pole,* which is the heavenly counterpart of the Earth's north pole. For observers in the northern hemisphere whose latitude is 40 degrees or greater, there is a group of constellations within 40 degrees of the north celestial pole (ncp) that never set; they belong to a region of the heavens called the *north circumpolar zone* (see fig. 4). The dancing Bears and their serpentine companion, Draco, are "circumpolar"; as are Cepheus, the king; Cassiopeia, his queen; a giraffe whose name, Camelopardalis, is longer than its neck; and the northernmost part only of Lynx, the

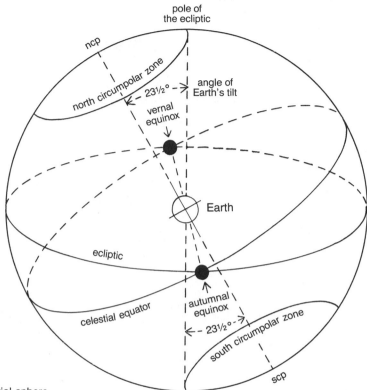

4. Great circles of the celestial sphere.

big cat that always manages to keep its tail above the water as it bathes in the ocean.* These configurations follow a circular path about the north celestial pole, and for most northern observers they never dip below the horizon. Seafarers of earlier ages appreciated this fact and utilized it as a navigational aid, although they did not know that the rotation of the Earth around an imaginary line running from pole to pole, the *polar axis,* is its true cause.

The circumpolar constellations are indeed appropriate as symbols of the sailors' rescue, in which these stellar pictures played so important a role. Many ancient peoples regarded the polar regions as the home of life-giving gods, a blend of Earth and sky (mortality and immortality); our early ancestors believed that the north polar skies rested upon a heavenly mountain called Mons Coelius—the Olympus of northern mythology—and that the two starry bears guarded this "seat of the gods and habitation of life" as it rose out of eternal seas that surrounded and supported the Earth. For many centuries prior to the Christian Era, however, no particular star marked the location of the north celestial pole: only in more recent centuries did this honor fall to Polaris, the North Star, which greeted our storm-tossed crew at the end of their ordeal.

---

* *A lizard (Lacerta) almost gets into the circumpolar picture but doesn't quite succeed, and Perseus the hero just manages to poke the peak of his cap above the 50° parallel of declination.*

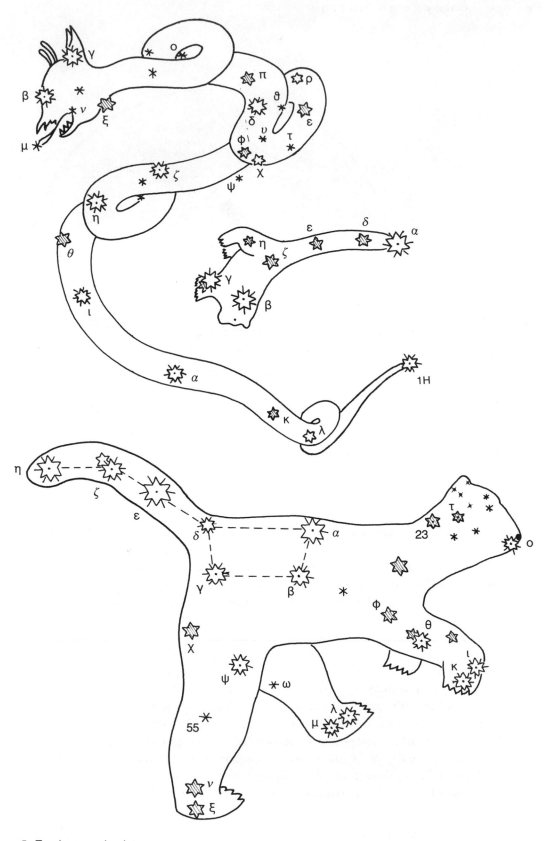

5. Two bears and a dragon.

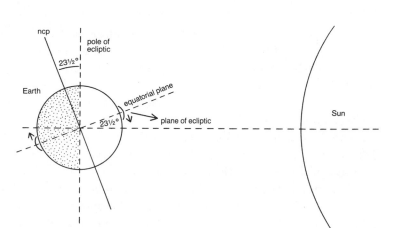

6. Earth, a spinning top.

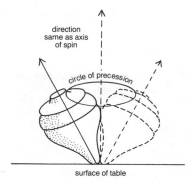

Precession

A phenomenon called precession, an additional motion of the Earth, slowly alters the direction of the polar axis, causing the gradual apparent shift away from the north celestial pole of the various "pole" stars that have marked the north and south celestial poles at various times. Like the Earth's rotation, precession is similar to the additional motion of a spinning top, whose axis slowly describes a circle as the top spins around this axis (see fig. 6). This is caused by the torque† exerted by the Sun on the bulges of the Earth at its equator.

Because of precession, the earth's axis describes a circle whose center is the *pole of the ecliptic;* completing one revolution every 25,725 years, the polar axis points in the approximate direction of the various stars that chance to lie near that circle of revolution. If one of these happens to be visible to the naked eye, it is generally called the "Pole Star." In 2830 B.C., Thuban, in Draco, was closest to the north celestial pole; for the possibly technically advanced peoples living fourteen thousand years from now, brilliant Vega, in Lyra, will be the Pole Star.

Long before the era of the "tall ships," Polaris had already occupied this useful position in the northern skies—an event that saved many lives, not only at sea but in our

† *A combination of equal forces acting in opposite directions but not along the same straight line.*

great deserts as well (for even in the middle latitudes, travelers still looked to the Pole Star as a guide). The northernmost celestial areas, therefore, were not only holy to peoples of earlier days in whose lore sacred mountains reached up to support the pole, but vital to those in subsequent centuries who undertook solitary and dangerous journeys and found their immediate salvation not in the ancient gods, but in a great natural compass of the heavens!

It would be a fitting culmination of this navigational story if we could state that Christopher Columbus had also followed the North Star, and that we therefore owe the discovery of the New World to Polaris' steadfastness. Such a thesis may in fact seem plausible, but it is incorrect, for Columbus made his fateful voyage in a relatively modern age. The compass, which had been discovered six generations before his time, had the undeniable advantage of guiding the sailor on his course in all kinds of weather and in the daytime as well, so that he was no longer dependent upon a clear night sky for his information. In his voyages, therefore, Columbus relied upon daily reckonings

7. *Christopher Columbus.*
Engraving based upon the oil painting by Carl Piloty (1826–86). On the night of October 11, 1492, Columbus consults his charts, as he first glimpses the coast of the New World. The original painting hangs in the Schack Gallery, Munich. (*The Bettmann Archive*)

made with a compass and maps, and a daily calculation of the distance covered.

However, he always checked out the direction of his compass needle against the light of Polaris. After leaving the Canary Islands, therefore, and again, far out in the ocean, he was greatly disturbed to notice that the needle pointed toward the northwest, rather than to the north pole. Columbus had discovered "variance"—a phenomenon occurring in certain regions and caused by a weak secondary magnetic field of the Earth which is superimposed on the stronger dipole field, which may be pictured as caused by a bar magnet lying along the north-south direction in the center of the Earth. The secondary field varies in an irregular fashion from point to point on the Earth.

"Unvarying" Polaris, however, held its true position.

THE NATURE OF STARS

> I paced the terrace, till the Bear had wheel'd
> Thro' a great arc his seven slow suns.
> Tennyson, *The Princess*

The information available to fifteenth-century navigators was relatively limited; but Columbus' voyages and the achievements of other daring pioneers in all fields of endeavor greatly expanded our horizons and generated more correct concepts of the Earth and the heavens. Over the centuries, this body of knowledge grew; advances in astronomy have been noteworthy during the past hundred years, with the pace of discovery increasing significantly in recent decades. This updating in our understanding of the universe has its reflection in a large and vastly more informed public, to whom science is now within easy reach, not only through the educational process and an outflow of many books and specialized publications, but also through the communications media, including newspapers and television. Today's reader is more or less familiar with references to the distant stars as "great Suns," that is, globes of superheated gases radiating energy produced by continuous thermonuclear fusion.

In the past, however, the stars were celebrated as tiny twinkling mysteries in folklore and in the nursery rhymes of childhood, and facts about the universe were not always presented to schoolchildren along with the three R's. One of the authors recalls an especially frustrating experience of her own childhood. An avid astronomy fan at age ten, she became involved in conversation about the Sun with an eight-year-old boy whose world was obviously Aristotelian. When she remarked, *"The Sun is a star,"* he burst into derisive laughter. "She must be an idiot!" thought he. "Doesn't everyone know that a star is a star and the Sun is *the Sun?"* In a state of virtual shock the poor child ran home to his mamma.

A Matter of Infinite Variety

In more recent times, scientists as well as schoolchildren have learned a great deal about the stars: they now know that while the Sun is indeed an average star and the stars are "Suns," they are not all alike, for they differ considerably in size, mass, and chemical structure. The gradual accumulation of facts, paralleling the advance of technology, has revealed that while many stars are similar to our own Sun, vast numbers also differ greatly from this "norm." In our celestial journey, we shall see that stellar objects exhibit great individuality; as with the inhabitants of a metropolis, numerous contrasting "personalities" are represented: some aging, some youthful; some gigantic, some minute; many stable and well-behaved; others eccentric and unpredictable. For this reason the stars have been grouped according to various classifications based on the quantities and the kinds of light they emit, which reveal their age, mass, chemical composition, temperature, and other details. The sifting out and orderly arrangement of this information help build an accurate picture of their evolution and of the dynamic processes within them.

Before we examine some of the ways in which stars are classified, however, it is essential for us to define the basic units of cosmic distance and to discuss one of the methods by which a star's distance is measured.

MEASUREMENT IN SPACE (Fig. 8)

(1) The astronomical unit, or AU: applied to relatively small regions like the Solar System; the "mean distance," or half the sum of the closest distance (perihelion) and the greatest distance (aphelion) of the Earth from the Sun; approximately 93 million miles

(2) The light year, or LY: the distance light travels in one year (at the rate of 186,000 miles per second), about 6 trillion miles

(3) The parsec, or pc: 206,265 astronomical units, or 3.26 light years

(4) The kiloparsec, or kpc: 1,000 parsecs

(5) The megaparsec, or Mpc: 1 million parsec, or 1,000 kiloparsecs

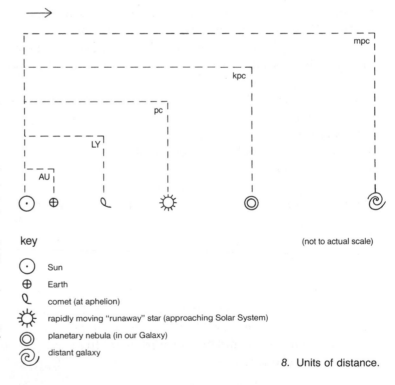

key

⊙   Sun

⊕   Earth

☌   comet (at aphelion)

☼   rapidly moving "runaway" star (approaching Solar System)

◎   planetary nebula (in our Galaxy)

🌀   distant galaxy

(not to actual scale)

8. Units of distance.

Stellar Parallax

Astronomers do not measure the distance of a star directly but, rather, its *parallax*, which is the semiannual angular displacement of the star as viewed from the Earth. The measurement of stellar parallax is, in principle, the application on a cosmic scale of triangulation, a method used for finding the distances of objects on the Earth that are out of immediate reach.

For example, suppose that an architect wishing to determine the acreage of a plot of land decides to measure the distance between two trees on opposite banks of a deep stream that he cannot cross (fig. 9a). As his baseline for carrying out this measurement, he lays off a line from A, the position of a tree on his side of the stream to a point C, some definite distance, say 186 feet, from A. He then uses his surveyor's instruments to sight the tree B, on the other side of the stream, and carefully determines the angle that his line of sight to that distant tree makes with his baseline, AC. At point C he again sets up his tripod, sights the tree B, and determines the angle that this second line of sight, CB, forms with AC. Having found these two angles in the triangle ABC and knowing that the line common to them, the baseline AC, is 186 feet, the surveyor uses tables of trigonometric functions to calculate the distance AB according to the rules of simple plane geometry.

An astronomer performs a similar task to determine the parallax of a star, which is formally defined as the angle subtended at the star by the astronomical unit, the AU. Here, an imaginary line, E_1E_2 (fig. 9b) corresponds to our architect's baseline; it is 186 million miles long, or twice the AU: the average diameter of the Earth's elliptical orbit around the Sun. The two "sighting points" are now diametrically opposite points of the orbit. Using the Earth as a space vehicle to go from E_1 to E_2, the astronomer takes photographs of the star from these points six months and 186 million miles apart. The parallax is then half the separation between the two photographic images of the star divided by the length of the telescope; the separation of the two images, obtained by superimposing the two photographic plates, must be expressed in the same units as the length of the telescope. An analogy may be useful here: Imagine a gigantic creature (with a head larger than the Earth's orbit) whose eyes are separated by 186 million miles. If, while gazing at a star, he closes first one eye and then the other, he detects a minute shift in the star's position against the stellar background field; half this shift is the parallax of the star. (The reader might try this on a small scale, holding out a pencil or his finger and noting how it shifts against the background, as he views it first with one eye, and then the other.)

Even with today's equipment, however, accurate parallactic measurement is complex. Before the trigonometric parallax can be determined with accuracy, the actual proper motion (p. 40) of the star in relation to the Solar System must also be taken into account through additional photographs; and other corrections must be made, allowing for the bending of starlight by the Earth's atmosphere, as well as the effects on the telescope and photographic plate of variations in temperature and photographic emulsion.

Trigonometric parallax is reliable only if the star is no farther away than 400 light years. An important method has been developed for measuring greater distances, based upon the star's apparent and absolute magnitudes, spectral type, and luminosity

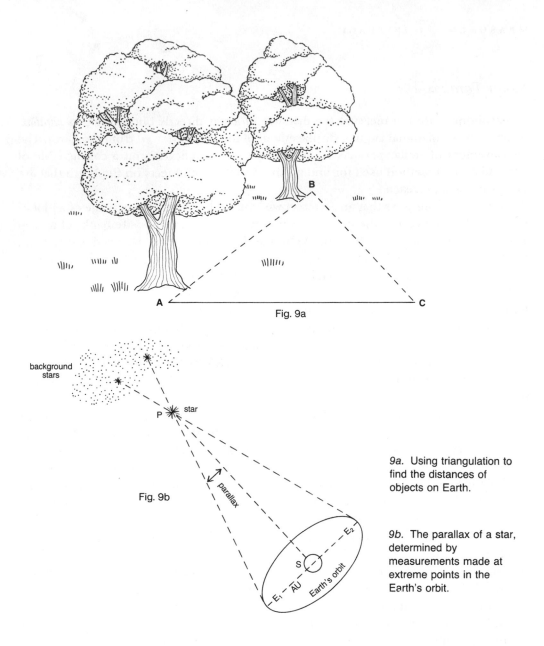

Fig. 9a

Fig. 9b

9a. Using triangulation to find the distances of objects on Earth.

9b. The parallax of a star, determined by measurements made at extreme points in the Earth's orbit.

class; we discuss these subjects in the next pages. F. W. Bessel in 1838 was the first astronomer to measure the parallax of a star, 61 Cygni (p. 279), and his calculation was considered one of the greatest scientific accomplishments of the nineteenth century.

Note in the illustration that the ellipse (very nearly a circle in actuality, but shown here in perspective) represents the orbit of the Earth, and the distance ES, the astronomical unit, is the base of the triangle SPE. The parallax is the angle of the triangle SPE at the position of the star P (the angle subtended at the star by the astronomical unit). Once this angle and the distance ES are known, the distance PS (or PE_1 or PE_2), all considered equal because of the great distance to the star, can be determined in the following way:

The parallax p''‡ is half the angle subtended by the baseline E_1E_2; the reciprocal of

‡ *Parallax expressed in seconds.*

this quantity ($^1/_{p''}$), expressed in seconds of arc, is the star's distance in parsecs. Thus, with a parallax of one tenth of a second of arc ($0''.1$), a star is 10 parsecs distant, or 32.6 light years. This very simple formula for a star's distance is therefore written as $r = ^1/_{p''}$ parsecs, where r is the distance and p'' is the star's parallax. Thus a star with a parallax of one second of arc is at a distance of 1 parsec.

As we have indicated, the distances of stars and the quantities and kinds of light they emit determine how we classify them; a star's "magnitude" is an important classification.

STELLAR MAGNITUDE

Although not nearly so ancient as the mythology of the Bears, the Dragon, and other archaic constellations, the system we use to classify the apparent brightnesses of their stars is very old. In 134 B.C. the Greek astronomer, mathematician, and geographer Hipparchus of Rhodes, an untiring observer of the heavens, noticed a brilliant "new" star in the constellation of Scorpius, situated where no star had previously been visible. Such a blazing object must have held great significance for him, since he regarded the brightest stars as the most important.

Accordingly, Hipparchus began to list his observations, using instruments of his own invention involving graduated circles to specify the coordinates of eight hundred stars, which he then classified on the basis of their importance, or "magnitude." He ranked those which appeared most brilliant as stars of the "first magnitude" (first in importance), the somewhat less brilliant as those of the "second magnitude," still fainter stars as "third magnitude," and so on. He thus produced the first catalogue of naked-eye stars, rating them according to their apparent brightnesses.

While carrying out this work, Hipparchus created the first planisphere, or star chart, showing their positions according to a definite system of coordinates; in doing this he discovered the precession of the earth's axis (which we have discussed). He is therefore known as the "father of systemic astronomy."

In the year A.D. 180 Ptolemy (p. 4), who owed his theories to Hipparchus, extended the latter's work by assigning 1,028 naked-eye stars to six magnitude categories or steps in his own great work the *Almagest*. First-magnitude stars were therefore the brightest, second-magnitude stars like Polaris were less bright, etc., so that the higher the magnitude number the fainter the star. The term "magnitude" gradually lost its original meaning of "importance" in astronomical literature; as a technical term, it came to represent the brightness, on an arbitrarily agreed-upon scale, rather than the importance, of a star.

The Pogson Magnitude Scale

In 1854, the British astronomer Norman R. Pogson introduced a precise standard for measuring magnitude (fig. 10) which was immediately accepted by the scientific world. In this scheme, an average sixth-magnitude star was chosen from a widely recognized catalogue (the *Bonner Durchmüsterung),* and the first magnitude was assigned to any star exactly one hundred times brighter. Therefore, on the Pogson scale, a first-magnitude star is exactly one hundred times brighter than the sixth magnitude. Stars more than one hundred times brighter than the sixth magnitude, and therefore even brighter than the first magnitude, are assigned numbers less than one (including positive fractions down to zero and negative numbers). A star of any given magnitude on this scale sends us very nearly 2.512* times more light per second than one of the next-higher magnitude. The brightness interval between any two adjacent magnitude steps appears equal to the unaided eye because of a limitation in its powers of perception, but actually the brightness increases or decreases by the fixed factor (2.512) mentioned above. The interval is multiple, rather than additive. The important thing to note here is that five steps in magnitude represent exactly a hundredfold increase or decrease in brightness. Thus a fifth-magnitude star is exactly ten thousand times as bright as a fifteenth-magnitude star, which in turn is one hundred times as bright as a twentieth-magnitude star but is 1 million times *fainter* than a zero-magnitude star.

To avoid confusion over the basically simple concept of magnitude, one should remember its original meaning, importance. Sixth in importance ranks much lower than first in importance. Therefore a higher magnitude indicates a fainter star, and a lower one a brighter star. For those first learning the concept of stellar magnitude, we suggest memorizing the following slogan:

HIGHER MAGNITUDE: FAINTER STAR
LOWER MAGNITUDE: BRIGHTER STAR

The lowest stellar magnitude belongs to our sun, at −26.78. Excluding the Moon, Venus, and other planetary objects, the next-ranking star is Sirius, with a magnitude of −1.4. Then follow Alpha Centauri (the closest star to our Sun), with a magnitude of −0.3, and Vega, with a *positive* magnitude of 0.04, and Capella, magnitude 0.05. Polaris, as we have noted, has a more modest magnitude of 2.0, but as a second-magnitude star, it is bright enough for its important navigational function of guiding sailors at sea.

In setting up a magnitude standard for all other stars, astronomers originally chose several circumpolar stars, the so-called "Mount Wilson Polar Sequence," because they are always visible to the northern observatories. Recently, however, improved and much more precise methods of measuring the brightnesses of stars through photoelectric photometry led to an enlargement of the standard sequence to include stars scattered throughout the heavens whose brightnesses have been very accurately established by these new methods.

Apparent Magnitude ("m")

This is the measure of a star's *apparent brightness:* the amount per second of the star's radiant energy that strikes a square centimeter of a surface held at right angles to a line from the star to the surface.

The apparent brightness (and hence the apparent magnitude) of a star depend upon two things: its intrinsic brightness and its distance from us. The intrinsic, or true,

* *The fifth root of 100.*

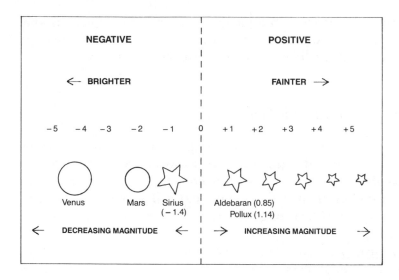

10. Stellar magnitude.

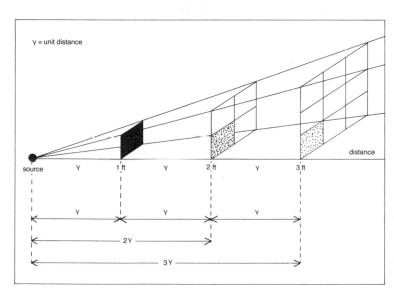

11. The inverse-square law of light propagation. The drop in brightness is indicated by the decreasing blackness of a unit square at uniformly increasing distance.

brightness of a star, which is called its *luminosity*, is the total energy it actually emits per second in all directions. When two stars of equal luminosity are at greatly varying distances from us, the closer star appears much brighter than the more distant, in accordance with the distribution of the light over ever increasing surface areas as it travels from its source out into space. The unvarying characteristic of this distribution is that the fixed amount of energy in a given cone of light spreads out in such a way that the quantity passing through a unit area decreases with the square of the distance from the source, a property of the propagation of light called the *inverse-square law* (fig. 11).

This phenomenon may be likened to the hospitality of an embarrassed hostess who, forced to save jam when unexpected guests arrive, spreads the same amount on more and more pieces of toast, so that each individual slice becomes more thinly spread and tastes less sweet!

Absolute Magnitude ("M"), a Modern Concept

Owing to the effect of distance on the brightness and therefore the apparent magnitude of a star, astronomers introduced *absolute magnitude,* a concept that enables them to make a true comparison of the stars' luminosities by hypothetically placing all the stars (and the Sun) at a standard distance. With the distance factor thus removed, the stars' apparent brightnesses become their true brightnesses, i.e., their luminosities; all stars of equal luminosity then appear equally bright.

In selecting this standard, astronomers might have put all celestial objects at the Sun's distance, but the stellar brightnesses would then be very large and unwieldy. It was preferable to "relocate" all the stars, including the Sun, so that they appear neither too "far" nor too "close." By international agreement, 10 parsecs (32.6 LY) was chosen as the most workable distance for this standard of comparison.

The apparent brightness of a star placed at 10 parsecs is called its *absolute brightness,* and its apparent magnitude at this distance is called its absolute magnitude: M. To see what this means, we consider how bright the Sun would appear at 10 parsecs, which is exactly ten times 206,265 astronomical units, or roughly 2 million astronomical units. Owing to the inverse-square law, the Sun's brightness would then be reduced by a factor of 4 trillion (four followed by twelve zeros). We then determine its absolute magnitude, remembering that each factor of a hundred in brightness means five magnitude steps. Since in 4 trillion there are slightly more than six such groups of five-magnitude steps (about 31.5 steps), the apparent magnitude of the Sun at 10 parsecs would be 31.5 minus 26.78 (its negative magnitude at 1 AU), or +4.7. This quantity is called the absolute magnitude of the Sun (M $\odot$). We may follow this procedure to obtain the absolute magnitude of any star whose distance is known or measurable.

Bringing all the stars to a standard distance is only a game astronomers play. Picture how life on Earth would be affected if such a mammoth (and divine) engineering feat were possible! We begin with the Sun itself, now enjoying its magnitude rating of −26.78. Demoted to a modest +4.7 at the standard distance of 10 parsecs, it would no longer be perceived as a warm, life-giving body; in fact, we would scarcely be able to see it at all with the naked eye except when the skies were very clear, and then only as a faint star (assuming that advanced technology could protect us from an atmosphere of solid ice: in which case, we would have an endless night for stargazing).

However, if the other stars were relocated at 10 parsecs, many of them would be seen as blazing orbs, lighting up the night sky with their brilliance; Rigel, for example, at M −7, would appear fifteen times as bright as Jupiter and Venus at their most brilliant.† Incidentally, a star that is *actually* 10 parsecs from us has the same apparent and absolute magnitudes. There is no major star at exactly this distance, but a few are fairly

---

† *Since there are really trillions of stars in the universe, if we brought all this superheated mass in to 10 parsecs, we would of course cause the universe to contract in a vast implosion and return to its primal state, ending life on Earth—which illustrates the importance of setting reasonable limits upon one's imagination.*

close to the standard, as shown by the small differences between their apparent and absolute magnitudes:

	Pc	App. m	Abs. m
Pollux (Gemini)	10.7	+1.2	+1.0
Arcturus (Boötes)	11	−0.1	−0.2
Vega (Lyra)	8.1	+0.04	+0.5

If a star is farther away than 10 parsecs, its apparent magnitude is greater than its absolute magnitude, because bringing it in to 10 parsecs increases its brightness (and therefore decreases its magnitude)! When the star is closer than 10 parsecs, the reverse holds true: its apparent magnitude is smaller than its absolute magnitude.

Bolometric Magnitude

Standard photographic instruments do not give us a complete picture of the different kinds of radiation emitted by a star. Radiation with wavelengths greater than those at the red end of the spectral scale, that is, infrared and radio waves, has played a prominent role in the development of infrared astronomy and radio astronomy. With the more recent measurements of ultraviolet radiation and X rays, using artificial satellites, ultraviolet astronomy and X-ray astronomy have also grown, as has gamma-ray astronomy even more recently.

Bolometric magnitude, which takes account of a star's ultraviolet, infrared, and radio waves as well as the visible light, was therefore introduced in the development of modern astronomy. Because it includes the contributions of all invisible radiation, it is always smaller than a star's visual magnitude.

The bolometric magnitude of the Sun is arbitrarily placed equal to its absolute magnitude (+4.8). The bolometric magnitudes of stars in general are then obtained by a mathematical correction based on the differences between their surface temperatures and that of the Sun.

Having agreed upon 10 parsecs as the distance for determining absolute magnitude, scientists also chose the Sun's intrinsic brightness (luminosity) as a standard for all stellar luminosities. Its luminosity (3.90×10^{33} ergs per second‡) is therefore taken as the astronomical unit of luminosity, which we designate $L \odot$.

The symbol M, absolute magnitude, should not be confused with the Latin script symbol $\mathscr{M}$, which we use for mass.

Magnitudes are no longer determined with the naked eye, with its natural limitations, as in Ptolemy's time, but are measured with highly sophisticated photoelectronic instruments and with large telescopes in conjunction with photographic equipment. Two types of photographic plates are used: those especially treated to imitate the human eye by reacting more strongly to red and yellow light waves, from which we obtain the so-called visual magnitude; and "normal" blue-sensitive plates, which yield the "photographic" magnitude.

‡ *The erg is the physicist's unit of energy; one ordinary calorie, which is the heat required to increase the temperature of one gram of water by one degree equals about 42 million ergs.*

$$
\begin{array}{ll}
L & = \text{luminosity} \\
m & = \text{apparent magnitude} \\
M & = \text{absolute magnitude} \\
\mathcal{M} & = \text{mass} \\
\odot & = \text{"of the sun"}
\end{array}
$$

12. *Frequently used symbols.*

For practical purposes, however, all magnitudes given in this volume, whether apparent or absolute, are of the "visual" type, and references to "magnitude" imply "apparent magnitude."

STELLAR SPECTRA

When we observe the stars, whether with binoculars or through a telescope, we are not only aware of the great differences in their apparent brightnesses, but also of their many contrasting colors. Careful investigation has revealed that these widely differing hues are related primarily to the temperatures of the gases in the stars' outer layers. The atoms of each basic chemical element in a star emit their own characteristic *photons*, or quanta of energy, producing light of particular wavelengths, or colors; but the extremes in stellar temperature limit the ability of some of these gaseous elements to reveal their identities through their spectral patterns. The stars' temperatures, however, can always be directly inferred from their colors: we find, simply, that the redder stars are cooler, and the bluer, hotter.

Astronomers have developed an interesting technique for analyzing the constituent colors in a beam of starlight, i.e., its *spectrum*. This method utilizes a phenomenon called *dispersion*, which occurs when light passes from a rare medium (a vacuum or a gas) into a dense medium such as water or glass. The various colors travel at different speeds in a dense medium, with blue traveling more slowly than red, so that the various colors are spread out and thus separated from each other as they pass through the medium.

If the light is directed through a prism and then focused onto a white surface by a lens, the various colors in the beam are imaged as a band whose color changes gradually from violet, the shortest visible wavelength, at one end, to red, the longest visible wavelength, at the other. Blue, green, yellow, and orange lie in between, to form a continuous array of color. The rainbow is a familiar example of such a "color spectrum" in nature; this spectacle is created through dispersion of the Sun's rays by drops of moisture in the atmosphere.

To study a star's spectrum, astronomers use a remarkable but simple instrument called the *spectroscope,* which in addition to its basic constituent, a glass prism, consists of two telescopes so arranged as to give a sharply defined image of the star's spectrum with its band of colors. This spectrum is of the utmost value in determining the various characteristics of stars: their temperature, chemical composition, orbital motion, etc. Indeed, so remarkable is the spectroscope that scientists have learned more from it about matter, energy, and the universe than from any other single instrument in the history of science. Never in technology has man obtained so much from so little!

When used with a camera, the spectroscope becomes a *spectrograph* and projects and records spectral bands on a photographic plate. The stellar "fingerprints" may be then be studied, compared, and analyzed at length. From this kind of analysis, three important types of spectra can be obtained, depending upon the physical conditions prevailing in the stellar medium. A dense hot gas like that in the interior of a star (or a solid, like an incandescent ball of steel or copper) produces what we call a *continuous* spectrum, in which, as the name implies, the colors are spread out continuously from red all the way to violet without a break. This effect is caused by the crowding together of the atoms in the light source so that they cannot vibrate independently of each other and reveal their intrinsic characteristics through the radiation they emit.

The situation is analogous to a passenger standing in a crowded bus, streetcar, or subway car, trying to jot down a message. As more and more people pour in at each stop, his arms become increasingly constricted until it is impossible to move pen across paper and the important message is no longer legible. Such "rush hour" conditions prevailing in the stellar (or other) medium have a similar effect upon its atoms, preventing their independent vibrations, so that they cannot produce an intelligible message concerning their identities. Thus a continuous spectrum reveals nothing about the chemical nature of the source.

The second type of spectrum, called the *bright-line emission spectrum,* is characterized by sharp lines that originate in a highly rarefied gas, in which the atoms can move about freely and vibrate independently of each other, thus "telegraphing" us the codes of their individual structures. In the laboratory, this type of spectrum is produced by sending an electrical discharge through a tube containing gas under very low pressure. The color of the light obtained depends on the kind of gas that is used, as some of us know from experience with red neon lights, blue mercury lamps, and the yellow flame of sodium—all examples of emission spectra. The spectrum in each of these cases is not a continuous array of colors, but consists of isolated colors concentrated in discrete, fairly sharp bright lines which alternate with dark spaces.

The Absorption Spectrum

Greatest in importance to astronomy of the three types of spectra, the *absorption spectrum* is characterized by fine dark lines, produced by atoms in the relatively cool stellar atmospheres as they *absorb* radiation from the hot photospheres (surfaces). These faint stellar absorption lines were first discovered in 1815 in Munich by the optician and telescope builder Joseph Fraunhofer (1787–1826), who plotted more than seven hundred of them while observing the Sun with a spectroscope that projected the image on a screen.* The young physicist saw that the Sun's light does not produce an ordinary continuous spectrum, but instead, a bright continuous background crossed by thousands of dark lines, which we now call *Fraunhofer lines* in his honor. In 1859, the explanation for this phenomenon was discovered by Gustav Robert Kirchoff (1824–

*** WARNING: NEVER LOOK DIRECTLY AT THE SUN** *through binoculars, telescope, or any optical instrument: permanent blindness will result!*

87), of Heidelberg, while experimenting with sodium vapor placed between his spectroscope and a beam of sunlight. The sodium vapor produced two dark lines because its atoms had absorbed the corresponding wavelengths from the sunbeam. Kirchoff's explanation of the origin of these absorption lines led to the spectral analysis of the chemical composition of the Sun (and all other stars), and marked the beginning of the science of astrophysics.

A Fraunhofer-line spectrum, or absorption spectrum is produced in the laboratory by surrounding a hot incandescent source of light (corresponding to a star's interior) with a cool gas (representing the cooler stellar atmosphere). The continuous spectrum of the hot, central source of light is then crossed by dark lines in precisely the positions that the lines of the emission spectrum of the more rarefied gas would occupy were this cooler gas isolated and emitting its own light. In fact, the absorption spectrum is rather like a reversed bright-line emission spectrum in the sense that a photographic negative may be compared to its positive. Whereas the emission spectrum is made up of bright lines separated by dark spaces, the absorption spectrum is composed of dark lines, in exactly the same positions, separated by bright spaces. The appearance of such Fraunhofer lines in a star's spectrum depends on the nature of the atom (or atoms) producing them and the temperature of the star. The atomic process that produces the absorption spectrum is a scattering of the light of a given frequency (or wavelength) rather than an absorption.

Spectral Classification

In astronomy, therefore, the importance of emission and absorption spectra is their usefulness in analyzing the stellar gases both for their chemical content and their temperature. This method of analysis led to the classification of stars according to definite spectral groups characterized by the appearance of different sets of absorption lines in the stars' spectra. The work of spectral classification was begun in 1885 by Edward Charles Pickering (1846–1919), director of the Harvard College Observatory and indefatigable surveyor of the heavens; and it was continued very successfully by his associate Annie J. Cannon (1863–1941). Pickering's Henry Draper Catalogue, compiled at first under his direction, in memory of Henry Draper, a pioneer in stellar spectroscopy, was completed in 1924 (after Pickering's death) and classifies 255,000 stars according to their spectra. The major classifications, which have been given the letters O, B, A, F, G, K, and M, are each in turn divided into ten subdivisions. Listed in sequential order, the main classes and their subdivisions show every gradation in stellar type from young, hot *blue giants* to aging, cool, and distended *red giants* and *supergiants*. These class letters read like a code, from which one can quickly recognize the general characteristics of a given star.

For example, a star that belongs to spectral class O5 is midway between classes O and B in type: extremely hot, blue-white in color, quite luminous, and massive. Because most stars have an abundance of hydrogen, we normally expect this basic element to produce its own characteristic pattern of dark absorption lines in the spectra of most stars. However, the extremely high temperatures (20,000–35,000° K.) of massive blue-white O and B stars "erase" the hydrogen lines from their spectra through *ionization*; in this process, a hydrogen atom absorbs such an energetic photon (p. 20) that, temporarily, its electron is torn completely out of the atom, which is thus *ionized*. In this ionized state, the hydrogen atom, having lost its sole electron, cannot produce an absorption line. Therefore, O and B stars show ionized helium (helium has one electron left even when ionized), doubly and triply ionized oxygen and nitrogen, and other highly ionized atoms, but they exhibit little or no evidence of their large amounts of hydrogen; whereas the relatively cooler A stars (10,000° K.) display strong hydrogen lines produced by the *nonionized,* or *neutral,* hydrogen atoms (fig. 13).

In our discussion of the circumpolar constellations, we state, for example, that Dubhe, in the Big Dipper, is a K star (often called "Arcturian," after Arcturus (α Boötis), the most famous K star); this classification means it is orange, somewhat cooler than our yellow Sun, and its spectrum shows the presence of some heavier metals. The Sun itself is a "middle-aged" class G star, its temperature an "average" 5700° K.; like those of the K stars, its hydrogen lines are relatively weak, even though it has a large abundance of hydrogen. Our navigational star, Polaris, is in class F8: the late subclass number indicates that it is almost like a class G star. In stars much cooler than the Sun, the hydrogen atoms can produce absorption lines in the stars' spectra only when colliding with photons of radiation that the electrons in the atoms absorb in being lifted to higher energy levels. The red giants of class M, for example, with temperatures averaging only 2,500–3,000° K., have much weaker hydrogen lines than type A, or "Sirian," stars (10,000° K.), which exhibit strong hydrogen lines, because very few photons in the cooler, red stars are energetic enough to excite the hydrogen atoms.

With the discovery of some new and unusual stellar objects, including the deep red "carbon" stars, three minor groups, N, R, and S (related to classes K and M), joined the seven main classifications; and an additional subclass called *Wolf-Rayet* was later added on to specify certain extremely hot and turbulent O stars encased in expanding gaseous shells (sometimes referred to as W stars, although they actually belong to class O).

The spectral class letters O, B, A, F, G, K, M, R, N, and S can be speedily learned by a

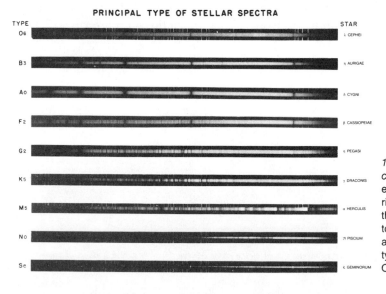

13. *Table of spectral classes.* The spectrum of each star (listed to the right) is typical of one of the principal types from O to S. Note the dark absorption lines in the type-A spectrum. (Hale Observatory)

mnemonic† familiar to all astronomy students and suggested by Henry Norris Russell (1877–1957), famous for his work on stellar spectra and the *Hertzsprung-Russell diagram* (p. 87). The stellar command is *"Oh, Be A Fine Girl (Guy), Kiss Me Right Now, Sweetheart!"*

A further classification of the stars, the *MKK system,* was introduced by W. Morgan, P. Keenan, and E. Kellman, assigning a luminosity class to each spectral type and using Roman numerals, as follows:

Ia	Bright supergiants
Ib	Faint supergiants
II	Bright giants
III	Normal giants
IV	Subgiants
V	Main-sequence (normal) stars and dwarfs
VI	Subdwarfs

These numerals are appended to the spectral-class letters. For example, Phecda (γ Ursae Majoris), the star in the left corner of the Dipper's bowl, has the spectral and MKK classification A0 V, denoting a class-A, white, main-sequence star (similar to Sirius) with a probable temperature of 10,000° K., its MKK class number V indicating a luminosity 50–100 $L \odot$.

Further examples of the usage of spectral classes and MKK numbers are the "runaway" star, Lalande 21 185 (p. 41), spectral class M2 V, a red dwarf; Tania Borealis (λ), in the Bear's left hind foot (p. 43), class A2 IV, a white (Sirian) subgiant; and Tania Australis (μ), M0 III, a normal red giant.

In Chapter IV (p. 85), "The Hunter," we give a detailed account of Betelgeuse, a famous red star, and the relationship of its light spectrum to its evolutionary development.

For the moment, however, we resume our story about two celestial Bears as we take temporary leave of modern science and its teachings. Once more we step magically into the past, not as "journalists" witnessing a primeval scene, but as "tourists" in classical Greece, sharing the ancients' wonder and awe as they adorn the star patterns with their fantasies.

---

† *From the Greek* mnemonikos, *pertaining to memory, root of the name Mnemosyne, the Greek goddess of memory and mother of the Muses by Zeus; her help is recommended for those studying the stars!*

TWO

Ursa Major and Ursa Minor

The bear that sees star setting after star
In the blue brine, descends not to the deep.
W. C. Bryant

THE MYTHOLOGY

Artemis (known also as Diana), moon goddess and mistress of the hunt, surrounded herself with a band of beautiful nymphs who always accompanied her on the chase. Among these hunting companions was an especially lovely maiden named Callisto; like the others, she took a vow of chastity on joining Artemis' band.

Zeus, or Jupiter, king of all the gods and husband of Hera (Juno), had a weakness for mortal women that often aroused the jealous ire of his queenly wife. On one of his frequent visits to the Earth, he happened to pass the woodland cove where the lissome Callisto, having put off her huntress' garb, lay soundly asleep. Falling instantly in love with the beautiful girl, Zeus disguised himself as Artemis' brother, Apollo, and then overwhelmed the unsuspecting Callisto, becoming her lover. She bore him a son, named Arcas (after the Greek *arktos,* or "bear").

Zeus, now perceiving that he would have to protect Callisto from the wrath of his slighted wife, Hera, as well as from the vengeful rage of Artemis, who brooked no desertion from her ranks, let alone a violation of the sacred vows of chastity which bound her followers together, turned his sweetheart Callisto into a bear.

One day, when Callisto's son Arcas had grown to manhood and mastered the skill of bow and arrow, he saw a great bear in the forest. The creature was in fact his unhappy mother, constantly forced to flee other beasts, with which she felt no affinity, and pursued by the very hunters in whose company she was once included. At the sight of her son she paused in joy, but Arcas, ignorant of his mother's transformation, drew his bow and took aim at her. At this moment, Zeus intervened and changed Arcas into a little bear, so that he could recognize Callisto. The godly source of all their troubles then transported mother and son to the heavens, allowing them a happier residence in

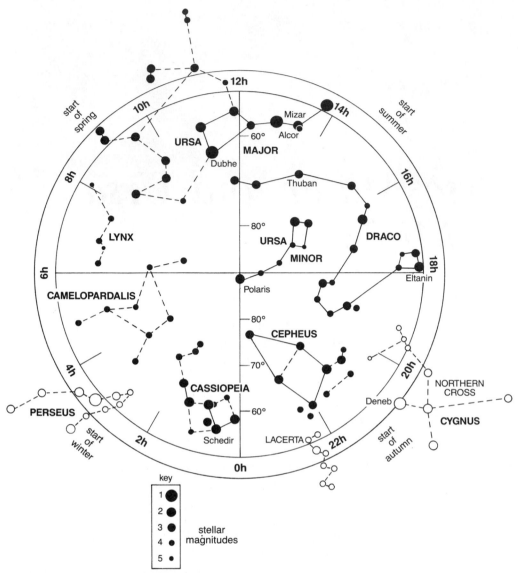

14. *North circumpolar sky* as it looks after sunset at the seasons approximately indicated. Hold the chart with the current season at the top and over your left shoulder at about 9:00 P.M. standard time.

the region of the north pole. Thenceforth, they have been known as the Greater and Lesser Bears.*

Hera, though, was far from satisfied with this turn of events; for in their new stellar domain the Bears brightened the heavens and, it was said, lit up the very pole that they now guarded. In protest against this unexpected honor to the "miscreants" (however innocent) who symbolized the indignity she had endured from her unfaithful husband Zeus, and resentful of the Bears' rivalry of her own brilliance, Hera pleaded with the ocean god never to permit Callisto and Arcas to bathe themselves in its immortal waters. The proud goddess' unkind wish was granted, and mother and son were forever

---

* *Somehow in this transformation, Arcas changed his gender to that of a she-bear, Latin* ursa *(bear; masc. =* ursus*).*

fated to circle the north celestial pole, never descending to join the other constellations in a pleasant ocean bath. In more recent times, this stern dictum has apparently suffered a slight violation: for points of observation south of the 41st parallel of declination, the Great Bear, at least, is allowed a partial dip into the waters.

As Camões would have it,

We saw the bears, despite of Juno, lave
Their tardy bodies in the boreal wave.

The Greater Bear, with its Big Dipper, is one of the most ancient constellations. According to Richard Allen, Ursa Major "always has been the best known of the stellar groups, appearing in every extended reference to the heavens in the legends, parchments, tablets, and stones of remotest times." It was frequently mentioned as a bear among both Greek and Roman writers, and found expression in the story we have told of Callisto's transformation, which originated in Hesiod's time.

Fainter and smaller than Ursa Major but more strategically situated, the Lesser Bear was not recognized by the Greeks until 600 B.C., about two hundred years after the time of Homer; therefore it may be said that Arcas' ascension to the heavens didn't become part of the mythological tradition until then. In other words, he was somewhat late in following his mother to the north pole! The constellation had been well known to the Phoenicians, however, as a navigational guide, and at this time Thales indicated to certain Greek mariners that the close proximity of the "Phoenician Bear" to the pole gave this smaller constellation an even greater usefulness than Ursa Major.

Ancient legend tells us that the two bears will live eternally in their abode among the everlasting stars.

An impressive and poetic narrative!

Modern science, however, tells another story, no less fascinating than the ancient myths and not lacking its own share of mystery and drama: no tale of immortality but, rather, one of birth, growth, dynamic change and ultimate death; a description of great events covering unimaginable time spans. Not an account of fanciful life histories with a cast ranging from god to demon and earthly hero to villain but, instead, an atomic detective story, whose clues lie in nature's laws and whose suspenseful plot, unfolding in the tremendous gulf of space, can be eagerly followed from star to star!

STAR LOCATION

The vocabulary of stargazing includes the following basic terms, which are generally used in locating objects on sky maps and atlases:

ZENITH: the highest point in the sky relative to the observer; the directly overhead point.

NADIR: its counterpart on the (hidden) "underside" of the celestial sphere, 180° from the zenith.

THE CELESTIAL MERIDIAN: a great circle passing through the north celestial pole and the observer's zenith and intersecting the celestial equator. When the observer faces to his south, he looks toward his celestial meridian, which extends from his zenith to his nadir (and passes through the south celestial pole).

CULMINATION: When an object reaches its highest point above the horizon, it is on the observer's meridian and is said to culminate. RA, or *right ascension,* and δ, or *declination,* correspond to longitude and latitude on the Earth. Not to be confused, however, with celestial latitude and longitude, which are measured from and along the ecliptic (p. 6).

The right ascension of a celestial body may be compared to the terrestrial longitude of a point on the Earth. It is measured eastwardly in hours, minutes, and seconds from the vernal equinox and is the angle between the great circle from the north celestial pole through the vernal equinox and the great circle from the ncp through the celestial body. One hour on this angular scale equals 15 degrees.

Declination, measured in degrees, gives the north–south location of a star with reference to the celestial equator (p. 6). Objects directly on the celestial equator therefore have a declination of 0; those "above," or north of, the equator (positive declination) are given plus signs followed by the appropriate number of degrees, ranging up to the maximum of +90 degrees, at the north celestial pole, and conversely, stars "below," or south of, the celestial equator have negative declinations, extending to −90 degrees, at the south celestial pole.

On star charts (except circumpolar charts), right ascension, α, is shown across the top and bottom, increasing from right to left (moving eastward) until 24 h (i.e., 0 h) is reached; declination δ, is indicated on the sides. The position of a star on the chart is given by the intersection of two lines: a vertical line corresponding to its right ascension, and a horizontal line giving its declination. On circumpolar charts the right ascension is given along the outer circle of the chart, and the two lines of declination shown intersect at the poles.

HOW TO USE THE CHARTS

For the north circumpolar zone (fig. 14), turn the chart (p. 26) so that the current season is at the top. Facing north at 8 P.M. standard time, hold the chart directly above your head; since directions in the heavens mirror those on the Earth, the bottom of this map corresponds to your northern horizon, the chart delineating a circular area of the sky. The center of that circle, or north celestial pole, marked roughly by Polaris, lies between your zenith and the northern horizon. (Only if you live at the north pole is it directly overhead!) The middle-latitude constellations and those in the north that are not circumpolar lie between your zenith and the southern horizon.

On spring evenings, when it is closest to your zenith, the Big Dipper is high in the sky and "upside down," and δ (Megrez), for example, is seen east-southeast, or to the "right" and somewhat "above" (i.e., *south* of) α (Dubhe). On autumn evenings, if you are at approximately 40 degrees north latitude, Ursa Major is closest to your northern horizon and "right side up"; it then lies low on the horizon, and the Bear's legs disappear below it. This also occurs just before dawn in late spring (not optimum times for viewing, however).

On segment maps and "blowup" charts of mid-latitude constellations, you face to the south. Because of the mirror-image effect of "reversed" directions when gazing upward to the skies, rather than downward upon the Earth, east will then be to your left and west to your right. An easy way to avoid confusion when visualizing celestial directions is to picture yourself lying supine upon a man-sized map of the U.S.A. (looking up at the sky)! With your feet to the south, your head to the north, and arms outstretched, the east coast will be designated by your left hand, and the western plains, your right. (If your head is to the south of the terrain map, the reverse holds true.)

In our text, we often describe the relative positions of stars and other celestial objects in terms of the approximate number of degrees separating them. If you hold a penny at arm's length, 1/2 degree is about equal to the area of sky that it will cover, the apparent diameter, in fact, of the full moon (31 minutes 5 seconds). Thus 1 degree is slightly less than the mean angle subtended by two full moons at the eye of an observer on the Earth.

URSA MINOR

POLARIS (α) *Alpha Ursae Minoris*
(RA 1 h 49 m   δ 89° 02')   m 2

We now have some of the basic knowledge necessary for an orderly and intelligent exploration of the constellations. Our heavenly safari gets underway with the star that enabled the sailors to orient themselves after the storm. The easiest way to locate Polaris is to imagine a line drawn between the two stars at the pouring end of the bowl of the Big Dipper: Merak (β Ursae Majoris) and Dubhe (α); and to extend the line northward about five times the distance between them. At the point that marks the tip of the Lesser Bear's tail, we find the North Star (see fig. 15). It lies almost exactly on the Earth's axis of rotation, about 1 degree from the north celestial pole.

History and Mythology

Known in early Greece as *Phoenice,* owing to the previous origin of its constellation, this strategically situated star acquired its present name, Stella Polaris, at a later date, when precession had rotated the north pole in its direction. Polaris was called upon for help by storm-tossed sailors—and the land bound—as *Stella Maris* ("Star of the Sea"); and nations to the south called it *Tramontane,* a name also applied to the whole of Ursa Minor, because of its location above the mythical Mons Coelius (p. 7). Other names for Polaris that allude to the guidance of sailors are the tenth-century Anglo-Saxon *Scipsteorra* (the ship star); the German *Angel Stern* (pivot star), and the Latin *Navigatoria;*

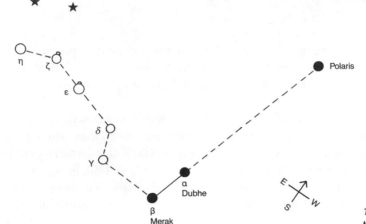

15. Locating Polaris from the "pointers."

similarly, the "Steering Star" of early English navigators, and in a later England, the "leading star," or "Loadstar" (lodestar). It acquired a scientific name, Cynosura, in the seventeenth and eighteenth centuries, by which it is still known.

To the Chinese, Polaris was the "Great Imperial Ruler of Heaven," its court the other circumpolar stars, with the whole region forming the "Purple Subtle Enclosure." According to legend, it was first used in navigation by the Chinese emperor Hong Ti (Hwang Ti), who claimed to be the grandson of Noah!

Cynosura, or Polaris, was also a religious object long ago. The peoples of India knew the star as *Grahadhāra*, "Pivot of the Planets," representing the god Dhruva. The Arab peoples also observed it for religious reasons: they called it *Al Kaukab al Shamāliyy* ("The Star of the North"), and it helped them to orientate themselves so that they could pray facing the Holy City of Mecca. For dwellers of the desert, Kaukab was not only a compass, but also a convenient medicine: they believed that prolonged staring at the star cures ophthalmia, an inflammation of the eyeball! The Arabic name *Kaukab* is similar to the Hebrew *kochab,* also meaning "star," which was the messianic name given to Simeon Bar Kochba, or "Son of a Star," who led the second revolt of the Jewish people against the Roman emperor Hadrian, A.D. 132–35.

The Pole Star, therefore, has played an important role in every human culture. Lettered Alpha, and one of the two brightest stars in Ursa Minor, its physical properties include a mass equal to that of 10 Suns and great luminosity (1,600 $L \odot$); however, because of its great distance from us—652 light years (200 pc)—Polaris is only of the second apparent magnitude. Despite its much greater luminosity, its spectral type (F8 Ib) is rather similar to that of the Sun.

In 1779, Sir William Herschel, the great British astronomer and telescope builder, discovered that Polaris is a double star, with a very faint, ninth-magnitude companion. More recently, a variation in its *radial velocity,* i.e., the speed with which it moves along our line of sight (17 kilometers—about 10½ miles—per second in approach) led to the discovery that the primary, that is, the more luminous visible object, is a spectroscopic double, which means that in addition to its faint telescopic companion, it also has a stellar component too close for detection except by the analysis of its light rays with the *spectroscope.* (We describe other important applications of this instrument and the doors of knowledge it has opened for astronomers in a later chapter.)

Such triple star systems occur frequently. Very often, the telescope first reveals that a well-known naked-eye star has a faint unseen companion; further investigation with the spectroscope indicates that the primary member of the pair consists of two stars that are very close to each other and in rapid revolution around a common point (the center of mass) between them.

Mass may be defined as a measure of the inertial properties in a body, i.e., the resistance of a body to a change in its state of motion. We determine the mass of a body by measuring the acceleration induced in it by a known external force or by measuring its gravitational interaction with some other body. Since it is impossible to subject a star to a force of our own choosing, we can obtain its mass only by observing the orbit of some object in the star's gravitational field. We can do this if the star is a component of a binary; only for such systems can the masses of the two components be obtained directly.

We discuss many binary systems in detail as we encounter them on our stellar journey, but here we consider only their orbital characteristics. From these, by applying *Kepler's third law of planetary motion,* as derived from Newton's law of gravitation, we can find the masses of the two components. Kepler (p. 4), who inherited Tycho Brahe's observational material, spent many years before his computations led him to his three laws of planetary motion. The third, or "harmonic," law, which states that the squares of the sidereal periods of the planets are proportional to the cubes of their mean distances from the Sun, tells us that the farther a planet is from the Sun, the slower is its speed of revolution. This formula applies to a binary system also since the two stars move around a common center of mass, as do the Sun and any planet. By measuring the position of the center of mass relative to the two stars and then determining the true orbit of one star with respect to the other (so that the mean distance can be found) we then, by measuring the period of the binary system, determine the mass of each star separately.

In a visual binary such as Polaris together with its faint telescopic companion, the position of the center of mass can be found and the true orbit can be obtained from the observed apparent orbit. The spectroscopic components of the bright primary of Polaris have an average distance from each other of 290 million miles (about three times the distance between Earth and Sun, or 3 AU), and they revolve about their common center of mass every 30.5 years. The telescopic secondary companion is much farther out from the close components of the primary: nearly 2,000 astronomical units, or 186 billion miles, and it requires thousands of years to circle them, in accordance with Kepler's third law.

Polaris is technically classified as a Cepheid pulsating variable. This type of star, first discovered within the circumpolar constellation of Cepheus (p. 70), is used as a yardstick to measure the distances of spiral nebulae and star clusters.

KOCHAB (β) *Beta Ursae Minoris*
(RA 14 h 51 m   δ +74° 22')   m 2.06

Occupying the western corner of the Little Dipper, this orange, metal-rich star is of the second magnitude but far less luminous than Polaris. Nevertheless, because it is only a hundred light years distant from us—less than one sixth as far away as Polaris—it

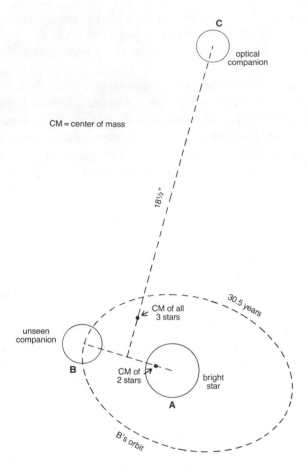

CM = center of mass

16. *Hypothetical model of the Polaris system.* The line from one component of a binary to the other must always pass through the center of mass of the system as the two stars revolve about each other.

appears nearly as bright. At the time of the conquest of Troy (1184 B.C.), the north celestial pole was closer to Kochab. Because of this, the star was also known as "Guardian of the Pole." We have mentioned that the Arab peoples called Polaris *Al Kaukab,* and the Israelites' name for the North Star was *Kochab;* in a confusing duplication of names, however, Beta Ursae Minoris has taken over the designation Kochab, and so is the only star with a Hebraic, rather than an Arabic, name.

PHERKAD *Major and Minor* (γ) *Gamma Ursae Minoris 1 and 2*
(RA 15 h 21 m   δ +72° 01')   m 3.01

Its name is derived from the Arabic *Aḥfā' al Farḳadain,* or the "Dim One of the Two Calves" (2nd-magnitude β is the brighter calf). Marking the Little Dipper's southern corner, about 18 degrees from the north celestial pole, this slightly variable star is a *wide double:* it consists of two stars *visually* close together but in reality so many light-years apart that they do not merge optically to appear as a single point of light (as in a *close double*), but may be seen individually even with the naked eye. Pherkad Major, the primary gamma star, is of the third magnitude. Pherkad Minor, the much fainter companion, like Kochab, is orange and metal-rich. Gamma 2 belongs to a stellar class called *variable*, because it does not emit a steady and unvarying amount of radiant energy; furthermore, because its peculiar variability doesn't conform to any regular cycle of periodicity, some scientists suspect that it belongs to an unusual, *dwarf Cepheid* class. Variability in stars is a subject worthy of lengthy exploration; we return to it in detail later on.

Kochab and Pherkad were designated by ancient poets as symbols of constancy, and like Polaris, they faithfully guided the night traveler; known also as the "Guardians of the Pole," β and γ were used to tell time down through the ages by country dwellers as well as navigators.

URSA MAJOR

The Big Dipper of Ursa Major

The pattern of the Greater Bear is somewhat difficult to discern as a whole, but it contains a familiar group of stars that constitute one of the most striking figures in the heavens. Our Big Dipper, the European "Wagon" or English "Plough," has a long history of titles. In ancient Egypt the seven bright stars of the Big Dipper were seen as a "Bull's Thigh," but to the Chinese they were the "Government" and also the "North-

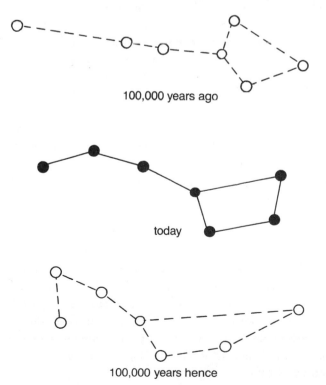

100,000 years ago

today

100,000 years hence

17. *The Big Dipper in the distant past, present, and remote future.*

ern Measure." In India these stars were called *Saptar Shayar,* or "Seven Anchorites," while the Israelites, like nations of our time, perceived them as a wagon, or *Agalah.* Both the ancient Hebrew and Arab peoples also saw them as a coffin or bier with mourners, while the Romans had various titles for our Big Dipper stars: *Currus* and *Plaustrum Magnum,* or "Great Cart"; but they also visualized the stars as threshing oxen circling the threshing floor of the pole!

Other names are the Italian *Carro,* or cart, the Scandinavian *Karls Wagn* ("Wagon of the Great God Thor"); the English "Charles' Wain," in which the English kings Charles I and Charles II took over the starry title from the ancient divinity; and the German *Himmels Wagen* ("Chariot of Heaven"), which refers to Elijah's chariot.

ζ MIZAR *(Zeta Ursae Majoris)*　m 2.4, with
ALCOR *(#80 Ursae Majoris)*　m 4.0
(RA 13 h 22 m　δ +55° 11')

The Discovery of Double Stars

Mizar (from the Arabic *mi'zar* or "girdle"), marking the middle of the Big Dipper's handle, is designated zeta, sixth letter of the Greek alphabet; nevertheless, it is a star of "firsts," its story a landmark in the history of astronomy.

In the year 1650, the Italian astronomer Giovanni Battista Riccioli aimed his simple Galilean telescope at this second-magnitude star and, to his surprise, observed a fainter companion, of about the fourth magnitude, 14 seconds of arc from it. Not to be confused with Alcor (a star optically close to Mizar but discernible with the naked eye), Riccioli's discovery had never been seen before. His early telescopic achievement was destined to be repeated innumerable times in subsequent years with regard to other stars, as it became apparent that the majority of stars exist not in majestic solitude but in good stellar company, sharing their motions through the heavens with companions. Many, if not most, stars, astronomers learned, come in twos, threes, or fours and often belong to wonderfully balanced, intricate star systems.

More than a century after Riccioli's discovery, Sir William Herschel published three catalogues of 800 double stars each. His son Sir John, continuing his father's work, compiled a catalogue listing more than 10,000 multiple-star systems as viewed from the southern skies. Meanwhile, in the northern hemisphere, another father-and-son team, Wilhelm and Otto Struve, produced a catalogue of their own observations of 3,110 double stars. Both of these works are used today.

Mizar is called a "visual double," because at its distance of 58.7 light years (18 pc), its components are sufficiently far apart (or close enough to us in space) to be seen separately through the average telescope. They are hydrogen-rich and white in hue, young on the stellar scale of time. The secondary component, Mizar B, has a peculiar metal line spectrum. Mizar A, seventy times as luminous as the Sun (L 70 ⊙), is separated from component B by 14.4 seconds of arc, or 390 astronomical units, a distance nearly five times the average diameter of the orbit of our outermost planet, Pluto. It therefore takes the Mizars A and B many thousands of years to complete one revolution as they circle their center of mass.

Spectroscopic Binaries

The *spectroscope,* an instrument developed to analyze light, is a fundamental astronomical tool that has played an extremely useful role in the detection of those binary stars whose components are too close to be seen separately through the telescope. When the orbital plane of such a binary is tilted toward our line of sight or is seen nearly edge on, a doubling of the spectral lines is recorded by the spectroscope as one star approaches us and the other recedes. This doubling is caused by a phenomenon called the Doppler effect of light, which we examine in detail later on.

Polaris A, as we have seen, is a good example of a "spectroscopic" binary. Similarly, the primary component of Mizar was leading a secret life, undetected, in this case, until 1889, when Pickering (p. 22) noted the shift of Mizar A's spectral lines in his spectroscope and disclosed its duality. Subsequent measurements revealed "stellar twins," each three times as massive as the Sun (3 $\mathcal{M}$ ⊙) and each 35 times as luminous (35 L ⊙). Since they are only 18 million miles apart, their period of revolution is 20.5386 days.

In 1908, E. B. Frost, professor at Dartmouth College (translator of an important work of the late-nineteenth century, *Spectral Analysis of the Stars*), discovered that Mizar B is also a spectroscopic binary; because its two components are very close to one another, its duality is detectable only through spectral analysis. The period of revolution for B is 182.33 days. For a time, therefore, the entire Mizar system was regarded as quadruple, that is, two spectroscopic binaries in revolution about a common center of mass: four stars altogether!

However, matters were further complicated by the more recent discovery of a third component of Mizar B at a considerable distance from the B twins and circling them every 1,350 days. This extra star in the Mizar system was detected only because of certain irregularities or a wobbling in the B stars' motions; it cannot be detected with a spectroscope, because its spectral lines blend with those of the other stars.

This type of binary component, one that reveals its presence in the effect it has upon the movements of nearby stars, is called *astrometric.*

Mizar, the first visual binary discovered, was also one of the first stellar objects photographed. This occurred at Harvard University in 1857; the photographer was G. P. Bond, who used the early process of the daguerreotype. (Seven years earlier, at the same observatory, his father, W. C. Bond, had made the first photograph of a star: Vega, in the constellation of Lyra.)

The Horse and His Rider

Centuries before the important discoveries of telescopic or visual double stars and spectroscopic binaries, however, the ancients had already observed a much fainter star very "close" to Mizar. The Arabs named this little companion Alcor, and are said to have used it as a test of good eyesight. It enjoyed various other names, including the Arabic *Suha,* or "Forgotten One"; the Chinese *Foo Sing,* or "Supporting Star"; and many titles also giving it a "horsey" connotation, such as the Latin *Eques Stellula,* or "Little Starry Horseman"; the German *Hans Duemken,* or "Hans the Thumbkin," a legendary wagoner who at his own request spends eternity driving his team (the other

Big Dipper stars) across the sky; the more recent "Horse and His Rider" (Mizar and Alcor); and the English name for this asterism, or group of stars, "Jack on the Middle Horse": the rider's steed, Mizar, being flanked by two stars of similar brightness, Alkaid and Alioth, the three together forming the handle of our Big Dipper, or the team that draws the Wagon or Plough.

It is now thought that Alcor is not actually a part of the Mizar system, does not share its proper motion, and appears "close" to it only because they are both in our line of sight; with an angular separation of 11.8 minutes of arc, they are one quarter of a light-year apart. This type of *apparent* double star, much rarer, incidentally, than the true binary system, is known as an *optical double*.

However, as you may already suspect, the story of Mizar and Alcor doesn't stop here! When examined with the spectroscope, Alcor was found to consist of two very close components. This modern discovery revealed that "the Rider," popularly regarded as Mizar's small companion, is actually in itself a dual object, a spectroscopic binary having fifteen times the luminosity of our Sun; another reminder of our own star's dwarflike status compared even to some of the faintest stars visible to the naked eye!

Of Interest While Viewing Alcor

Suitable for field glasses or the small telescope is the eighth-magnitude star "Sidus Ludovicianum," forming a flat triangle with Mizar and Alcor. This faint star was named after Ludwig V of Hesse-Darmstadt by a subject who mistook it for a new planet! It is interesting that "Sidus L.," despite its apparent faintness, is also a spectroscopic binary, with the total luminosity of fifteen Suns.

ALIOTH (ε) *Epsilon Ursae Majoris*
(RA 12 h 52 m   δ +56° 14')   m 1.79

The brightest star in the Big Dipper, Alioth is one of the aforementioned "horses" adjacent to Mizar ("Jack's Middle Horse"). Its Arabic proper name is said to have first been mentioned in an astronomical landmark of thirteenth-century Spain, the Alfonsine Tables ("Introduction," p. xx). Various authorities had conflicting notions of the origin of the name "Alioth," possibly from *alyat,* the fat tail of the eastern sheep, and more remotely as *aliare,* from *al hawar,* "the white poplar tree," meaning "intensely bright."

With a luminosity of 85 Suns, Alioth lives up to this description. It is a fairly close star, at a distance of 61.9 light years (19 pc). Its peculiar spectrum, which contains abnormally strong lines of chromium and europium, is of especial interest to scientists, because these heavy elements are rare in the universe and therefore not expected in any abundance in stars.

Variations in Alioth's luminosity are attributed to changes in its magnetic field. Like its neighbor Mizar, it is a spectroscopic double; its period is 4.15 years.

ALCAID (η) *(Eta Ursae Majoris)*
(RA 13 h 46 m   δ +49° 34')   m 1.9

Sometimes called Benetnasch, this star marks the end of the Big Dipper's handle or the tip of the Bear's tail and derives its Arabic name from *Ka'id Banat al Na'ash,* "Governor of the Daughters of the Bier," that is, the chief mourner. The other "daughters" were pictured as Alioth and Mizar, with the bowl of the Dipper seen as a bier. The Chinese called Alcaid *Yaou Kwang,* a "revolving light." Although 228 light years distant, it is one of the brightest Big Dipper stars, belonging to spectral type B3, which indicates a very young, white, and luminous object. Alcaid is approaching us at eleven kilometers (6.8 miles) per second, and it is believed that it may be variable.

The Bowl of the Big Dipper

The "Pointers"

DUBHE (α) *Alpha Ursae Majoris*
(RA 11 h 01 m   δ +62° 01')   m 1.9

Dubhe (also called Dubb), the northernmost "pointer," is appropriately named *Thahr al Dubb al Akbar,* or "Back of the Great Bear," and like ε (Alioth), was first listed in the Alfonsine Tables, of medieval Spain. In ancient times the Egyptians used Dubhe as an orientation point for their temples of worship and identified it with the goddess Isis. Although lettered α, it is slightly fainter than ε at the present time. Its magnitude differs in various listings, suggesting that its luminosity may have diminished slightly in recent years.

Dubhe's history resembles that of the binary Mizar with respect to its secrets of duality. It has, first of all, a seventh-magnitude companion, which is just beyond the perception of the naked eye but a very easy and attractive object for the simplest telescope; this white (F8) companion contrasts with the orange (K0 II or III) primary, creating the illusion of a bluish and gold double in most telescopes. Their separation of 6.3 minutes of arc indicates a distance between the two stars of at least 12,000 astronomical units; they therefore form a "wide" gravitational pair and do not complete one orbit about their common center of mass in less than six hundred thousand years. Thus a determination of their period of revolution cannot be based upon observation.

In 1889 S. W. Burnham, of Lick Observatory, also made a discovery echoing the "Mizar story," namely that the primary of Dubhe is a close visual binary: i.e., two components moderately distant from one another but visible as separate stars only with a powerful telescope. With a total mass of 3 $\mathcal{M} \odot$ and luminosity 145 $L \odot$, their orbital period is 44.66 years. This familiar pattern of secret duplicity was continued when, in 1959, R. M. Petrie announced that analysis of α's distant, bluish companion revealed this star as a *spectroscopic* binary. Its close components, which take six days to complete one revolution, cannot be seen individually with the telescope, but as in other cases involving spectroscopic binaries, are discernible through the telltale doubling of the spectral lines.

To sum up the Dubhe system: component A, the primary, is a close visual binary (two stars moderately separated from one another), circled at a great distance by component B (the companion), a spectroscopic binary (again two stars, but these very close to, or nearly touching, each other); the whole, therefore, four stars: a quadruple system.

MERAK (β) *Beta Ursae Majoris*
(RA 10 h 59 m   δ +56° 39')   m 2.37

Named *Al Marakh,* or "Loin of the Bear," by the Arabic peoples, this white-hot "Sirian" star, 80 light years distant, radiates with the luminosity of 65 Suns. Along with Dubhe, it points the way to Polaris (p. 29).

PHECDA (γ) *Gamma Ursae Majoris*
(RA 11 h 51 m   δ +53° 58')   m 2.4

Also called Phad, this star's name is derived from the Arabic *Al Falidh,* meaning "The thigh"; it marks the lower left (SE) corner of the Dipper's bowl. Its spectrum (A0 V) is like that of Sirius and also that of β Ursae Majoris, both white-hot stars. A single star, 90 light years distant, its luminosity is 75 $L \odot$. Phecda is approaching us at a velocity of about 7.5 miles per second.

MEGREZ (δ) *Delta Ursae Majoris*
(RA 12 h 13 m   δ +57° 19')   m 3.6

Marking the juncture of "bowl" and "handle," this somewhat fainter Dipper star derives its Arabic name from *Al Maghrez,* "The Root of the Tail." The Chinese gave it the more dignified title of *Tien Kuen,* or "Heavenly Authority." The Hindu *Vishnu-Dharma* says that it rules the other stars of the Bear; because this is not true at the present time, it raises the suspicion (as in the case of Dubhe), that δ, which is slightly variable, may have increased in magnitude, that is, become fainter, over the centuries. The present luminosity is 20 $L \odot$, and its distance a moderate 65 light years.

Megrez is a single star; therefore, owing to orbital stability, a planetary system is possible. Since it is situated more or less centrally in the Dipper, inhabitants of a planet of Megrez would see numerous brilliant stars scattered throughout the heavens, many of them members of their own cluster of stars. These would be interspersed with others that are not in our Dipper, some nearer to Megrez, and many, more distant; and for the "Megrezians" they would "combine" optically to form many shapes and patterns: their own constellations! However, the other Dipper stars would not be seen as forming a particular dipper-like asterism; the Dipper that we view would have no significance for these hypothetical beings.

Furthermore, Alcaid, the star at the end of the handle, is farther away from the other Dipper stars than they are from our Sun! In fact, it has been suggested that our own Sun may be physically associated with the stars of the Great Dipper, because it shares their motion fairly closely (p. 39, †). However, we do not chart our Sun as a part of the Big Dipper; nor would the inhabitants of the planet of Megrez which we have postulated, or one of any other Dipper star, count their own sun as a part of that constellation; the Dipper, or Plough, that we see simply would not exist for them.

These facts illustrate the illusory, humanly created patterns of the constellations:

purely imaginary sky pictures, however useful they are for locating and committing to memory the numerous fascinating objects that fill our heavens. We may therefore enjoy the fantasy of their entertaining mythology without making the astrological error of imputing to such star patterns the power to influence our personal destinies!

The Origin of Star Clusters

We have learned that stars in general "prefer" a double life: the majority of the massive, luminous stellar objects that we can easily see with the naked eye are members of double, triple, and quadruple star systems, keeping close company within a stellar family. However, even those stars which are separated by vast distances often belong to groups of stars that have a common motion through space. The stars in a given "cluster" originated within the same gas/dust cloud, and the motion, already imparted to this cloud by the explosion of an old, dying star, carries the group of newly formed younger stars on a common space trip. The remaining gas and dust particles of this cloud are the remnants of the old exploded star plus aggregations of material from other sources. Called the Ursa Major moving cluster, the Big Dipper, with the exception of its end stars Dubhe and Alcaid, is a good example of such an open star cluster; 30 light years long by 18 light years wide, it is the nearest cluster of this type in the heavens, its center being 75 light years distant. The five inner stars, ζ, ϵ, δ, γ, and β, together with several additional fainter stars, are all moving in the direction of Sagittarius, to the southeast, at an average space velocity (velocity relative to our Solar System) of 9 miles per second. Dubhe and Alcaid, the two exceptions, are rapidly moving away from the others on a path of their own, however, because they were not formed from the same gas/dust cloud as the other Dipper stars; they belong to the present Dipper pattern by the sheer coincidence of their optical alignment with the others.†

In time, the proper motions of Dubhe and Alcaid will carry them somewhat away from their present positions in the Dipper; they will still be considered members of that constellation but will gradually alter its shape. Eventually they will leave it altogether. Their present neighbors in the Dipper, however, maintaining the same positions *relative to one another,* will continue their space journey in concert, like a flock of birds. This moving cluster as a whole will, of course, change its position relative to other groups of stars. Many thousands of years ago, long before the epoch of our tribal hunter (p. xv), the now familiar pattern of the Big Dipper had a very different appearance. A hundred thousand years from now, the independent movements of Dubhe and Alcaid will have again given it a new shape. Not only will the stars' positions relative to one another change (fig. 17), but their chemical composition will also be greatly altered in the course of long intervals of time (billions of years), thereby contributing to the altered appearance of the asterism we now call the "Dipper." The evolution of stars is a fascinating study; in our story of Betelgeuse (Chapter Four, p. 91), we give an account of its details.

One of the authors is reminded of a conversation she had with an astronomically naïve person, this time no small child, but an elderly gentleman. Bewildered by the

† The "Ursa Major stream"—*several hundred light years in diameter—is a general stream of stars that share the Ursa Major cluster motion fairly closely and may or may not be physically associated with it; our Sun is included.*

turbulence of today's world events, he indulged in a torrent of complaints, but then offered the author a consoling observation: "At least the stars don't change!" Several decades had passed since her encounter with the child who couldn't accept the Sun's starlike character. Having lost the uninhibited forthrightness of childhood, she attempted a tactful if slightly embarrassed reply: "Not in a human lifetime!"

Proper Motions of Stars

The west-to-east rotation of the Earth causes all the heavenly bodies (except those in the circumpolar areas) to rise in the east and move across the skies in an east–west pattern. This apparent east–west movement of the stars was mistaken by our remote ancestors for a true motion imparted to them by the turning of a "celestial sphere" (p. 4). With the acceptance of the heliocentric universe, after the time of Galileo, it was understood that the stars' rising and setting is caused by the Earth's rotation, and that the Earth's motion around the Sun imparts to the stars a semiannual shift called the *parallactic motion.* However, most people still believed that stars were "fixed" relative to each other and did not have motions of their own.

Ultimately, it was learned that the concept of "fixed stars" was incorrect. Indeed, all the stars move through space and thus alter their positions with reference to each other, but very slowly as seen from the Earth. Totally independent of the "apparent" rising and setting movement of stars caused by the Earth's rotation and unrelated to the annual parallactic shift caused by the Earth's revolution around the sun, these *intrinsic* motions or velocities affect the stars' angular positions relative to the Earth. We call the resultant changes in their apparent positions their *proper motions.*

Astronomers define "proper motion" as the apparent angular displacement of a star relative to the Sun, measured in seconds of arc per year. Symbolized (μ''), this apparent displacement arises only from the star's *transverse* motion, i.e., its motion *across* our line of sight (perpendicular to that *along* the line of sight), or, to put it very simply, crosswise motion as we view it (fig. 18).

In the case of faint stars whose angular changes in position are generally small, their *absolute* proper motions must be determined by first recording their *relative* proper

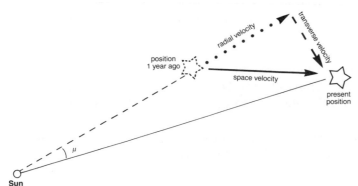

18. *Annual motion of a star relative to the Sun.*

motions, i.e., their motions or displacements in comparison to the faint stars in the same fields that have relatively much smaller positional changes. This is accomplished by making a photographic record of a given star field over a period of time, let us say once every five or ten years, and then comparing the stars' changes of position. In this process, a useful instrument for "spotting" stars that have altered their relative apparent positions during this time is the *blink microscope,* in which two photographic plates are aligned in the microscope so that their images are superimposed upon the retina of the eye, creating an effect whereby the stars with relatively large apparent motions appear to jump back and forth between their earlier and later positions. Thus the relative proper motion of any star can be measured by comparing different photographs of its star field taken over sufficiently long intervals of time. The absolute proper motion is then obtained by noting the absolute proper motion of the bright stars in the same field.

"Runaway" Stars

Proper motion depends not only on a star's actual velocity through space but also on its distance from us. Stars situated relatively close to the Sun tend to have large proper motions. Therefore, the determination is of great value in locating our closer neighbors in space by picking out large proper motions.

An excellent example of one nearby star showing very large proper motion is designated #1830 in the Groombridge catalogue (which lists stars of special interest). **GROOMBRIDGE 1830** is a yellow subdwarf just beyond the naked-eye limits of perception (m 6.5), and 1/7 L ☉, situated 16 degrees below Phecda, of the Big Dipper (RA 11 h 50 m, δ +38° 05'). It has a peculiar, metal-deficient spectrum; its metallic-atom content is only one thirtieth that of the Sun. This fact tells us that it was formed early in the history of the Galaxy, long before our Sun was born. At 28 light years, it is one of the nearest known population-II stars of the galactic halo, a vast ellipsoidal, almost spherical region surrounding the entire galaxy. With a proper motion of 7.04 seconds (the third-fastest known) and presently approaching us at 61 miles per second, it will be closest in the year 9900. Groombridge 1830 takes 511 years to change its position by 1 degree of arc, or 265 years to cross a line equal to the Sun's apparent diameter. At this rate, in 185,000 years it will circle the entire "celestial sphere."‡ Its actual velocity is 215 miles per second as it streaks through space on its way to the southern constellation of *Lupus,* whose wolfish company it will join in about a hundred thousand years! Keeping pace with Groombridge is a faint variable companion, of magnitude range 8.5 to 12, that may be a flare star.

Another "runaway" star in the same general area of the Greater Bear (RA 11 h, 6 m, δ 36° 18'), north of the Bear's left hind foot and 4° northwest from (ν) Nu, is numbered "21 185" in Lalande's catalogue. **LALANDE 21 185** is a red dwarf (spectral type M2 V), with a tiny mass of 0.3 ℳ ☉; at 8.3 light years, or 24.9 parsecs, it is closer to us than our stellar neighbor Sirius. With a proper motion of 4.78 seconds of arc per year, a space velocity of 62 miles per second, and a radial velocity of 52 miles per second in ap-

‡ *We use this term figuratively, rather than in the ancient—and literal—sense!*

proach, Lalande 21 185 doesn't "run" as fast as Groombridge 1830; nevertheless, it ranks eighth in the "Stellar Handicap"—not a poor finish, considering the stupendous number of competing entries!

Like so many other stars of Ursa Major, it reveals a tale of duplicity: Lalande 21 185 is an astrometric binary, the close companion detectable only through a wobble, or irregularity, that it produces in the motion of the primary. The unseen companion is of especial interest, because its mass is calculated at only 0.01 $\mathcal{M}\odot$; with such a small mass, it may be no star at all, but a large planet! For the bewildered residents of a planet circling a runaway star, however, the constellations would be rather unstable, changing their patterns every few centuries!

Stars Outside the Big Dipper

In addition to the "runaways," we find very interesting stellar objects in the Bear's head and feet. Our first stop on this anatomical trip is the third-magnitude star **MUS-CIDA,** (o) **OMICRON** Ursae Majoris (RA 8 h 26 m   δ +60° 53'), 150 light years distant, luminosity 85 $L \odot$, which marks the Bear's nose and is also part of the obsolete *asterism* (minor star group) known as Al Thiba, or "The Gazelle," the latter including several faint stars in the Bear's head. *Muscida,* a medieval word meaning "muzzle," appropriately indicates the star's location. A very faint (m-15) companion was discovered at a separation of 320 astronomical units, a dwarf of 1/600 $L \odot$, suitable only for great telescopes. Before departing from the head, we mention the long-period binary (about 700 years) (σ) **SIGMA II** (#13), m 4.78, which together with **SIGMA I** and (ρ) Rho constitutes the Bear's ears.

Two third-magnitude stars jointly form the Bear's forepaw. **TALITHA,** (ι) **IOTA** (#9) **Ursae Majoris,** the more northerly of these (RA 08 h, 56 m, δ +48° 14'), is only 50 light years distant. It has a very faint (10-m) companion of reddish hue at about 70 astronomical units separation. Large telescopes, however, reveal the companion as a close visual binary that in itself consists of two red dwarfs, i.e., ancient stars of low mass and temperature (spectral class dM1), each 1/100 $L \odot$, separated from one another by 11 astronomical units, a distance slightly greater than that of Saturn from the Sun; the whole (Talitha and its gravitational companion) therefore constitute a triple system.

Talitha's neighbor **AL KAPRAH,** or (κ) **KAPPA** Ursae Majoris (RA 9 h 0 m   δ +47° 21'), m 3.68, is a close visual binary (separation 0.3"), 250 $L \odot$, whose two fourth-magnitude components are probably as far apart as our planet Uranus and the Sun, with a period of revolution of about fifty-eight years. Talitha's dual nature is revealed only by powerful telescopes. The two stars, Talitha and Al Kaprah, present a good example of optical deception in space: they are only 1 degree apart, yet Al Kaprah, at 300 light years, is six times the distance of Talitha. Thus, while both stars mark the Bear's paw, our own Sun is more entitled to call Talitha "neighbor" than is the distant Kappa!

Several degrees northeast of ι and κ, indicating the Bear's foreleg, and sixty-five light years from us, the third-magnitude star (θ) **THETA** Ursae Majoris (RA 9 h 30 m,   δ +52° 50') is sixteen times as luminous as our Sun. Like so many other stars in the Greater Bear, it has a faint (m/13.5) companion: a dwarf star at a separation of 85 astronomical units, seven hundred times less luminous than the Sun, therefore visible only with the most powerful telescopes.

Higher up on the foreleg, marking the Bear's "armpit" and, visually speaking, Theta's next-door neighbor (at about 3° to its northeast), is the faint (m-4) star (φ) Phi (#30) Ursae Majoris, a binary whose components are about 13 astronomical units apart (a distance somewhat greater than that of Jupiter from the Sun), the period of revolution being about a century.

TANIA BOREALIS (λ) **LAMBDA** and *TANIA AUSTRALIS* (μ) **MU** together mark the Bear's "left" (more "forward"—i.e., western) hind foot, their names deriving from the Arabic *Al Kafzah al Thānīyah,* or "The Second Spring" (of the Gazelle as it escapes from Leo the Lion)! The Chinese name is *Chung Tae,* meaning the "Middle Dignitary," illustrating how differently the same sky pictures can be interpreted by varying national cultures.

One hundred and fifty light years distant, λ, the more northerly of these neighbors (RA 10 h 14 m δ 43° 04'), is a single white (A2 IV) star of magnitude 3.45; in contrast, μ (RA 10 h 19 m δ +41° 45') is a *red giant* (spectral type M0 III), 105 light years distant from us in space and about 50 $L \odot$. The term red giant indicates a very old, diffuse, and greatly distended star that has nearly completed its evolutionary cycle. We will say much more about the fascinating subject of red giants and the much more luminous, distended red supergiants in later chapters. Mu has an interesting dual nature: the spectroscope reveals two objects very nearly in contact, only 14.5 million miles from one another, circling their common point of mass every 230.089 days.

ALULA BOREALIS **and** *ALULA AUSTRALIS* (ν) Nu and (ξ) Xi Ursae Majoris, 1.6 degrees apart, mark the Bear's right, or hindmost, foot, and together these third-magnitude stars delineate the southernmost part of Ursa Major. Along with the two Tanias, they enclose the minor constellation of Leo Minor, or the "Little Lion," and with a bit of imagination on our part, it appears as though the Bear is hopping over him!

Alula Australis, the southernmost star of Ursa Major (RA 11 h 16 m δ 31° 49'), is worth mentioning because it marks another milestone in the history of astronomy. One of the nearest binaries, its two fourth-magnitude solar-type components, which are only 26 light years from us (at an average separation of 2"), were first discovered by Sir William Herschel in 1780. Forty-eight years later, it was the first binary for which the orbit was computed (this the work of Felix Savary), showing a period of 59.74 years.

Spectroscopic examination later disclosed that each of its components is doubled: Xi A's components are only 38 million miles in separation, and their period of revolution is 669.17 days; Xi B was found to have a nearly circular orbit, its components completing one revolution around their center of mass every 3.98 days. Thus, once again, we find a quadruple star system within the realm of the Bear.

Proceeding northward from the two Alulas, we arrive at the point where the Bear's hind legs cross, marked by third-magnitude *(ψ) PSI,* about 10 degrees below the Dipper's bowl, forming the apex of a long triangle with β and γ; δ and γ also point the way to ψ, making it easy to locate. Psi is a yellow giant, of spectral class K, about eighty-five times the Sun's luminosity and 130 light years distant. Compare the brightness of ψ to δ (Megrez), in the Dipper, which is somewhat fainter.

Contact Binaries

In the year 1933, a new type of binary was discovered in Ursa Major. *UX URSAE MAJORIS,* in the handle of the Big Dipper midway between Mizar and Alcaid, was found to have the shortest period of revolution then known, 4 hours 43 minutes. With a magnitude range of 12.7 to 13.8, it consists of two small but dense components, each one half the diameter of the Sun, but twenty-five and fifteen times its density, respectively, showing periodic increases and decreases in light owing to alternating eclipses of one component by the other. The primary eclipse, i.e., that in which the greatest dimming of light occurs, lasts about forty minutes. This type of variable star is called a *dwarf eclipsing binary.* The components were at one time "average" stars like the Sun, but they are now gradually evolving into compact *white dwarfs* as they approach this final stage in the life of all stars similar to the Sun in mass and chemical composition. Separated by only 1.2 million miles, they are affecting one another strongly; this transfer of material from one star to the other will result in a nova-like eruption in the UX system, perhaps in the near future, particularly because of great similarity between its light curve and that of famous DQ Herculis, the brilliant nova of 1934.

Right on the tip of the Bear's nose we find another erratic variable, a dwarf nova of the cataclysmic type. *SU URSAE MAJORIS* is very far away from us, 1,200 light-years in fact, or 368 parsecs, about fifteen times more distant than the average Big Dipper star. In other words, our Sun is much closer to the stars of the Dipper than any of the major stars of that constellation are to SU.

At 200 light years, however, and therefore much closer to our Sun than SU, we find the dwarf eclipsing binary lettered *W,* situated 2.2 degrees north-northwest from ϕ (p. 43). The components of W are similar in stellar type to our Sun (late F); the secondary is 60 percent as massive as the primary. Enclosed in the same gaseous envelope, they are separated by only 1 million miles from center to center, and because of their strong gravitational tidal action, they are egg-shaped. These tiny "Siamese twins" of the heavens take only eight hours to complete one revolution, during which two eclipses occur, one of them total. Like the close components of UX, those of W are believed to be the forerunners of the explosive SS Cygni stars (see p. 282), a type of nova-like contact binary found in the constellation of the Swan.

Symbiotic Partners

About 2 degrees east-northeast of (χ) Chi, on the Bear's hindquarters, and some 8 degrees to the east-northeast of ψ, lying more or less on a line with these stars (and south-southeast of γ), we find the variable *BE URSAE MAJORIS* (RA 11 h 55 m δ +49° 13′)   m 15, also known as *PG1155+492* (a designation type for radio stars) because of its ultraviolet excess and its unusual spectrum. Approximately 150 parsecs distant, BE is a close (but detached*) binary which, because of its extraordinary properties, was under scrutiny by astronomers B. Margon (University of Washington) and R. Downes and J. Katz (University of California) during March and April of 1981 at Kitt Peak

---

* *I.e., the components are not exchanging mass.*

Observatory. BE Ursae Majoris consists of one of the hottest known white dwarfs (at least 100,000° K.), with a probable mass of 0.7 $\mathcal{M}\odot$, and a cool, class-M red main-sequence star of 0.4 solar mass, their period of revolution just 2.29 days.

The atoms that constitute the gaseous material in the upper layers of the cool red star are ionized (i.e., stripped of some of their electrons—see p. 23) by photons of intense ultraviolet radiation emitted by the white dwarf, which results in a spectrum for BE that varies between emission and absorption (p. 21) on a time scale that apparently corresponds to its light variations (the latter, however, are not due to the eclipses of the system). The spectral variations occur because radiation from the white dwarf is absorbed by the facing photosphere of the red dwarf and then reprocessed by it, producing a strong emission spectrum as the red component reappears after an eclipse and we observe its irradiated face. Although there seems to have been little of the mass transfer between these components that we find in cataclysmic variables, as typified by SU Ursae Majoris (above), U Geminorum (p. 151) and SS Cygni (p. 282), Margon believes that BE Ursae Majoris may be a possible forerunner of this type of binary. Further studies are awaited with interest.

A Pulsating Red Star

In 1853, a year before he developed the modern magnitude scale, Pogson discovered the first long-period variable known in Ursa Major, R Ursae, 7 degrees north-northwest from Dubhe. R is a pulsating red giant similar to the famous variable "Mira the Wonderful," in Cetus ("The Whale"), which was the first variable star of its kind discovered. R Ursae Majoris is very remote, about 1,350 light years; therefore, although its luminosity is 250 times that of the Sun, its magnitude range is high: 7.5–13; thus it cannot be viewed with the naked eye. Strangely, its reddish color is strongest when R is faintest; when the star reaches its smallest magnitude (and is therefore brightest), the red color fades.

Galaxies

We have seen how stars tend to form multiple gravitational systems, closely bound "families" of two, three, or more components. These families, along with various "single" stars, also group together in clusters like the Ursa Major moving cluster, "local communities" that originate in the same gas/dust clouds and have "common motions."

There is, however, a much higher order of organization among stars, which may be likened to a "city-state" or a "nation." We call this supersystem a *galaxy*. The Sun and the Solar System, together with all the stars we can perceive with the unaided eye, plus billions of telescopic stars, are loosely bound together into such a system, which we call "the Galaxy." Containing every conceivable type of celestial object, this huge stellar conglomerate revolves slowly about its own center of mass. Stars in the direction of the central area of the Galaxy are "packed" more closely together than those at the periphery. They also are much older, belonging to an ancient generation of stars astronomers call *population II*. The majority of stars (including our Sun) that lie in the less concentrated areas of the Galaxy surrounding this central core were born out of the debris of exploded population II stars and constitute a younger generation, designated *population I.*

The ancient, concentrated *central core* of our Galaxy is globular in shape, but because we view the Galaxy edge on, from our position in it, it appears as a long, misty band. This great luminous path, which we call the Milky Way, (fig. 19) forms an extended gradual arc as it stretches northward across the heavens all the way from Canis Major and Auriga to Cassiopeia and Cepheus, there dividing into two branches: one passing through Cygnus and Aquila and winding southward again through Sagittarius; the other, rather broken up, extending through Lyra and Ophiuchus, eventually rejoining the first branch in Scorpius and Norma, far to the south. The Milky Way then sweeps onward through Lupus, Centaurus, and Crux (the Southern Cross), where it is most brilliant. Later on in our exploration of the constellations, we will encounter portions of the heavens very richly filled with Milky Way star fields.

Ancients named this immense encompassing band the "Circle of the Galaxy" and also the "River of Heaven." Its glowing mystery was bound up with numerous myths, among them the belief that Juno created the white path when she spilled some of her milk into the sky while nursing the infant Hercules. In fact, the word "galaxy" is a reflection of this ancient tale, being derived from the Greek *galaxias,* or *galaktikos,* which means "milky." Nevertheless, in accounting for the true nature of the "galaxios," as the Greeks called the Galaxy, the philosophers Pythagoras (ca. 580–500 B.C.) and Democritus (460–370 B.C.) rejected the fanciful explanation of Juno's maternal difficulties. Both guessed instead that the Milky Way is really composed of a vast number of stars that merge optically to appear as a misty band because of their great distances. Centuries later, Galileo's telescope confirmed this clever hypothesis.

Scientists in the past century or so correctly conceived of the star system to which our Sun belongs as lenticular but erroneously believed that the Sun itself occupies a central position. We now know that the Sun lies near the periphery of this vast system; the central region of the Galaxy, some 30,000 light years distant, appears to us as the band of the Milky Way. From this central galactic core originate two winding stellar "arms," giving the system a spiral form. (Our Sun is near the inner edge of the middle galactic arm.)

The hierarchy of stellar organization in the universe does not stop here. About a hundred "globular clusters" circle the core of our Galaxy like a "halo." Each of these is a system unto itself, containing hundreds of thousands or even millions of stars. Somewhat more distant than the globular clusters but still within the "suburbs" of our galactic community are two irregularly shaped structures called the Greater and Lesser Magellanic Clouds, (pp. 389, 392) after Ferdinand Magellan, the first European explorer to observe them, in the southern hemisphere. These great star clouds (each a galaxy in its own right) contain billions of suns of every conceivable type. Much farther away in space, at distances conveniently expressed in megaparsecs, we find vast numbers of super star systems, huge galaxies like our own, some even exceeding its dimensions.

Most of these distant galaxies, or "island universes" as they used to be called, can be seen only with the telescope, as tiny patches of "mist," and the details of their varying forms and stellar content can be resolved only with the most powerful optical instru-

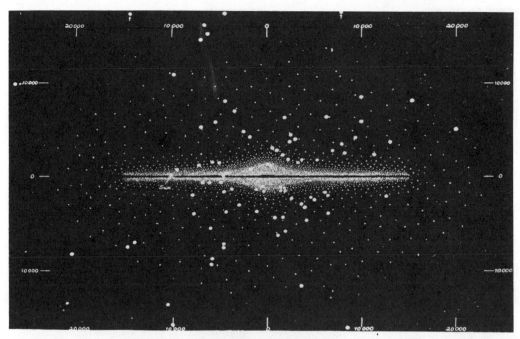

19. *Diagram of the Milky Way system* edge on, showing globular clusters, stars, and the location of the Sun. (*Yerkes Observatory photograph, University of Chicago*)

ments. For this reason, the astronomers who first viewed them, using the earliest telescopes, imagined these objects to be gaseous nebulae, clouds of gas and dust. Strangely enough, this misconception led to the compilation of the earliest and one of the most useful catalogues of nebulae and galaxies that we possess, especially for amateur observers in the northern hemisphere.

Charles Messier (1730–1817), an astronomical observer working at the Hôtel de Cluny Tower Observatory, in Paris, under the guidance of the astronomer Joseph Nicolas Delisle, set out to make a comprehensive record of all the comets he could discover. Annoyed with the "interference" of many misty "spots" he tended to encounter in this search, Messier decided to make a compilation of them so that other comet seekers might avoid these patches of light and save time in the all-important search for true comets. His revised listing of 103 "nebulae" was issued in 1784 and became known as the Messier Catalogue. It was ultimately recognized that the objects he had compiled were much more important than the comets he had so arduously sought. Without realizing it, Messier had observed and recorded the greatest and the most all-inclusive objects that we know of in our universe: its galaxies.

William Herschel continued Messier's work and compiled his own listing of 2,500 nebulae, and his son John expanded this in his General Catalogue of Nebulae (1864) to include 5,079 objects. This catalogue was in turn revised and enlarged to include 7,840 clusters, nebulae, and galaxies by J. L. E. Dreyer in 1888, which he listed by number in his New General Catalogue: NGC; this was followed by two supplements, the Index Catalogues: IC. Although it was suspected that many of the "nebulae" were really "island universes," these early cataloguers did not know the exact nature of the celes-

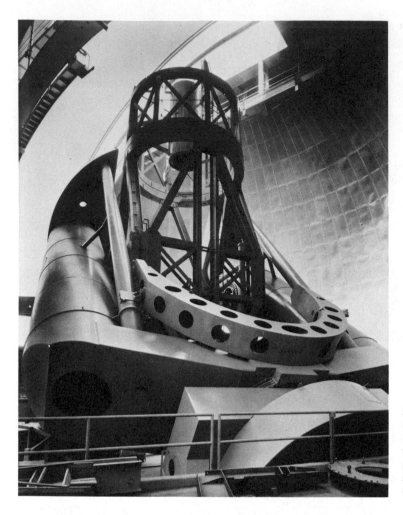

20. *The 200-inch Hale
telescope* on Palomar
Mountain; one of the
world's largest telescopes,
pointing to the zenith;
seen from the south. (*Hale
Observatories*)

tial objects they had listed. Nevertheless, the Messier and NGC listings are widely used today.

Galaxies display a variety of structures, believed to be related to their age and their evolution; our own spiral-shaped system is one of the more frequently found types. A number of interesting "spirals" can be viewed within the boundaries of Ursa Major, all, of course, at great distances from the stars of the Bear. The billions of stellar objects that belong to these galaxies evolve, live out, and end their existence millions of light years from our Galaxy, separated by the tremendous mileage of intergalactic space from such "nearby" stars as Dubhe, Merak, and company.

In pointing out a few spirals optically situated "behind" the star fields of the Bear, we caution the amateur viewer not to expect too much from a small telescope; even a fairly large instrument cannot reveal all the beauty and detail shown in long-exposure photographs made by the world's greatest observatories.

GALAXY M101 (Messier number), or *NGC 5457* (RA 14 h 01 m δ +54° 35′), lying east of Mizar and northeast of Alcaid, and forming a slightly flattened triangle with these two Big Dipper stars, is one of the oldest types of spirals, its arms "loosely wound" from its central core. Of the tenth magnitude, it appears on the meridian for northern observers on June 7 and is suitable for small (10- or 12-inch) mounted telescopes.

The unsuspecting Messier described it as a "nebula without star"; its mass, however, 250 billion $\mathcal{M}\odot$, is 10 percent greater than that of our Milky Way; thus M101 represents the combined luminosity of two hundred fifty billion stars like the Sun. Its spiral arms contain multitudes of hot, young, hydrogen-rich stars, and their light gives it a very blue appearance. This remote system is close to 15 million light years in deep space, and its diameter is approximately 90,000 light years across, one of the largest of its type. In its knotty arms are many condensations that were erroneously given separate NGC numbers; however, several small galactic companions do accompany it. Thus the familiar pattern of binaries—stars that have gravitational companions—is continued into the equally sociable galactic hierarchy, wherein major star systems travel in the company of smaller ones.

Three notable supernovae lie in M101. The first of these old exploded stars, which reached only the twelfth magnitude (a considerable luminosity, however, considering the distance of the galaxy), was discovered in 1909 and is now called SS Ursae Majoris; the second, very faint, appeared in 1951 (see fig. 21); and an eleventh-magnitude supernova was discovered in 1970. All these occurred in stellar condensations at the galaxy's circumference.

GALAXY M81, NGC 3031 (RA 09 h 52 m δ +69° 18′), lying northwest of Dubhe and just behind the Bear's ears, is a middle-aged (Sb) spiral similar to our own Galaxy and to the famous Andromeda galaxy (p. 225). It displays a bright, glowing nucleus with very faint spiral arms, which assume greater definition on modern long-exposure photographs (fig. 22). About 7 million light years, or 2.1 megaparsecs, distant, it is one of the densest galaxies known, in contrast to the distended M101. The longest diameter of M81 is about 36,000 light years; its mass is about 250 billion $\mathcal{M}\odot$, somewhat greater

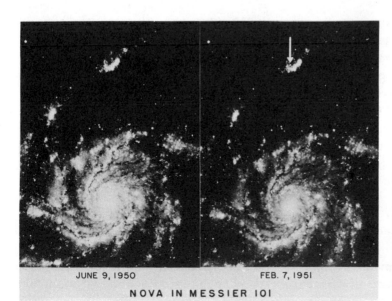

JUNE 9, 1950   FEB. 7, 1951

NOVA IN MESSIER 101

21. *M101 (NGC 5457),* a striking example of a spiral galaxy, photographed with the 200-inch Hale telescope before and after the appearance of the supernova of 1951 (arrow). (*Hale Observatories*)

22. *M81 (NGC 3031), a spiral galaxy in Ursa Major.* One of the densest known galaxies; with its bright center and delicate, faint arms, it has a cameo-like appearance in this photograph made with the 200-inch telescope at Palomar Mountain. (*Hale Observatories*)

than that of the Milky Way; and its luminosity is 250 billion $L \odot$. M81 contains many red and yellow giants, middle-aged and old stars, which give it a yellow color. A variety of other types of stars have been identified in M81, including blue and red variables, cepheids (see p. 225), and novae: explosive stars whose eruptions are more limited than those of the cataclysmic supernovae.

GALAXY M82, NGC 3034 (RA 09 h 52 m   δ +69° 56′), 38′ north of M81, is a so-called "peculiar" radio galaxy, irregular in structure. Best seen on April 4, it looks like a curved splash of light in smaller telescopes, but long-exposure photographs, made with great reflectors, reveal an elongated, highly patterned, bright nebulous mass covered with dark patches and rifts. The 200-inch Palomar telescope achieves an even wispier effect, which resembles a feathery bird in flight, and, we think, might earn it the name of "Dove" galaxy. (fig. 23).

Not so dove-like, however, were the events that led to the present appearance of M82 and its strong radio emission. Approximately 1.5 million years ago, an extremely energetic expulsion of materials from its center occurred, causing the formation of a massive system of filaments which are still expanding outwardly at a rate of about 600 miles per second. The exact nature of this colossal explosion is a mystery, but some

23. M82 (NGC 3034), the "Dove" Galaxy in Ursa Major; a peculiar radio galaxy that owes its strange appearance to a powerful explosion in its center 1.5 million years ago. (200-inch Palomar telescope photograph from Hale Observatories)

theorists suggest that it was initially *isotropic* (moving equally in all directions), but as the debris collided with the outer gas in the galactic core it was forced to move poleward. In this case, relativistic (extremely high-speed) electrons spiraled in the magnetic field, producing the strong radio emission and "continuous" optical emission (p. 21) recorded in the halo of the galaxy. James Fowler in 1963 postulated, however, that the initial explosive mechanism involved the gravitational collapse of a massive ensemble of gas.

"M82 may be the first recognized case of a high-energy explosion in the central region of a galaxy, and may explain many extra-galactic radio sources," suggest R. R. Lynds and A. A. Sandage (May 1963 *Astrophysical Journal*).

Galaxy Clusters

We have noted the tendency of stars to associate with companions and to form multiple systems of various sizes; we have also observed how all these star systems are bound together in great conglomerates called galaxies. In an apparent universal plan characterized by stellar systems of ever increasing dimensions, galaxies also combine gravitationally to form their own supersystems; each of these may contain anywhere from a dozen to thousands of galactic members. These clusters of galaxies have been roughly divided into two classes: regular and irregular. The regular are round and symmetrical, with dense central regions; they contain a minimum of one thousand members, mostly the *elliptical* type of galaxy.

The irregular, as implied, have no definite shape and little central concentration; they are of every type of galaxy. Our own Galaxy and some seventeen galactic "neighbors," within a region 3 million light years in diameter, form a small irregular-galaxy cluster called the Local Group. Some of the best-known members of the Local Group

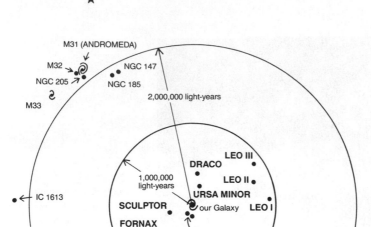

24. *The Local Group of galaxies.*

are the great M31, in Andromeda; M33, in Triangulum; NGC 6822, in Sagittarius; and the Magellanic Clouds, in Dorado and Tucana.

The M81/M82 Galaxy Group

M81 and M82 form the nucleus of a small galactic group, which, at about 7 million light years, may be the closest to our own Local Group. *GALAXY NGC 3077,* lying 45 minutes east-southeast from M81, is a tiny companion of the eleventh magnitude; with a diameter of approximately 6,000 light years, it is considered a "dwarf" among galaxies. Another dwarf nearby to M81 (about 1.4° to its south-southwest) is NGC 2976, an elliptical system of mottled appearance. There are several other faint galaxies in the northwest regions of Ursa Major, including *NGC 2685* (fig. 25), notable for its encircling outer ring (R.A. 08 h 52 m, δ +58° 59'), a few degrees southeast of Omicron.

Other Galaxies in Ursa Major

One of the best examples of a very regular and symmetrical "middle-aged" spiral galaxy is *NGC 2841,* lying 1.71° west-southwest from the star Theta, in the Bear's foreleg (RA 09 h 19 m   δ +51° 12'). Of the tenth magnitude, it must be viewed with a mounted telescope no smaller than 10–12 inches.

Certain spirals display a band of glowing material that extends across the nucleus; the arms of this type extend from the opposite ends of the band. These are known as *barred spirals.* An example in the Bear is the eleventh-magnitude object *NGC 3992* (fig.

25. *NGC 2685, a ringed spiral galaxy* in Ursa Major. Classed as a barred spiral, this galaxy has a strange, elongated shape reminiscent of the famed Spindle Galaxy, in Sextans; the dark dust lanes of its unusual series of encircling rings are clearly visible, overlying the bright central galaxy. (*200-inch telescope at Palomar Mountain, Hale Observatories*)

26), located in the field of (γ) *Phecda* (40′ to its southeast and appearing on the meridian May 6 (RA 11 h 55 m   δ +53° 39′).

A True Nebula

When Charles Messier drew up his list of "fuzzy" obscurations, he gave them the Latin name for "clouds," i.e., *nebulae*. In their catalogues, the Herschels continued this nomenclature for all these hazy objects, although they suspected that many of them were really condensations of great numbers of stars. This of course proved to be true, and eventually such stellar conglomerates came to be known as "galaxies." However, many of the original "nebulae" did not resolve into star clouds with higher magnification, but are true aggregates of interstellar gas and dust.

Like galaxies, the nebulae were classified mainly into two groups: "dark" (wherein relatively dense clouds of solid grains absorb the surrounding starlight) and "bright" (particularly dense dust clouds in the region of hot, young stars that illuminate them).

26. *NGC 3992, a barred spiral galaxy* in Ursa Major, showing typical Theta structure with graceful, well-formed spiral arms, photographed with the Shane 120-inch reflector, at Lick Observatory.

We will have more to say about bright and dark nebulae in the chapter on Orion and its Great Nebula.

There is a third type of nebula, first discovered in the constellation of Lyra by the French astronomer Antoine Darquier in 1779. Although called a *planetary nebula* (because its green color reminded Herschel of the planet Uranus), it is actually a large shell of expanding gas that has been ejected from a central star approaching its final, white-dwarf stage (see Chapter VIII, "Three Tragic Myths," the Ring Nebula).

In 1781, two years after the discovery of the famous Ring Nebula, the astronomer P. Mechain turned his telescope toward a region southeast from the "pointer" Merak in the bowl of the Dipper (RA 11 h 12 m, $\delta + 55°$ 18'). There he was greeted by a great "owl": a spherical but rather dim and hazy starless object. Sir John Herschel saw it as a "large, uniform nebulous disk"; its owl-like appearance was later noted by Lord Rosse, at Birr Castle in 1848, who first measured its dimensions. One of the largest "planetaries," *M97,* or *NGC 3587* (fig. 27), has a diameter of about 3 light years (150"); its distance has been variously estimated, averaging about 3,000 light years. The extremely hot, compact central star has a surface temperature of perhaps 85,000° K., with

27. *M97 (NGC 3587), the Owl Nebula.* In this photograph made with the 60-inch telescope on Mount Wilson, the central star and two dark areas (the "eyes" of the Owl) are clearly visible. The structure of the expanding shell is very complicated, and intense radiation of the central star keeps it in a state of turbulence. (*Hale Observatories*)

a tiny radius only 4 percent that of the Sun. The entire nebula has a mass of about .10–.15 $\mathcal{M}\odot$. Despite its large size, the Owl is of about the eleventh magnitude, its central star of the fourteenth. A 12-inch mounted telescope is therefore required for viewing it. In the same field, 48 minutes northwest, may be seen the edge-on spiral *NGC 3556,* also discovered by Mechain.

A Modern Discovery: The Lens in Space

Among the most intriguing mysteries in our universe are the extremely distant, high-energy "starlike" objects called *"quasars"* (for "quasi-stellar") or simply, "QSOs." For their apparent size, QSOs emit abnormally strong radio waves, and for this reason they are of great interest to radio astronomers. Data concerning the velocities of recession which their spectra reveal are also vital to a correct estimate of the rate of expansion of the universe. QSOs are not rare: about one of these puzzling celestial bodies is present in every 30 square degrees of sky patch the size of the Big Dipper's bowl.

Early in 1979, the astronomers D. Walsh, R. F. Carswell and R. J. Weynam were collaborating at Kitt Peak Observatory, the University of Arizona, in a search for optical counterparts of the QSOs which were then being recorded by the great radio telescope in a survey at Jodrell Bank, Manchester, England. Under study was a pair of QSOs known since the 1950s as the "Twins," situated several degrees northwest of the fourth-magnitude star (φ) Phi (p. 43), and slightly southwest of fifth-magnitude Fl. #17. The three astronomers made a spectrogram which revealed a pair of blue QSOs with nearly identical spectra, 6 seconds of arc apart, receding at 70.7 percent of the speed of light, apparent magnitude +17: twenty thousand times fainter than the most obscure naked-eye star! The "Twins"† are identified by ($\alpha 0907\ \delta + 561$), a designation based upon their coordinates.

After extensive analysis, the Kitt Peak astronomers concluded that these QSOs are not really "twins" at all but, rather, two images of *the same object,* and that the original quasar is behind an intervening galaxy that acts as a gigantic gravitational "lens in space," splitting the QSO's image in two. This thesis, however, was rejected by other astronomers, who believed that 0907+561 A and B, the northern and southern images, are two objects in rotation about a common center of mass: a "binary quasar"!

On the nights of April 20 to 22, 1979, three weeks after the revealing spectrogram was made at Kitt Peak, the newly completed multimirror telescope, or "MMT," of the Smithsonian Astrophysical Observatory, at Mount Hopkins, Arizona, was activated. This instrument consists of six 1.8-meter telescopes joined in a single structure, the equivalent of one single-mirror telescope with a diameter of 4.5 meters! The scientists mounted a *spectrograph* (p. 21) at the focal plane of the MMT, the first time, in fact, that this procedure had been used with it. The high-quality result revealed even greater similarities between the Twins' spectra and led astronomers Walsh, Carswell, and Weynam to propose formally that A and B are gravitational-lens images of a single object.

Soon another highly developed and powerful telescopic instrument entered the drama of the Twins. On June 23 and 24 of 1979, the so-called Very Large Array, or VLA, at the National Radio Astrophysics Observatories, near Socorro, New Mexico, was used to obtain new data. The VLA normally consists of twenty-seven antennas each 25 meters in diameter; in effect, it has a total diameter of 27 kilometers and is two hundred times more powerful than any other radio telescope! At the time, only fourteen antennas were operating, but they still yielded an incredibly fine resolution: .8 second of arc. The observation showed two blobs or emissions, out to the northeast of the northern image, which were absent in the southern. Furthermore, analysis of the space between A and B revealed no galaxy that might operate as a lens; this was not conclusive, however, because elliptical galaxies are known to be weak sources of radio waves.

On the basis of these data, scientists of the Massachusetts Institute of Technology concluded that the "Twins," true to their name, are actually two separate quasars. In this controversy, optical astronomers tended to favor the gravitational-lens theory, while radio astronomers supported the "binary quasar" hypothesis, a more conventional explanation of the Twins' similarity. Its proponents postulated a cloud of matter (or gas) which they believed had surrounded these quasars from the time of their

† *(Don't try to find them with your telescope.)*

formation and thus produced identical absorption lines in their spectra.

An important breakthrough in the controversy was made by Alan N. Stockton, working at the University of Hawaii at Manoa, who announced that he had located the "missing galaxy" in the form of a bit of jetlike "fuzz," very faint, and almost coincident with Twin B. The galactic lens, he found, is one second of arc north of B and is the brightest member of a cluster of sixty galaxies lying halfway to the QSO, all of them playing a role in the formation of the twin images. Soon afterward, a group of astronomers at the Hale observatories (Palomar Mountain / Mount Wilson, CalTech) corroborated Stockton's discovery; they used the 200-inch telescope in combination with yet another new tool, the *charge-coupled device*, or "CCD," which is slated to replace the photographic plate as the astronomer's chief recording medium. The CCD is superior because it overcomes "saturation" on long exposures and can record images of vastly different brightnesses, still allowing the relative brightness of each object to be determined. A two-hour exposure of the Twins, using the CCD, revealed that the southern image, B, is elongated to its north by one second of arc, while A is circular. Furthermore, the Hale astronomers suggested that B is really two superimposed images of the original quasar, so close to each other that they cannot be resolved and we therefore see their combined light. Therefore B, although covered by the deflecting galaxy, is almost as bright as A, because the lens in space actually produces three images: A, the northern, and two overlapping southern images: B and C! The real quasar is a single optical point with two extended blobs of radiation to its northeast. The elliptical galaxy, or "lens," is displaced three seconds of arc south of the visual line connecting the original QSO and the Solar System. Thus our story concerning many "single" starlike objects in Ursa Major that are revealed as multiple systems completely reverses itself in the case of 0907+561: two, and even three, images prove to be one object!

The gravitational deflection of light waves in the vicinity of massive objects, which gives the lens effect, accords with Einstein's general theory of relativity, showing that gravity is a distortion of spacetime geometry. As far back as 1937, in fact, Fritz Zwicky, of CalTech, concluded that such a lens in space is possible and would provide the simplest and most accurate method for detecting nebular masses, although Einstein, in the same year, stated that there was "no great chance of observing the phenomena." Until the discovery of quasars, twenty-five years later, the space-lens effect had remained unexplored. However, modern technology made possible its ultimate revelation in the drama of the Twins. Still more recently, a second galactic lens, PG 1115+08, showing at least three images, was found in the southern part of the constellation Leo, not far from the ecliptic; as newer and more powerful instruments are turned toward the heavens, it is anticipated that similar discoveries will follow. Such gravitational lenses are of great importance, because their study enables scientists to obtain a more accurate picture of distances in the universe. What is even more significant, the reality of the lens in space demonstrates that general relativity can be applied to phenomena millions of light years away, as initially—and correctly—surmised by Einstein.

THREE

Other Polar Pictures

DRACO

The "Dragon" is a very ancient constellation, adapted by the Greeks to their story of the Titans' revolt against the Olympian gods. The Titans, who sought to free certain of their brothers imprisoned by Zeus in Tartarus, were giants with feet like serpent tails; almost invincible, they could not be overcome by any god. They seized rocks and firebrands and hurled them upward from the mountaintops; in the course of this fierce battle, Athena, goddess of wisdom, snatched one of the Titans' serpentine feet and whirled him to the sky, where he became entangled in the axis of the heavens and remains to this day. Draco's many other identities include the biblical snake that tempted Eve; a great dragon worshiped by the Babylonians; Hea, third god in the Assyrian triad; and the Euphratean dragon Tiamat, symbol of primeval chaos. On the other hand, the Chinese saw the constellation as *Tsi Kung,* or "Palace of the Heavenly Emperor."

THUBAN (α) *Alpha Draconis*
(RA 14 h 03 m   δ +64° 37′)   m 3.64

We begin our exploration of Draco "tail first," with this famous third-magnitude star. Situated at the beginning of the draconic appendage and about 214 light years distant, Thuban, a white star of class A0 III, has a luminosity 135 $L_\odot$; about 2,830 years ago, Thuban was near the north celestial pole (see *Precession,* p. 9). The Egyptians apparently constructed the descending passage of the Great Pyramid of Khufu, at Gizeh, to point directly at this star when it would pass 3° 34′ *below* the pole, i.e. about 3,475 B.C., some 645 years before its closest approach to the pole, so that the interior chambers of the pyramid served as an ancient observatory. There is some disparity, however, between the foregoing date and 2,600 B.C., the approximate date that marks the beginning of the IVth dynasty and the reign of Khufu. Owing to the margin of error of some 800 years that exists in these dates, we cannot be sure that Khufu's pyramid builders used Thuban as a point of orientation until the problem can be solved through a plausible theory that reconciles astronomical and archaeological data.

Like many stars we have already mentioned, Thuban consists of two components detectable only with the spectroscope; they are 20 million miles apart, less than a quarter of the distance between the Earth and the Sun. The brighter star shows a peculiar silicon spectrum. Through the years, observers of Thuban have reported apparent magnitudes ranging from 2 through 4, so that in the past it was suspected of variability, although not at the present time.

While exploring the realm of the Dragon's tail, we note the two rather faint stars at its end: (κ) Kappa and (λ) Lambda. The latter is an orange star, its proper name, Giauzar, having been derived from the Persian *ghauzar,* or "poison place," because it was incorrectly believed to mark one of the points, or nodes, where the Moon crosses the ecliptic, the apparent path of the Sun. These lunar nodes were regarded as poisonous, because they were thought to be situated in the Dragon's head and tail. (A good example of the bizarre conclusions one may reach when the reality of the constellations is taken too literally!) According to some catalogues, the tail of Draco does not end with Giauzar but extends nearly 12 degrees of declination farther north, to the faint star numbered "1" by the cataloguer Hevelius, lying about 9 degrees from Polaris. Over the centuries, in fact, the great Dragon has yielded up portions of its anatomy: as an example, the ancient Chaldeans are said to have visualized it with horns and claws; and with wings, these covering the area now occupied by Ursa Minor. In 600 B.C., however, Thales gave Draco's wings to the Lesser Bear, which he suggested as a new guidepost for Greek mariners, as we have mentioned.

Because Draco has always been a long constellation, even more extensive in ancient times than today, various parts of it were imagined as separate figures (asterisms) by different peoples, the observers interpreting these stellar images in terms of their own lives. To early desert dwellers, the Serpent's head resembled a herd of camels! Seen as a dragon's head, however, the star pattern of Beta, Gamma, Nu, and Xi has also been variously depicted, sometimes with the brighter stars β and γ as the eyes, and ξ as the lower jaw, but more recently with the head turned in profile, β marking an eye, γ an ear, and the faint star ν at the corner of the mouth; or else with the two brighter stars on the top of the head. The Dragon's head, however, when pictured face on, has much greater symmetry during winter and early spring than at other times, with γ (Eltanin) as Draco's long, orange nose, β and ξ as the eyes, and the faint star ν defining the peak of the head. When Draco is viewed in the late summer and early fall, this face is reversed, showing ν as a tiny, short nose, and the bright Eltanin as a hornlike structure on top of its head! In support of these conceptions, we offer our illustration (fig. 28).

RASTABAN (β) *Beta Draconis*
(RA 17 h 29 m   δ +52° 20')   m 3.0

The Dragon's eye is a yellow star (G2 II), like the Sun. Its Arabic name is derived from *Al Ras al Thu'ban,* meaning "The Dragon's Head." Although solar in spectrum,

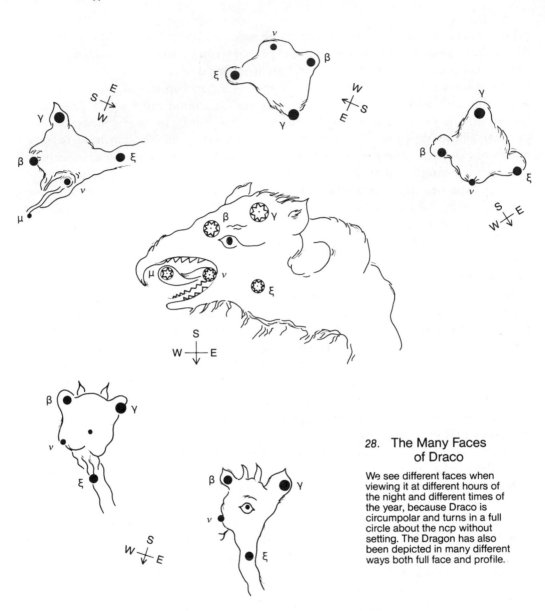

28. The Many Faces of Draco

We see different faces when viewing it at different hours of the night and different times of the year, because Draco is circumpolar and turns in a full circle about the ncp without setting. The Dragon has also been depicted in many different ways both full face and profile.

Rastaban, with a luminosity 600 $L \odot$, far outshines our tiny home star. About 300 light years from the Solar System, β has a very distant companion, of the eleventh magnitude, thought to be gravitationally related.

ELTANIN (γ) *Gamma Draconis*
(RA 17 h 55 m δ +51° 30') m 2.4

The most prominent of the Arabian "camels," originally known as Draco's "other eye," but the top of its head in modern charts, is second-magnitude star γ, named Etamin or Eltanin, from *Al Ras al Tinnin,* which translates as "The Dragon's Head." Eltanin is 110 light years distant, with a luminosity 145 $L \odot$; its late spectrum, of K5 III, gives it an unmistakable orange hue.

The Discovery of the Aberration of Light

Legends in astronomy sometimes tend to repeat themselves in slightly varied forms. Often, an astronomer, searching for one truth, finds another of equal or even far greater importance. We have told the story of the comet seeker Messier, who quite inadvertently and unknowingly compiled the first list of galaxies, which he erroneously imagined to be distracting blobs of haze. Similarly, in 1669, Robert Hooke observed a shift in the position of Eltanin which he thought to be parallactic. He really had observed the aberration of starlight, but this phenomenon was not correctly understood until 1725, when James Bradley, who had also been trying to measure Eltanin's parallactic shift, discovered that the apparent position of this star (which he had recorded) differed from its true position by about 40 seconds of arc, because of the Earth's motion around the Sun, combined with the velocity of light. Bradley's discovery enabled him to deduce a value for the speed of light (confirming the first such calculation, made fifty years earlier).

Aberration is most noticeable when the true direction to the star's position is at right angles to the direction of the Earth's orbital motion; the observer's telescope must then be tilted in the direction of the Earth's motion through a very small angle with respect to the true direction of the star if the star's image is to lie at the center of the field of the telescope (see fig. 29). The tangent of this angle is given by the ratio of the speed of the Earth to the speed of light. For stars whose directions from us are parallel to the Earth's motion, the aberration vanishes. For stars in intermediate directions, aberration can be calculated by taking into account only that part of the Earth's motion which is at right angles to the direction of the stars. The aberration is constant for the circumpolar stars, because their directions are very nearly always at right angles to the plane of the ecliptic and hence to the Earth's motion.

The aberration of light ultimately became the first of three basic proofs that the Earth is indeed a "world in motion," the other two being the parallactic shift of stars, discovered by Bessell at a later date with more sensitive instruments, and the Doppler effect (see Chapter IV). Aberration was destined to form one of the arguments in support of Einstein's special theory of relativity.

Although several faint stars are visible with the telescope or binoculars in the same field as Eltanin, it is a single star. Classed as metal-rich, or "Arcturian" (K5 III), which is one of the most numerous spectral types in our Galaxy, it is about 110 light years distant from us, with a luminosity of 145 $L \odot$.

Some Minor Stars in the Dragon

The faint star (v) *NU,* one of the four marking Draco's head, is a fine double star for small telescopes. Of approximately fifth magnitude, both of these white stars (class A5)

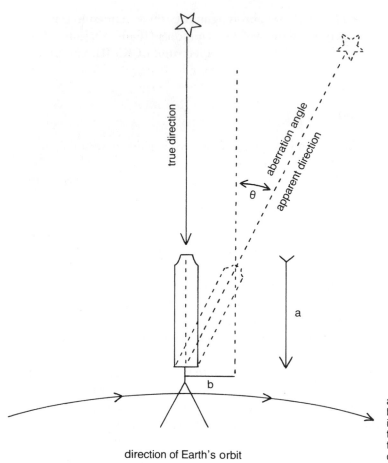

29. The aberration of starlight due to the Earth's annual motion.

Key

a = length of telescope
b = displacement of telescope
in a unit of time. The light
travels the distance "a" in the
time the telescope moves the
distance "b" (not to scale).

show very strong metallic spectral lines. The primary is also a spectroscopic binary, with a period of revolution of 38.5 days. About 5 degrees west-southwest of ν and forming a triangle with β and ν is fifth-magnitude (μ) *MU,* a long-period binary circled by a distant dwarf companion. On modern charts where Draco's face is shown in profile, faint μ marks his nose or tongue; but the Arabs included it in their herd of camels.

Another Arabic "camel," or "facial" star of Draco, is the fourth-magnitude (3.76-m) (ξ) *XI,* currently known as Grumium (called *Genam* by Ptolemy and in Bayer's catalogue). It must now be evident that the Greek lettering sequence of the Draconian stars bears a very imperfect relationship to their magnitude sequence. ξ, an orange K2 III giant, appears to have a very faint (m-15) gravitational companion, now believed to be a white dwarf or extremely dense "degenerate" star (pp. 131–2).

Dwelling momentarily on the "camel" scene, we point out (ζ) *ZETA* and (η) *ETA* on the Dragon's third coil, stars that early Arabia saw as two hyenas or wolves waiting to pounce upon the Mother Camel's Foal, which is a very faint star centered in the "herd" (Draco's head) and "protected" by the larger camels. The star ζ, so wolf-like in character to the Arabs, was known by the Chinese as *Shang Pih,* the "Higher Minister":

similarly, they gave η the title of *Shang Tsae,* or "Minor Steward," demonstrating, we think, the importance of matters of government and rank to the ancient Chinese. In contrast, the Arabian desert dwellers were closely attuned to the natural surroundings of a nomadic existence, ever aware of the importance of their beast of burden, the camel, and the presence of wild animals that threatened the herds.

Another asterism in the long constellation of Draco is the second quadrangle of stars that form part of the first two coils of the Dragon. It comprises the stars (δ) Delta, (ε) Epsilon, (π) Pi and (ρ) Rho, along with the extremely faint (m-6.5) star (σ) Sigma or Alsafi (the latter of interest because it is only 13 light years from us). This star picture differs greatly in the eyes of varying national cultures. Quite appetizingly, the Chinese people called it *Tien Choo,* or "Heaven's Kitchen." Nearby in the Dragon's coils are (φ) Phi and (χ) Chi, of magnitudes 4.2 and 3.7, respectively, a pair of Chinese "minor ministers" called *Shaou Pih* and *Kwei She*—perhaps on their way to dine in Heaven's Kitchen! Very appropriately, that section of the Dragon including Zeta, the Minor Ministers, and Heaven's Kitchen resembles a handsome "horn of plenty" (fig. 30). Lastly, we must not fail to mention the ministers' neighbor (ψ) Psi, named *Niu She,* or "Palace Governess," and also known as a "literary woman." Psi is a double star, the components 30 seconds apart, 4.3 and 5.2 in magnitude, pearly white and yellow.

(δ) DELTA DRACONIS
(RA 19 h 13 m   δ +67° 34')   m 3.2

Al Thais, meaning "The Goat," while only of the third magnitude, is the brightest star in "Heaven's Kitchen." About 120 light years distant, it has a late solar spectrum (G9 III) and is 75 times as luminous as the Sun.

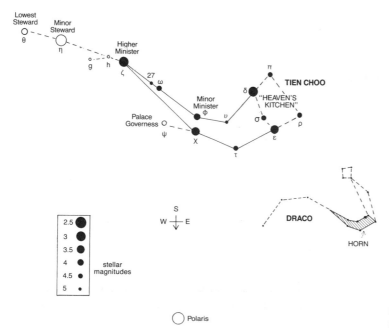

30. A Chinese "horn of plenty" in the Dragon.

Of Unusual Interest

Situated more or less in the neck of the Dragon, approximately 60 seconds from Draco's head and not far from the orange and blue contrast double (o) Omicron Draconis, is a pair of miniature red dwarfs numbered *Σ2398* (Otto Struve listing). They are only 11.3 light years distant, constituting one of the doubles closest to our Solar System. The tiny components have luminosities of .0027 and .0013 $L \odot$, and their masses are only .29 $\mathcal{M} \odot$ and .25 $\mathcal{M} \odot$, respectively. Of the ninth and about the tenth magnitudes, respectively, they can be viewed only telescopically.

NGC 6543: Midway between Delta and Zeta (RA 17 h 58 m, δ +66° 38′) is a small but very bright planetary nebula somewhat resembling the "face" of the Owl nebula, in Ursa Major. Its greenish color results from the "forbidden" spectral lines of doubly ionized oxygen, which we discuss in detail in connection with the famous "Ring" nebula, in Lyra (Chapter VIII).

The Schmidt 48-inch telescope at Palomar Observatory reveals a double shell of gas in NGC 6543: the present bright nebula, the product of a "recent" ejection, being enclosed in a very faint and diffuse gas shell, which means that the central star experienced a previous ejection of material in its distant past. This planetary nebula figures in astronomical history as the first to undergo spectroscopic analysis. On August 29, 1864, William Huggins made this important observation, which determined the gaseous nature of planetaries, previously thought to be clusters of unresolved stars.

If C. R. O'Dell's estimate of 3,200 light years for the distance of NGC 6543 is correct, then the diameter of the inner, bright shell is 20,000 astronomical units. The faint outer shell is then about 3.5 light years in diameter, a tremendous object when one considers that it takes light 3.5 years to travel through it! The ancient, central star is an extremely hot, O-type small star whose radius is one fourth the radius of the Sun, with a temperature of about 35,000° K. and a luminosity of 100 $L \odot$.

"Other" Galaxies in Draco

NGC 5985: a fine multiple-arm spiral

NGC 6643: many-armed spiral; looks like a splash

NGC 3147: more compact (younger) spiral

NGC 5907: edge on

The Story of Cassiopeia and Cepheus

Cassiopeia (meaning "cassia juice") was a beautiful black queen married to Cepheus (from *cepeus,* "gardener"), who was the Ethiopian king of Joppa, a city in the land of

Philistia (Phoenicia, later Palestine), northeast of Egypt. The royal pair was blessed with twenty sons, and two daughters named Aeropa and Andromeda. The latter was exceptionally attractive, having inherited Cassiopeia's renowned good looks. In those days, many gods of greater and lesser stature inhabited the Earth, all of them keeping close watch over mortal affairs, so that even the sea itself "had ears." Unmindful of this danger, the foolish queen boasted that both she and Andromeda were more beautiful than the fair Nereids, or sea nymphs.* These privileged spirits overheard Cassiopeia's insult and complained of it to their protector, Poseidon, god of the sea and brother of Zeus and Hades. Of a surly, quarrelsome nature, Poseidon was easily angered, and responded by sending a raging flood and a female sea monster† to devastate Philistia. King Cepheus quickly consulted with the powerful Oracle of Ammon, who advised him that he could assuage Poseidon's wrath and save Philistia only by sacrificing his daughter Andromeda to the monster.

Luckily, at this time Perseus, son of Zeus, was returning home to Greece from a successful quest in which he had destroyed the Gorgon Medusa, a creature so frightful that all who gazed upon her hideous face were turned to stone. Carrying the severed head in a magic wallet and wearing winged sandals which were the gift of three Stygian nymphs, Perseus flew over the coast of Philistia. Suddenly he was startled by a strange spectacle on the shore below: from a safe distance, a mob of angry subjects cried for the immediate sacrifice of the princess; the greatly distressed king and queen stood near the water's edge, while above them, chained to a rock and clad only in her royal jewelry, beautiful Andromeda awaited the deadly approach of the sea monster. Overwhelmed by the sacrificial victim's loveliness, Perseus approached her despairing parents and offered to rescue her, but only on his terms: she must become his wife and accompany him back to Greece. They quickly agreed, and Perseus beheaded the monster.

Although Andromeda insisted on honoring the marriage pact, Cassiopeia conspired with Poseidon's son Agenor, or, as some say, with Cepheus' brother Phineus, who interrupted the wedding festivities with armed troops in an attempt to kill Perseus and exercise his own, prior claim to the princess' hand. Greatly outnumbered, Perseus exposed the Gorgon's head, turning his attackers to stone.

At a much later date, Cepheus and seventeen of his sons were slain in Heracles'‡ successful expedition against Sparta. The archives of mythology are silent on Cassiopeia's earthly fate. After their deaths, however, Cepheus and Cassiopeia were transported to the heavens by Poseidon himself; but to punish Cassiopeia for her treachery, he seated the vain queen in a circumpolar area which rotates perpetually about the north celestial pole, causing her chair to turn upside down every winter! Ineffectual and helpless, Cepheus stands nearby, his position appropriately marked by faint stars of the third and fourth magnitudes.

---

* *Daughters of Nereus (a prophetic old man of the sea) and attendants on the sea goddess Thetis.*
† *Identified by some mythologists as Draco.*
‡ *Hercules.*

CASSIOPEIA

The principal stars of the Queen form a conspicuous *M* in the late fall and winter months, when her chair is above the pole, and a *W* in the summer, when below it; thus this starry letter serves as a convenient guidepost in the northern skies. For example, from Caph (β) Beta Cassiopeiae, at the west end of the *W*, we may locate Polaris, because Caph is very close to the *equinoctial colure,* the hour circle that intersects the ecliptic and the celestial equator at points 0 and 12, the vernal and autumnal equinoxes, respectively. By following this circle northward from Caph for a distance of about 30 degrees of declination, or somewhat more than twice the breadth of the *W* of Cassiopeia, we arrive at the pole star (see fig. 31). Lying to the south of Cassiopeia and delineating the eastern corners of the Great Square of Pegasus, the bright star Alpheratz and its southern neighbor Gamma Pegasi also mark this colure; together with Caph they have therefore been known as the "Three Guides." Centuries ago, owing to the precession of the Earth's axis, the three stars were exactly on this great circle, but in future years they will move farther away from it as their right ascension gradually increases.

The letters of Cassiopeia's principal stars, β, α, γ, δ, ε, and ι, form the "word" *B-A-G-D-E-I,* a traditional aid in memorizing their order.

Cassiopeia and Cepheus (fig. 32) are linked to the autumn constellations of Andromeda and Perseus (Chapter Twelve) by their adjacent location and, as indicated, by their common mythological story. Because the circumpolar constellations can be viewed all year long, we will return to them for occasional orientation in subsequent chapters.

The W of Cassiopeia

The Romans called her *Mulier Sedis,* "The Woman of the Chair." At the back of this "chair" and upper right (western) end of the *W*, we find **CAPH** (β) ***Cassiopeiae*** (RA 0 h .06 m   δ +58° 52')   m 2.25, a white subgiant of class F (F2 IV). Caph is 45 light-years distant, its luminosity 19 *L* ⊙, listed as a spectroscopic binary whose components are

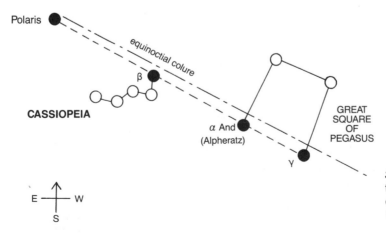

31. How to find Polaris from Caph and follow the colure to the Square of Pegasus.

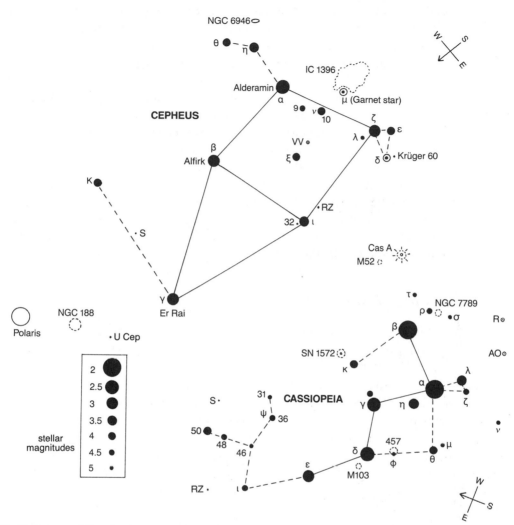

32. Cepheus and Cassiopeia.

quite close. The Persians, somewhat to the south, saw Cassiopeia's prominent *W* on their northern horizon, and they imaged it as a kneeling camel; to them, Caph was the camel's hump! Of interest is Caph's slight variability, because it has the lowest apparent magnitude (and is probably the closest member) of a rare class of short-period pulsating variables called Delta Scuti stars (after δ in the modern constellation Scutum, or "Sobieski's Shield.") These large, aging stars have rather complicated cycles of light variation. Epsilon in Cepheus also belongs to this class.

At the last trough of the *W* lies the star Schedar, or Schedir (from *Al Sadr* or "The Breast"), (α) *ALPHA Cassiopeiae* (RA 0 h 37 m   δ +56° 16′)   m 2.23. Schedar is an orange star of spectral class K, many times the Sun's size and mass, and about 230 *L* ⊙. Schedar is not a true double, but has three faint optical companions at varying distances. Southwest of α are the binaries (ζ) *ZETA* and (λ) *LAMBDA,* by tradition marking the queen's face. On the central hump of the *W* we next find (γ) *GAMMA,* which has no

Arabic name but was known to the Chinese as *Tsih,* or "Whip" (RA 0 h 53 m, δ +60° 27') m 2.40. It is of especial importance in the history of astronomy, because the bright emission lines of hydrogen in its spectrum were the first to be discovered in any star (by Father Secchi in 1866). As a B0 subgiant (100 $L \odot$), it would normally be expected to produce only dark absorption lines originating in helium atoms (p. 22). In addition to its spectral peculiarity, Gamma is also quite erratic in its general behavior, changing its magnitude, spectral lines, color, temperature, and diameter. In 1937, the time of its largest drop in temperature, it apparently ejected a great gaseous shell. There seems to be no indication, however, that it is a nova. Gamma Cassiopeiae also has a faint visual companion, but the latter is visible only in a large telescope.

Cassiopeia's "knee" (as the Arabic peoples saw her), the first trough of our *W,* is represented by **RUCHBAH** (δ) (RA 1 h 22 m δ +59° 59'), a normal "Sirian" star, 45 light years distant, whose very faint variation may be due to partial eclipse by a companion. Ruchbah marks still another "first": the use of the telescope (in 1669) for determining latitudes on the Earth (the star was used to measure an arc of the meridian).

Fainter Stars in and Near the W

EPSILON (ε) Cassiopeiae (RA 1 h 51 m δ+63° 25') m 3.38, marks the queen's leg and the beginning (or left-hand, northeastern end) of the *W.* Much more distant than the other stars of the *W* (520 LY), its high luminosity exceeds that of the Sun by a factor of a thousand. Beyond the bright *W,* but following along the same line of direction after δ and ε, we find the fainter star (ι) **IOTA** (RA 02 h 25 m δ+67° 11'), m 4.51, perhaps marking the hem of the queen's stiff robe. In the spring, when ι is uppermost, Cassiopeia suffers her greatest discomfort, her head hanging downward near the horizon. Iota is of special note as a fine triple star, its components representing the consecutive spectral classes A, F, and G; the primary a white (Sirian) star, with a variation in its spectral lines of chromium and europium, related to an intense magnetic field. Such rare variables are called magnetic-spectrum variables of the Alpha Canum Venaticorum type, after Cor Caroli, the brightest star in Canes Venatici and the classic example. Alioth (ε Ursae Majoris) is also included in this rare group of stars.

Optically near Schedar we find the beautiful contrast double (η) **ETA** (0 h 46 m δ +57° 33'); magnitudes of 3.5 and 7.2 make it an easy object for the small telescope or binoculars. Sir William Herschel discovered this binary in 1779. The primary is a solar-type star, and the companion may be a red dwarf or approaching that late stage in its evolution. East and slightly to the south of Schedar are the two stars known as **MARFAK:** (θ) **THETA** and faint (μ) **MU**: these were called *Al Marfik,* or "The Elbow," by the Arabians. Mu is another "runaway" star, of great proper motion.

SN 1572

Above the *W* and forming a trapezoid with α, β, and δ is the fourth-magnitude star (κ) **KAPPA,** of special interest because it lies in the star field of Tycho's Star, the famous supernova of 1572 (1.5° NNW of κ). Known also as the "Stranger" or "Pilgrim Star," it reached an apparent magnitude smaller than that of Venus, and was first seen by several independent observers; on November 11, Tycho Brahe noticed it, and not trusting his own eyes, called upon witnesses to confirm its miraculous appearance. He kept a valuable record of the star's light changes and its exact position; he even tried to determine a parallax for it but failed, concluding—important for his time—that it lay beyond the Moon and the planets, among the "fixed" stars. He recorded his observations in a book, *De Nova Stella* (*On the New Star*). One of only four known galactic supernovae and the most brilliant to appear in five hundred years, the "new star" was named in Tycho's honor. Its appearance started him on his career as an astronomer in much the same way that the nova of 134 B.C. inspired Hipparchus to commence work on his catalogue. Do not look for Tycho's Star in today's heavens, however: Over 10,000 light-years distant, this aged object was blown apart following its sudden colossal gravitational collapse, and its core must have been compressed to the great density of a neutron star.* No star has been seen in the position Tycho gave for it, although faint shreds of nebulosity, believed to be its remnant, were recorded by Palomar Observatory and confirmed by radio astronomers. Measurements show that the average velocity of expansion of this nebula is a tremendous 5,600 miles per second, but it has virtually no luminosity.

Some supernova eruptions may have gone unnoticed in the days when there were no telescopes. One such in our Milky Way was *CASSIOPEIA A* (23 h 21 m   δ +58° 32'), 11,000 light years distant and a strong radio source. Its appearance in the approximate year of 1680 was overlooked, because heavy nebulosity in the field where it was seen greatly reduced its brilliance. Radio astronomy is gradually uncovering remnants of such "hidden" supernovae.

Minor Stars of Interest

R: Far below the western end of the *W* (RA 23 h 56 m   δ +51° 07') is another first— discovered by Pogson in 1853: the first long-period variable (431 days). A red giant, it shows a visual magnitude range of 5 to 13, which is deceptively large as far as its total radiation is concerned, because it then emits only infrared waves, which are not visible to the human eye. For identification purposes, we mention that R has two eleventh-magnitude optical companions, one very close.

S: To the north, far above the *W* (RA 1 h 16 m   δ +72° 21'), discovered in 1861 at Bonn; fainter than R, and of interest as one of the coolest stars known (2,500°–1,900° K.).

RZ: (RA 02 h 44 m   δ +69° 26'): Farther out from Iota than these. An eclipsing variable, m 6, fading to m 7.8. Easy for small telescopes.

AO: "PEARCE'S STAR" (RA 0 h 15 m   δ +51° 09'): Look south of Alpha and Beta. A

---

* *A star in which the protons in all the atomic nuclei have been changed to neutrons by an influx of the highly compressed electrons. The neutrons then seep out of the nuclei to form a neutron gas. (See Chapter Four, p. 95).*

sixth-magnitude variable, one of the most massive and luminous binaries known. The components are class-O giants separated by only 15 million miles. Although 7,000 light years distant, their enormous luminosity brings them into faint naked-eye visibility.

Clusters

M52 (NGC 7654): Open cluster, 3,000 light years distant; follow Schedar and Caph in a straight line in the direction of the constellation Cepheus; rich and dense, containing many young stars. To its southwest, just inside the border of Cepheus, lies *NGC 7510*, an irregular star cluster of jewel-like beauty, containing thirty tenth-magnitude stars.

M103 (NGC 581): A fan-shaped cluster 1 degree east-northeast from Delta. Approximately 8,000 light years distant, it contains a red giant. Principal stars range from m 7 to m 9; easily observable with a 10-inch mounted telescope.

NGC 457: Southwest of Delta and next to the m-5 star Phi (φ). Phi may possibly belong to this cluster, and if so, at 9,300 light years, it would be one of the most luminous stars, exceeding even Rigel, in Orion.

NGC 7789: Lying between m-5 stars Rho (ρ) and Sigma (σ) Cassiopeiae, a beautiful cluster containing at least a thousand stars about 6,000 light years distant, possibly three thousand times the Sun's luminosity.

CEPHEUS

With arms outstretched and feet extended toward the pole, the king of Joppa sits on his starry throne. Known to the Chaldeans of 2300 B.C., the Greeks of 500–400 B.C., as well as the Hindus and the Arabs, Cepheus was associated with many legends, and the patterns of his stars were imaged in various ways. Such a plethora of interpretation for this faint, inconspicuous constellation may surprise us; but then we remember that the ancients enjoyed dark, unpolluted skies, and Cepheus' figure was viewed with clarity. On today's charts, he is shown as a lopsided quadrangle with a tall peak resembling, appropriately, a dunce's cap! (The king was certainly not too clever about handling his family affairs.) When the peak marking his feet (which point toward the pole) is above the quadrangle, the traditional figure of the king is inverted and his little head hangs downward, a fate he shares every six months with his vain queen.

The principal star of Cepheus' square is *ALDERAMIN* (α) *ALPHA CEPHEI* (RA 21 h 17 m   δ +62° 22′)   m 2.6, 52 light years distant, originally *Al Dhira al Yamin*, "The Right Arm" (more exactly, the king's right shoulder—to the viewer's left). In the spring, however, when the pole is above Cepheus, Alderamin marks the lower right-hand (southwestern) corner of the quadrangle. Of note are the star's unusually rapid rotation, which blurs its spectral lines, and its future destiny as the polestar, seventy-three thousand years from now.

ALFIRK (β), meaning "The Flock," lies northeast of α, and in the spring it marks the upper right-hand corner of the quadrangle (RA 21 h 28 m   δ +70° 20'). About 980 light years distant, m 3.15, β is a hot, blue-white giant. It belongs to a rather rare class of variables, having a period of only 4 hours 34 minutes and an exceedingly tiny magnitude range (0.04). An eighth-magnitude companion is present. Farther to the north, *ER RAI* (γ), or *GAMMA* (RA 23 h 37 m   δ +77° 21')   m 3.21, marks the peak of the dunce cap (and the king's knee); the name is derived from *Al Rai,* or "The Shepherd" (as the desert dwellers of Arabia once identified parts of Cepheus, pictured together with his dog and a flock of sheep). An orange subgiant, Gamma is 50 light years distant, and its luminosity is eleven times the Sun's. In only two thousand years, Gamma will be the pole star.

Proceeding southward to the upper right-hand (northeastern) corner of the "square" (22 h 47 m   δ +66° 17'), we find (ι) *IOTA,* m 3.7, with companion Σ *2948* (Fl. 32), a spectroscopic binary of class B9, its components of m 7 and m 8.5. Lying more or less centrally within the quadrangle (RA 22 h 02 m   δ +64° 23') is fourth-magnitude (ξ) *XI,* an attractive double. At the lower left (southeastern) corner is (ζ) *ZETA* (RA 22 h 09 m   δ +57° 57')   m 3.36, an orange supergiant at the great distance of 1,240 light years. Nearby are (δ) Delta and (ε) Epsilon, which together with (ζ) Zeta, and faint (λ) Lambda, to its northeast, delineate the king's head—distinguished by the right triangle of δ, ζ, and ε. King Cepheus sports two outstanding jewels in his crown: *EPSILON,* a fourth-magnitude variable of very short period (61 minutes) and exceedingly small magnitude range (0.03), of the rare Delta Scuti class (p. 264), and neighboring *DELTA,* possibly the most noteworthy star in the circumpolar zone because of its great value to astronomy in revealing the true dimensions of the universe. No visit to the heavens is complete without an account of Delta's history and its unusual character; we pause here for the details.

The Story of Delta Cephei and the Cepheid Variables

A deaf-mute of noble birth, John Goodricke of York was physically too frail to survive past his twenty-second year. Yet his acuteness of vision and talent for careful observation assured the severely handicapped young man an important place in the history of astronomy. When he was only eighteen years old, Goodricke's diligent studies of the night sky led him to a landmark discovery. On this occasion, he noticed that Algol, in the constellation of Perseus, was becoming fainter by a whole magnitude and then brightening up again at regular intervals. Goodricke guessed, correctly, that the cause of the star's changes in brightness was its periodic eclipse by a less luminous companion. Two years later, in 1784, the young astronomer directed his sharp gaze northward to Cepheus' faint head, where he discovered that Delta Cephei was becoming fainter and trading places in magnitude with Zeta and Epsilon. According to Goodricke's observations, these variations conformed to a cycle of 5 days 8 hours 37½ minutes, with a rise in brightness much steeper than its decline. He reported his findings on Algol and on Delta Cephei in a paper to the Royal Society, suggesting in remarkably prophetic words that "such enquiries may probably lead to some better knowledge of the fixed stars, especially of their constitution and the cause of their remarkable changes."

Stars, as we now know, are actually not "fixed" but move through space. In addition, many also vary periodically in luminosity; these fluctuations are *intrinsic,* owing to dynamic factors in their internal structure. In contrast to *nonintrinsic* variable stars, like Algol, whose light changes are caused by an eclipse, these *intrinsic* variables undergo *true* changes in their output of light, which, in turn, affect their spectra. As in stellar duplicity, our indispensable instrument the spectroscope shows its great value by also helping to solve the riddle of intrinsic variability. Thus, we learn that δ Cephei's spectrum changes with the same period† (5 days 8 hours 48 minutes) as does its brightness. When brightest (its apparent magnitude 3.6), δ Cephei belongs to spectral class F5; as it fades to magnitude 4.3, its spectrum gradually becomes that of a yellower, G3 star; as it brightens, it returns to its original, F5 class. Its surface temperature rises and falls with its spectral changes, varying from 5,500° K. at minimum brightness to about 6,750° K. at maximum brightness. The spectroscope also reveals that δ Cephei is a pulsating variable, with its atmosphere alternately approaching and receding from us with the above period. When the atmosphere is approaching us, the absorption lines in its spectrum are displaced toward the blue from their normal positions; the amount of this displacement gives us the speed of approach of the atmosphere. This is followed by displacement toward the red end of the spectrum. In accordance with the Doppler effect,‡ we deduce that the star is pulsating, with its atmosphere alternately approaching and receding from us. When the star is faintest, it seems to be moving away from us, and when it brightens, it seems to approach us. Obviously, the star as a whole cannot be moving through space in such an erratic manner, so it is evident that δ Cephei is undergoing regular periods of expansion and contraction. Because δ is actually approaching us, its overall velocity of approach combines with the velocity of expansion of its outer layers (with reference to its center) when it is in a state of expansion, to produce a total radial velocity of approach of about 23 miles per second. When it is contracting, the velocity of recession derived from its absorption lines is a mean between the star's speed of approach and the inward motion of its atmosphere, about 5 miles per second.

The astrophysicist Sir Arthur Stanley Eddington (1882–1944), outstanding British astronomer at Cambridge, made detailed mathematical researches on the subject of stellar pulsations, which he attributed to the opposing effects of gravity versus radiation and gas pressure in stars. Since such pulsations, however, would tend gradually to lose their energy and die out, researchers had to develop a more adequate explanation of the pulsations of stars like δ Cephei. We now know that the driving mechanism responsible for its behavior is *ionization:* the hydrogen and helium atoms in the star's outer zones become ionized as their electrons absorb a fraction of the radiation from its interior when the star is most highly compressed (and its internal temperature the highest); the temperature in these outer layers thus drops. These ionized atoms then release this trapped energy as they recapture their lost electrons when the star is expanding most rapidly and thus give an added energic boost to its expansion. Certain very specific internal conditions in addition to the foregoing must be fulfilled for the

† *Ten and a half minutes longer than Goodricke's surprisingly accurate calculation.*
‡ *When a source of light is receding from an observer, the color of the light from the source is redder than it normally would be, because the waves emitted by the source are stretched (increased wavelength), and when the source is approaching, the light is bluer, because the waves are crowded together (shorter wavelength)—the "accordion effect" (p. 96).*

star to pulsate in this manner: it must be quite massive and in the helium-burning phase of its evolution, therefore not a young star.

The Period-Luminosity Relation

Short-period pulsating stars of this type were named *Cepheid variables*, or "Cepheids," after δ Cephei, the classic example. Since young Goodricke first reported his work on variables nearly two centuries ago, many other Cepheids have been discovered in our galaxy and in far more distant parts of the universe. In 1912, Henrietta Leavitt (1868–1921), of Harvard Observatory, investigating the many Cepheid variables in the Lesser Magellanic Cloud and correlating their periods to their average apparent magnitudes, discovered a remarkable relationship: the longer the period of a variable the greater its apparent brightness (the smaller its apparent magnitude). Harlow Shapley (1885–1972), one of the first to suggest the pulsation theory as an explanation for Cepheid variability, showed that this relationship between period and *apparent* magnitude also holds true for the period and the *absolute* magnitude of a Cepheid, by determining the distance of such variables in our own Galaxy. Although his conclusions were correct, we now know that his data were misleading, because he did not realize that there are two populations of Cepheids, the younger, population I Cepheids being on the average 1.5 magnitudes more luminous than the older, population II Cepheids.

Nevertheless, Shapley's work enabled Edwin P. Hubble in 1924 to use the Cepheids in the Andromeda nebula and other galaxies to determine their distances and to establish that these so-called nebulae are true star systems, or galaxies. Shapley's oversight was later (in 1950) corrected by Walter Baade of Mt. Wilson and Mt. Palomar Observatories, in the course of his own investigations. Baade was studying variable stars in the Andromeda galaxy in an effort to determine its true distance. When he was unable to obtain photographs of the luminous RR Lyrae variables in the "nebula," he concluded that it must be much farther away than Shapley had estimated. This led him to reclassify the Cepheids in the Andromeda galaxy along with those in our own as population I stars and to revise the period-luminosity curve accordingly, to obtain the correct distance for the Andromeda galaxy: 2,200,000 light years. With the revised period-luminosity curves for population I and population II Cepheids one can now obtain accurate distances of the many globular clusters surrounding our Galaxy of extragalactic systems like the Magellanic Clouds, of large galaxies like the Andromeda and of those much more remote in space. Thus Goodricke's early discovery of the variability of δ Cephei opened doors of knowledge that led to the development of one of the most useful astronomical tools, the Cepheid "yardstick in space" (fig. 33).

Delta Cephei, like all other stars of its type, is a yellow giant, about 1,200 $L \odot$; its *photographic* absolute magnitude is −2.9 and its mean photographic apparent magnitude about +4.65.* By calculating the difference between the absolute and apparent magnitudes, we arrive at a distance of 1,031 light years. Before its variability was discovered, by Goodricke, δ was already known as a double star. The sixth-magnitude blue companion (spectral class B), 41 seconds (13,000 AU) distant from its primary, presents a good subject for the small telescope. Even this faint companion is 250 times

---

* *Not to be confused with the visual magnitudes.*

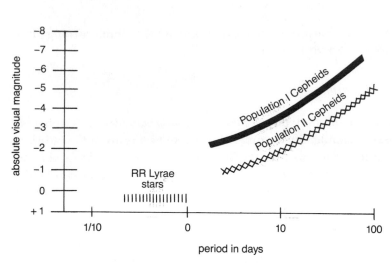

33. *A celestial yardstick:* the Cepheid period-luminosity curve for populations I and II.

as luminous as our Sun! A very faint optical companion was also discovered at 20.9 seconds from the primary by S. W. Burnham in 1878.

The Tiny Red Twins

Before leaving the star field of δ Cephei, we pause briefly to discuss the double star **KRÜGER 60,** about 43 minutes to δ's southeast (RA 22 h 26 m, δ +57° 27′). Only 13.1 light years distant, m 9.8 and 11.4, this wide pair is unusual because both of its components are red dwarfs of exceedingly low luminosity (.0013 and .0004 $L \odot$). The mass of the tinier dwarf is only .14 $\mathcal{M} \odot$, one of the smallest known. This midget star also shows irregular variability in the form of sudden flares which may double its luminosity in the course of a minute or two. It may be viewed with a small telescope, and there is always the exciting chance of witnessing a flare as the tiny speck of light marking Krg 60 B suddenly seems to "blow up" like a miniature nova and overtake its primary in apparent size and brightness.

Herschel's Garnet Star

Attached to the King's little head is his outstretched left arm, its elbow marked by a fourth-magnitude star (μ) *MU* (RA 21 h 43 m δ +58° 47′), noted for its deep red color, which may be compared to that of nearby Alderamin. Named by Herschel, its variability was discovered during his lifetime, and its light curve and spectrum have been thoroughly analyzed since then to reveal a pulsating red giant (M2 Ia) similar to famous Betelgeuse in Orion. Of interest are Mu's great luminosity (perhaps larger than 12,000 $L \odot$) and its diameter, several hundred times that of the Sun; however, Mu Cephei lies near the northern edge of the vast, dark, gaseous nebula IC 1396, an obscuration that hinders astronomers in calculating the Garnet Star's "vital statistics."

One very interesting footnote is the presence of rare water-vapor bands in the spectrum of this highly turbulent star (see "Convection," footnote, p. 92).

Other Variables of Interest

Lying approximately 18° due north of β Cephei and east of κ, in the northern part of Cepheus, *S CEPHEI* (RA 21 h 35 m   δ +78° 24′), a famous variable, forms an almost right triangle with Polaris and Gamma Cephei (the Pole Star is at the northern vertex). About 2,000 light years distant, S Cephei belongs to spectral class N, that rare group of very cool pulsating giants whose low temperatures allow the existence in their atmospheres of carbon molecules, which produce strong carbon bands in their spectra. As we have explained, temperature determines the color of a star; the coolest are the reddest, and S Cephei, a deep ruby red, is no exception. Its strong tint renders it an easy object even for a small telescope.

Farther north (and east) in Cepheus (only 8.5° from Polaris) we find the eclipsing variable *U CEPHEI* (RA 0 h 57 m   δ +81° 37′). Just out of naked-eye range, U is an easy object for the small telescope. Its components, a hot bluish primary and a diffuse yellow giant only 6.5 million miles apart, are exchanging matter. As a jet of gas passes from the older, expanded star to the younger, blue star, their relative mass slowly changes, so that their eclipse period (now 2 days 11 hours 49 minutes) is gradually increasing, a change that will become apparent in centuries as the seconds accumulate.

A very distant RR Lyrae pulsating variable, *RZ CEPHEI* is known as the fastest star in the heavens (about 400 mi/sec in transverse motion). At 2,000 light years, this ninth-magnitude star is not too easy to observe. Look for it approximately 2½ degrees south-southwest of Iota (exact position, RA 22 h 37 m   δ +64° 36′).

VV CEPHEI (RA 21 h 56 m   δ +63° 37′), about 1¼ degrees southwest of Xi and 3,000 light years distant, is an eclipsing variable, a huge binary whose components are a red-giant primary and a blue companion revolving about their common center of mass every 7,430 days. Both stars are also believed to be *intrinsically* variable, a fact that greatly complicates VV's light curve. The primary is one of the largest known stars, with an absolute magnitude of at least −5 (refer to the account of Epsilon Aurigae, p. 347).

Clusters and Galaxies

NGC 188 (RA 0 h 39 m   δ +85° 03′) is a globular cluster, extremely ancient (12–14 billion years old), containing only late-type stars. Look for it 5 degrees south of Polaris. *IC 1396*, lying south of μ, is a dark nebula, elongated, faint, and very large; it surrounds Σ 2816, a spectroscopic double of class O (blue and extremely hot).

NGC 6946 (RA 20 h 33 m   δ +59° 58′), lying 2 degrees southwest of η at the King's right wrist, is a type Sc very large spiral galaxy (Fig. 34), its arms well unwound, 11 megaparsecs† (35,860,000 LY) distant and remarkably productive of supernovae. Five have been discovered in it; the first, in 1917, led to an understanding of the true nature of spirals: gigantic star systems like our own Galaxy. Supernovae were subsequently

† *According to Sandage and Tammann.*

34. *NGC 6946, spiral galaxy in Cepheus;* one of the nearest galaxies beyond the Local Group; photograph shows rich field of foreground stars. (*Yerkes Observatory, University of Chicago*)

seen in 1939, 1948, 1968, and 1980, when a supernova appeared about 16 kiloparsecs (52,160 LY) from the nucleus of NGC 6946, outside the galaxy's visible spiral structure; this exploding star was about two magnitudes brighter (m 11.2 at maximum) than those previously discovered in NGC 6946.

Not to be confused with the population I and II star classifications, supernovae are also classed I or II; *type I supernovae* are somewhat more luminous than *type II,* which show greater variations in their light curves (and are associated with population-I stars). Classed as Type II according to its photometric and spectral characteristics, the NGC 6946 supernova of 1980 is one of the brightest of its type. It is also a radio source; this was detected by the VLA, the same multiantennaed instrument used for studying the "lens in space." The spectra of "bright" supernovae, both types I and II, are of especial interest for probing the interstellar medium which lies between our own galactic center and those of distant galaxies.

NGC 6939, a rich star cluster, lies 38 minutes to the northwest of 6946. In determining the magnitude of Supernova 1980, photographic plates were used to compare it to photoelectric standards in the cluster.

CAMELOPARDUS (Camelopardalis)

It is regrettable that the Giraffe's ten major stars are no brighter than than the fourth magnitude, for their outline really resembles this distinctive animal. A modern constellation, it was formed by Bartschius (Jakob Bartsch) in 1624; however, he saw it as a biblical camel, the one that brought Rebecca to Isaac. The imaginative (and sharp-eyed) Chinese pictured several smaller figures in that region of the sky, including the

35. *NGC 7510, a sparse open star cluster* in southeastern Cepheus near the border of Cassiopeia. In this exceptional photograph, its thirty tenth-magnitude stars resemble a sparkling diamond brooch. Taken with the 120-inch Shane reflector. (*Lick Observatory*)

usual assortment of ministers and guards, and a symbolic personage named *Yin Tih*, or "Unostentatious Virtue": an excellent description of Camelopardalis' qualities; for despite its modest faintness, the Giraffe contains some very interesting stars!

One of the newer classes added to the spectral tables, class R pertains to certain stars with spectra similar to those of classes K and M, in which neutral metals and OH and TiO‡ molecular bands dominate; however, the bands of carbon molecules dominate the spectra of class-R stars. They are all reddish giants. A few of them are variable; one of these, *S CAMELOPARDI*, (RA 5 h 36 m   δ +68° 46′), lying under the Giraffe's long neck, about 10 degrees east-northeast of α (and near spiral galaxy NGC 1961), was discovered in 1891. It completes its cycle in 326 days, and its brightness ranges between the eighth and tenth magnitudes.

‡ *Hydroxyl and titanium oxide.*

We usually think of *novae* as giant stars which are subject to a series of explosions late in their evolutionary cycles. But exceedingly rare dwarflike eruptive stars that belong to various spectral classes also exist. Erratic in their behavior, these so-called *dwarf novae* alternate between small-scale eruptions occurring at intervals of a few weeks, and periods of calm which may last for several months. The classic example of such an *irregular* variable is **Z CAMELOPARDI** (RA 8 h 20 m   δ+73° 17′), a very close "midget binary" with an orbital period of about 7 hours and a magnitude range of 9.9 to 14.3. Its eruptions may be caused by an exchange of material between the tiny components. Ordinary recurrent novae are believed to be red-giant/white-dwarf binaries, wherein a younger main-sequence star of a few $\mathcal{M} \odot$ evolved rapidly, and, as it expanded into the red-giant stage, lost most of its mass to a less massive companion and therefore became the present white dwarf. The companion, just approaching the red-giant stage, now returns hydrogen-rich material to the surface of the white dwarf, which the latter compresses, triggering the proton-proton nuclear reaction (p. 91) in this hydrogen, which ignites explosively and creates periodic nova-like effects.

The Cepheid That Stopped Pulsating

We spoke briefly of the older and less luminous population II Cepheids (p. 73), whose presence at first misled researchers trying to measure the distances of variables in our galaxy. One of these ancient Cepheids, discovered early in the century by the Russian astronomer L. Ceraski, is **RU CAMELOPARDI** (RA 7 h 16 m   δ +69° 46′), approximately 12 degrees east of L (the faint star lying below the bend of the Giraffe's long neck and east-northeast of α). When discovered, RU completed its cycle in about three weeks; classed as a type-R carbon star when at minimum and a class-K supergiant when at maximum, it had a magnitude range of 8.3 to 9.2. Fifty-eight years later, Canadian astronomers discovered that RU's light changes had virtually ceased; further

36. NGC 2403, a spiral galaxy in Camelopardalis. A young (type Sc) spiral only 8 million light years distant, containing many Cepheids and blue giants. The rough-hewn, crablike appearance of its long, "unwound" spiral arms and the small nucleus of the galaxy indicate its youthful stage. Photographed with the 200-inch reflector. (*Hale Observatories*)

investigation showed that this "stabilization" had occurred in less than three years (not even the wink of an eye in a star's life). RU has been constant since 1965, a unique case in stellar annals. Nevertheless, these facts are not as surprising as they may seem, because astrophysicists who study stellar evolution know that a massive star approaching its red-giant phase (p. 88) is only in the proper condition to pulsate a few times in its life history, the first of these instability phases lasting a few thousand years at most; after the star starts burning helium, it becomes unstable again for another few hundred thousand years, when it can be detected as a Cepheid. Unless some other factor is operating, we seem to have caught RU in the very act of passing out of an instability phase, and this presents an exciting opportunity for astronomers to witness the evolution of a supergiant.

Clusters and Galaxies

NGC 1502 (47[7] HERSCHEL)(RA 4 h 03 m   δ +62° 11'), a rather rich irregular cluster in our Galaxy, lying under the Giraffe's midriff; contains twenty-five stars, including doubles Σ 484 and Σ 485.

NGC 2403 (44[5])(RA 7 h 32 m   δ +65° 43'), m 8.8 (fig. 36), approximately 12° northwest of Omicron Ursae Majoris (the Bear's nose) and about 8 million light years distant, with a luminosity of 4 billion suns; a large spiral, of interest as the first galaxy outside our Local Group in which Cepheids were identified. Young and still evolving, it contains many blue giants in its arms.

NGC 2523 (RA 8 h 09 m   δ +73° 45'), a "middle-aged" barred spiral of unusual appearance.

IC 342: (RA 3 h 42 m   δ +67° 57'), an older spiral; very large *apparent* size because of relative closeness (approximately 2 million light years), but quite faint (m 12).

LYNX

Before departing from the circumpolar regions, we pause briefly within the borders of Lynx, a very faint constellation formed by Hevelius from stars at the forefeet of the Greater Bear, but extending for some 29 degrees all the way northwest to Auriga and Camelopardalis. In explaining the title, its inventor said that those who would examine the Lynx ought to be lynx-eyed. Nevertheless, for those with binoculars or small telescopes, it contains many beautiful objects.

Principal Stars

In the southeast corner of Lynx, about 18 degrees east of Castor and Pollux (the bright "twin" stars of Gemini), we find the great cat's brightest star (α) *ALPHA* (RA 9 h

18 m   δ +34° 36′) m 3.14, a red giant some 155 light years distant. Nearly 3 degrees north of α and slightly to its west is the white double star *F1. #38,* nearly of fourth magnitude and about 88 light years distant; its magnitude 4 and 5.9 components, only 3 seconds apart, are described by R. H. Allen as "white and lilac." About 7 degrees to the northwest of #38 is the fourth-magnitude binary designated as *#3679* in the Yale Catalogue of Bright Stars, but *"10 U Ma"* (Ursae Majoris) on some charts, an F5 star with a sixth-magnitude companion at 0.6 second. Another 8 degrees to the west-northwest, we find *#31 Lyncis,* magnitude 4.25, an orange giant (class K2 III), which is also a former star of Ursa Major, and was the Arabs' *Alsciaukat,* or *Al Shaukah,* "The Thorn."

Still farther northwest along the straggling line of stars in the figure of Lynx is the A-type subgiant *#21,* magnitude 4.6; the magnitude-4.5 class-A2 (white) main-sequence star *F1. #2* lies at the far northwestern end of the cat's tail, thus belonging to the circumpolar group of stars that appear to revolve about the north celestial pole. A few degrees to the east is the triple star *F1. #12,* Σ *948* (RA 6 h 41 m   δ +59° 30′)   m 4.87, an A2 white Sirian type star, whose broad spectral lines show rapid rotation; the 6.0- and 7.5-magnitude components are 1.8 and 8.5 seconds from the primary, respectively. The primary, magnitude 5.5, and the sixth-magnitude star form a binary with a period of revolution of about seven hundred years; with a 3-inch refractor, they appear as one star, but the more distant component provides a good test for the small telescope. All three components may be seen with 6-inch and larger refractors.

A distant Globular Cluster

Far to the southwest corner of Lynx, near the border of Auriga and about 7 degrees north of Castor, in Gemini, lies *NGC 2419* (RA 7 h 34 m   δ +39° 0′)   m 11.5, the most distant globular cluster of our Galaxy, possibly 182,000 light years from the Sun and even farther from the galactic center. Sometimes called the "Intergalactic Wanderer," it is classed by some astronomers as an object outside our Galaxy. With an apparent diameter of 2 minutes, the Wanderer is visible as a fuzzy spot in small instruments, resolvable only in very large telescopes, which reveal it as a very luminous object, rich in stars.

We have found the north circumpolar zone rich in the traditions of mythology and astronomy. Contained within its boundaries is a wealth of information about the countless Sun-like objects of which our universe is formed. It has served us well as a basic introduction to the mysteries and adventures lying ahead.

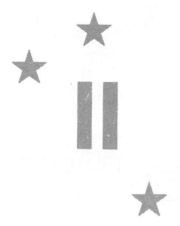

Winter

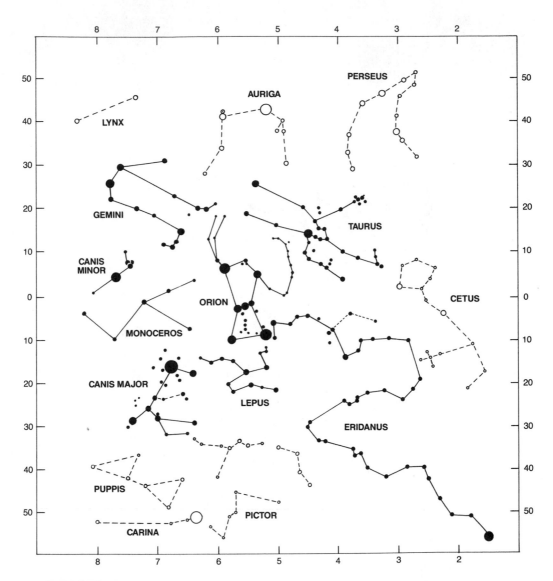

37. THE WINTER SKIES

FOUR

The Hunter

Man Against Beast: The Starry Contenders

A silvery thread in the heavens, Eridanus, greatest of all rivers, winds slowly through vast fanciful forestlands. On its grassy banks stands a solitary, impressive figure dressed in hunting attire. The man's powerful arms are upraised; his right hand clutches a stout club, and the left holds aloft a shield made from the skin of a lion. His calm, steadfast gaze is directed at the fearsome visage of a deadly opponent: horns lowered, angry red eye glaring with the lust of combat, a great white Bull threatens to charge the hunter.

Through unending reaches of time and with inexhaustible patience, the "Great Hunter" Orion awaits his challenger, Taurus. Nearby crouch two faithful dogs who always accompany the hero on his exploits: Canis Major, the larger and more aggressive, and Canis Minor, smaller and more timid. At the Hunter's feet a little woodland Hare quietly watches the confrontation. Lepus the Hare is afraid neither of the man nor his dogs, for it knows that their attention is focused entirely on the charging Bull.

Attracted by the excitement, other forest dwellers occupy ringside seats near this fierce contest. Although Orion is a popular personality, Taurus, too, has his fans and supporters. These include a strange and wonderful mythical creature who finds his way through the crowd and settles down between the hunting dogs, close to Orion's side. Monoceros, fabled beast of single-horned fame, anxiously awaits the "decision." Will Taurus overcome the dreaded Hunter?

This stellar audience is augmented by two young men whose deeds are celebrated in the ancient world of sport: Castor, great horseman, and his immortal brother Pollux, champion pugilist. Gemini, the famous twins, rest from their own busy conquests while they observe the unfolding battle.

Despite the determination of Orion and Taurus, however, the earthly spectator soon notes that an element of reality is missing from their promised fight: it will never take place, for the distance between man and beast cannot be closed! Belligerent stance unchanged, each maintains a respectable and safe distance from the other while quietly rising and setting at the appointed time in the celestial mid-latitude. However stirring and suspenseful, this action-packed scene is but the frozen, ghostly echo of a flesh-and-blood drama said to have once been enacted on the Earth. Those shadowy events, if ever they did occur, belong to a past so distant that their reality lies at the farthest border of man's memory. Precise details as well as names of "dramatis personae" have long since been filled in by his creative and unfailing imagination.

ORION

The Mythology

His name is derived from *Uru-anna,* meaning "The Light of Heaven."

He was a man of gigantic stature, unrivaled good looks, and a hunting prowess celebrated throughout the ancient world. On a visit to the isle of Chios, he fell in love with Merope, the beautiful granddaughter of the god Dionysos. Merope's father, Oenopion, promised her hand in marriage to Orion on the condition that he free their island from the dangerous wild beasts troubling the inhabitants.

Orion exercised his famous hunting skills with great success, but Oenopion invented excuses to evade the nuptial agreement by assigning Orion further tasks. This led the desperate Hunter to down a flask of the unwilling father-in-law's most potent wine; succumbing to its effects, Orion then forced his intended bride to consummate the marriage bond prematurely. For this misdeed, he paid with his eyes: Oenopion arranged to have Orion reduced by further drink to a state of total helplessness and then blinded him.

Instructed by an oracle to seek the Sun's morning rays if he would see again, Orion set out for the east with his sailors; attracted by the sound of hammering, he docked at the island of Lemnos, where Hephaestus (Vulcan), the gods' blacksmith, ruled over the one-eyed race of cyclopes. Responding to Orion's pleas for help, Hephaestus gave him Cedalion, an apprentice, who sat on the giant's shoulders and guided him to the farthest ocean; there Orion appealed to the Sun god, Helios, for the restoration of his sight. His plea granted and the optical wonder accomplished, the handsome patient enjoyed a love affair with Helios' sister Eos, goddess of the dawn.

In his vengeful pursuit of Oenopion, Orion eventually went to Crete, where he met Artemis (Diana), goddess of the Moon and an ardent huntress. Rejoicing in their common love of the sport, Diana and Orion now accompanied each other on the chase. This newfound camaraderie became a source of concern to Apollo, "destroyer and healer," famed god* and brother of Diana, for he feared that she might be next on Orion's list of amorous conquests.†

Unfortunately, the mighty and godlike Orion had a human weakness that led to his undoing: he bragged injudiciously about his achievements. He claimed no animal existed he could not conquer. Apollo, slyly awaiting an opportunity to discredit the Hunter, repeated his foolish boasts to Mother Earth. Piqued at Orion's rash words, and perhaps fearful that he would totally wreck the Earth's ecology by destroying all the

---

* *Eventual successor to Helios as sun god.*

† *His concern may well have reflected his own guilt in such matters, for Apollo had a record of aggressive seduction that far overshadowed the limited love life of the young Hunter, who seems to have been more a victim of circumstance than a freewheeling and ardent womanizer!*

predators (as he had on the isle of Chios) she decided to halt his exploits by producing a monstrous scorpion under his feet, a creature with such heavy armor it could not be slain. For the first—and the last—time in his life, Orion was defenseless in the presence of a deadly beast. The scorpion promptly stung the Great Hunter and he died.

In anguish over the loss of her friend, Diana requested the gods that she be permitted to immortalize him among the stars. Her wish granted, she placed him in the mid-latitude skies; as a reward for conquering such a mighty hero, the scorpion was also placed among the immortals, where he may be seen in the southern skies as the constellation Scorpio, but not at the same time as Orion, for the antagonists were tactfully positioned in opposite quadrants of the heavens so that Orion should never be reminded of his ignoble defeat.‡ To relieve the Hunter of idleness in his starry abode, the gods thoughtfully positioned a charging Bull named Taurus nearby. As he faces his prey with shield and upraised club, Orion's striking pictorial clarity seems to account for his ancient "recognition" as a constellation (see fig 38). The poet Homer, who lived about 800 B.C., refers to the "Mighty Orion" in his famous epic *The Iliad;* Orion's Belt is also mentioned in the Old Testament at a time when most star groupings were still known by the names of the objects and animals with which they had long been associated (Intro., xviii). We may infer that primitive identification of the great stellar figure as a "hunter" was preserved through folklore and later elaborated upon by the Greeks, who attached to it the story of their Orion.

Not only easily recognizable, the Hunter is also extremely useful for scientific exploration. Most of the heavenly phenomena valuable to this instruction are well represented within his vast boundaries. The Great Hunter, arms outstretched, invites us to examine these astronomical wonders; gratefully, we accept his hospitality and begin our safari to Orion with the study of a famous red supergiant.

BETELGEUSE (α) *Alpha Orionis*
(RA 5 h 53 m   δ +7° 24')   m 0.2–1.8

Its name derives from *Ibṭ al Jauzah,* the "Armpit of the Central One." The second-brightest star in Orion, but designated α, it occupies the position of the Hunter's right (northeast) armpit (viewer's left), the arm that holds aloft a club as he prepares to smash Taurus. Betelgeuse constituted the fourth *nakshatra,* or ancient lunar mansion of India, named *Ārdrā,* or "Moist," depicted as a gem with the storm god as its presiding divinity, and influential in worship. The star was also known as "Arm" to the Hindus, Persians, and Copts.

As if enraged at Orion's taurine enemy, Betelgeuse glows with a somber red color. The naked-eye observer sees a warm, bright point of light tinted with a slightly roseate or coppery cast. This intriguing ruddiness is far more significant, however, than the imaginative ancients could have dreamed. Our modern photographic instruments have taken a closer look at the great star and revealed to us the amazing secrets of its evolution and its chemistry implied by its hue and brightness.

‡ *There are two versions of this story; Orion's death is often attributed to Diana herself, whom Apollo tricked into firing an arrow at the Hunter's head as he swam away in the ocean to escape the scorpion.*

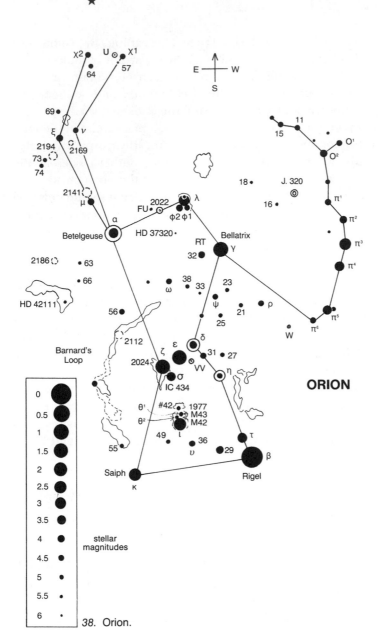

38. Orion.

Magnitude and Color of Betelgeuse

With a mean apparent magnitude of +0.1 and an absolute magnitude of approximately −5.9, Betelgeuse is visually about the tenth-brightest star. Its roseate tint is classified as 1.85 on the *color-index scale,* a standard of measurement of a star's color based on the difference between its blue-sensitive photographic magnitude and its red-sensitive visual magnitude. The eye and the camera respond equally to the color white, and absolute white therefore becomes zero on the color-index scale. Any bluer tinge is a minus number on this scale of measurement, and tints from yellow to red have plus numbers: the redder the star, the higher the number. When accustomed to the use of this color index, we need only glance at Betelgeuse's rating of +1.85 to know that this large star is blushing rather deeply, which in turn reveals its structural and biographical secrets!

A Gigantic Star

Betelgeuse, a *red supergiant,* has a high luminosity and a relatively low surface temperature, and is exceedingly large. For a striking contrast, its mean diameter of 640 million miles may be compared to that of our Sun: only 865,400 miles. If Betelgeuse were at the center of the Solar System, its rim, even at its smallest, would extend beyond the orbit of our planetary neighbor Mars, which is at a mean distance of 141 million miles from the Sun. Earth, only 93 million miles out, would be completely swallowed up! Because of its great size, Betelgeuse is about twelve thousand times as luminous as the Sun (12,000 $L \odot$). Because this supergiant is so huge, its gaseous material is extremely rarefied: only one one-thousandth the density of the Earth's atmosphere! In stellar society, Betelgeuse's hugeness may be regarded as unusual. The majority of stars are very much smaller, even dwarfish, like the Sun, and to find a single star approximately Betelgeuse's size, such as Antares, in the zodiacal constellation of Scorpius, we must go out to a distance of 130 parsecs, or 424 light years from the Sun. At *180 parsecs,* or *590 light years,* Betelgeuse, receding from us at the rate of about 13 miles per second, is the largest visible star. Theoretically, Betelgeuse's disk should be discernible from the Earth; only the Earth's varying atmosphere prevents detection of Betelgeuse's disk with large telescopes. In 1975 at Kitt Peak, however, a computer called the Interactive Picture Processing System (IPPS) was used to analyze photographs made with the 159-inch Mayall reflector and an image intensifier. The results indicate that this aging supergiant has a blotchy face! Made visible for the first time by this process, immense star spots are revealed, each almost the size of the Earth's orbit! Astronomers believe that *convection* (rapid mixing of gases in the star's interior—see p. 92), and its weak surface gravitational pull, which allows this agitated gas to break through the surface layer, produce these mottled and cloudy effects on Betelgeuse's disk.

The Hertzsprung-Russell Diagram

Two investigators, the Dutch astronomer Ejnar Hertzsprung and the American Henry Norris Russell, constructed a diagram that is of fundamental importance in studying the evolution and physical characteristics of the stars. Hertzsprung first began this work in 1905, and Russell (who became known as the "dean of American astronomers") later carried it on. In this diagram, the absolute visual stellar magnitudes are plotted along the vertical axis against the stellar spectral classes (surface temperatures or colors may also be used) O, B, A, F, G, K, M, N, and S (p. 22), as designated across the bottom of the "H-R" diagram. When large numbers of stars chosen at random are shown by their respective points on the diagram, they form definite branches according to their temperature, chemistry, and evolutionary stages. The majority lie on a long, curved branch running from the upper left corner of the diagram down to its lower right corner. This heavily populated stellar "street" is known as the *main sequence.* The most luminous and massive blue and blue-white main-sequence stars, those in classes O and B, are in its upper left-hand corner; yellow and orange F and G stars, including our Sun, which are of smaller mass and lower temperature, lie in the central part of this avenue; and the cool, small *red dwarfs* are clustered in the bottom right corner. Other very small stars, the *white dwarfs* (their story to be told later on) have left the main sequence altogether and are grouped in the lower left-hand corner of the H-R diagram (see fig 39).

As a general rule, a position on the main sequence indicates a definite mass, chemical composition, and "homogeneity" (even mixing of the various gas molecules); it represents the state of a "normal" star throughout a great part of its life span. When a star begins to leave the main sequence, however, the event marks the beginning of a change in its fortune, for it is the harbinger of stellar aging; as the star moves farther away from the main sequence, it begins to exhibit chemical changes, inhomogeneity, and changes in surface temperature, luminosity, and size.

Extending away from the main sequence and across to the right-hand side of the diagram is another branch, occupied by the *red giants,* a particular category of large, massive, and aged stars that evolved from earlier stages on the main sequence. With initial mass greater than that of the Sun, they expanded to a giant size in their later years, acquiring relatively low temperatures and reddish hues. Because of their great size, however, these stars are very luminous. Parallel to this giant branch, but situated above it, a position on the graph that indicates a still higher luminosity than that associated with the red giants, is a much smaller group of very large red stars. Called the *supergiant* branch, it is a very exclusive stellar club whose aging members suffer the

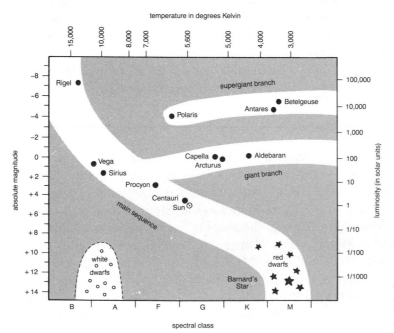

39. Betelgeuse, Rigel, and other major stars on the Hertzsprung-Russell diagram.

symptom of redness, indicating low temperatures (circulatory problems, perhaps!) combined with tremendous, overly inflated girth.

Betelgeuse has taken its place in this company. As its position on the diagram indicates, it belongs to spectral class M0 Ia. Although in former times no one was certain of the exact history of this gargantuan star, we now know that it started life as a blue-white giant of class B about 100 million years ago but, after a few million years, turned far away from its former position on the main sequence as it became chemically inhomogeneous. Evolutionary theory tells us that it then expanded rather quickly into a supergiant. This departure from the main sequence occurred much more rapidly than it would have if its mass had been much smaller, like that of our Sun, wherein chemical changes proceed at a very gradual pace. Millions of years ago, as a blue-white giant, the young Betelgeuse exhausted its hydrogen fuel supply (and thereby changed its chemistry and continues to do so) far more swiftly than the Sun. In contrast to the better-behaved stars it left behind, Betelgeuse is now exhibiting the vagaries of an advancing stellar age.

Variability

One of the most interesting fields in modern astronomy is the study of stars whose luminosities fluctuate owing to their unusual internal conditions. The variations in some of these stars are totally unpredictable, while others have mean periods that change continuously; the fluctuations in still others follow a regular cyclic schedule. All these stars are known as *intrinsic variables* (p. 72). Still another type of variable star, the *nonintrinsic,* eclipsing binary, owes the apparent changes in its brightness to a companion star that passes in front of it periodically, cutting off some of the light as we receive it. Both types of variable are well represented in the constellation of Orion and will be discussed as encountered.

Over a period of time, Betelgeuse does not glow steadily. Owing to its pulsations, its magnitude ranges from 0.3 (when it is biggest in size and brightest) to $+1.1$ (after it contracts again); for this giant is having difficulty deciding what size it would like to maintain! When it expands to its maximum size, its diameter is 920 times that of our Sun, and when it contracts, its minimum diameter is 550 times that of our Sun. Correspondingly, its luminosity increases when it expands and decreases when it shrinks by about 60 percent. This behavior is also quite erratic in another sense: the changes do not occur with regularity. Officially, the period is 2,070 days—between five and six years—but Betelgeuse doesn't actually stick to this schedule. Two different pulsations may be involved: a periodic fluctuation every two hundred days superimposed on the longer, six-year oscillation. According to definition, this marks it as an intrinsic *semiregular variable,* the type whose mean period changes continuously.

A star is stable because of an exact balance between two forces: gravity, which exerts an inward pull on the star's mass, and the compensating outward thrust of gas pressure created by the internal heat and radiant energy. The pulsations of Betelgeuse may be attributed to changes in the state of ionization of its atoms near its surface or to changes in its energy-generating mechanism, possibly related to continuing contraction of the stellar core at its late evolutionary stage. Current theory suggests, however, that temporary ionization of the hydrogen and helium in the outer regions of the star, causing collapse of the stellar materials, and subsequent rising temperatures after completion of this process, also leads to regular expansions and contractions. In this succession of events, reminiscent of the continuous contractive and expansive movements of an accordion, radiation is absorbed by the rapid ionization of hydrogen and helium atoms in its outer layers. As soon as this ionization sets in and the radiation is momentarily trapped by the electrons ejected from the atoms, the star's temperature drops and its outer layers contract. This contraction causes a subsequent rise in temperature, starting at the bottom of the star's ionized outer layers as the electrons are recaptured (forced back into the atoms) and reemit the absorbed energy. The star thus expands and the ionization process sets in again.

The pulsations of the supergiant Betelgeuse may remind us of the movements of an accordion, one that entertains us with celestial serenades. In this indulgence, we hasten to note that the concept of stars providing music is by no means a new one in the world of astronomy! We are merely following the example set by a great name of antiquity: Pythagoras, Greek geometer and mathematician who lived during the sixth century B.C. and was the discoverer of the numerical properties of music. He founded a school of philosophy which taught that all nature is governed by these mathematical relationships, and he advanced a "theory" that the motions of the planets, stars, and therefore the celestial sphere follow the rules of an exact harmony. He and his followers taught that this phenomenon resulted in the production of distinctive musical tones (which, alas, only Pythagoras himself could hear). In the spirit of the Pythagoreans, we therefore recommend close listening with an ear of fantasy as Betelgeuse adds its voice to that heavenly symphony, serenading us on its "variability accordion."

The Mass-Luminosity Relationship

Mass, as we mentioned, is the measure of a body's *inertia,* i.e., the resistance of a body to change in its state of motion (p. 31). The more massive it is, the greater the force required to alter the direction or magnitude of its velocity. Mass also is the source of the gravity of a body; this increases as more atoms are brought together in a given volume. A star's luminosity is chiefly determined by its mass (which often bears a relationship to its size but is not to be confused with it). An increase in the mass of a star results in a much greater increase in its luminosity. The mathematical rule here is that the luminosity increases approximately as the fourth power of its mass. For example, if the star were to have twice the Sun's mass ($2 \mathcal{M} \odot$), its luminosity would be about sixteen times the Sun's luminosity ($16 L \odot$).

The more massive a star is, therefore, the greater is its luminosity, in accordance with the foregoing power law. This relationship is best shown by a graph in which absolute magnitude is plotted against mass (fig. 40). Stars with the same internal structure as the Sun, and which either belong to the main sequence on the H-R diagram or originated as main-sequence stars, obey the same mass-luminosity relationship. Points representing them on this graph, called the *mass-luminosity diagram,* fall on a well-defined curve that enables us to obtain the masses of main-sequence stars from their luminosities!

For a variable star like Betelgeuse, we take its mean, or average, luminosity (12,000 $L \odot$) in plotting its position on the mass-luminosity curve. (Its luminosity varies continuously, but this stems from intermittent "bottling up" and subsequent release of its

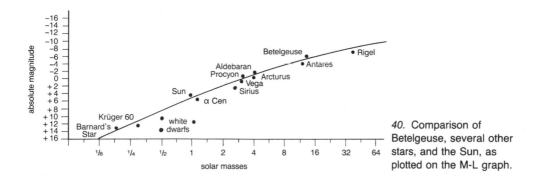

40. Comparison of Betelgeuse, several other stars, and the Sun, as plotted on the M-L graph.

energy as described, and is not related to any change in its mass, which remains more or less constant).

Great Mass: A Short Life

In terms of evolution, as we have noted, Betelgeuse is advanced. It has become unstable owing to the change in its chemical composition from light to heavy elements like iron. Nevertheless, as seen in human terms, it can look back upon a long life, beginning on the main sequence 100 million years ago as a blue-white giant. For a star, however, 100 million years is not a long time! From that viewpoint, therefore, Betelgeuse has enjoyed relatively few years of vigor, aging very quickly in comparison to our Sun, which had its remote beginnings five *billion* years ago and is at present only "middle-aged" from an evolutionary point of view.

Mass also determines the rate of evolution of a star. Changes in chemical abundance proceed very gradually in low-mass stars like the Sun, which therefore take far longer to leave the main sequence and reach the "giant" stage of their development. The life spans of the low-mass stars are thus reckoned in the billions, rather than mere millions, of years; as a rule of nature, they outlive their colleagues of greater mass many times over. Because of its very great mass, Betelgeuse has burned up its fuel reserves 10,000 times faster than the thriftier Sun, and its life span is thus more than 120 times shorter (the Sun's total life span may reach 12 billion years, while Betelgeuse, 100 million years old, is already nearing its final evolutionary stages).

Evolution of a Red Supergiant

Betelgeuse, like all stars, originated in the gradual condensation of a globe of gas and dust, with the release of gravitational energy. When this great accumulation of matter, undergoing continuous contraction, owing to the force of gravity, ultimately contracted to the point where its central temperature reached the critical value of 10 million degrees, the hydrogen thermonuclear reaction was triggered deep within the star. From that moment on, the fusion of hydrogen into helium began, causing the constant thermo-nuclear release of energy. During these early stages, energy was generated by the *proton-proton chain reaction,* in which hydrogen protons are fused into helium (four protons for each helium nucleus), according to the following set of reactions:

THE PROTON-PROTON CHAIN REACTION

H^1 + H^1 → H^2 + e^+ (positron) + ν (neutrino).

H^2 + H^1 → He^3 + radiation.

He^3 + He^3 → He^4 + 2 H^1 + radiation.

When a temperature of 20 million degrees was reached at the center of Betelgeuse's homogeneous convective core,* energy was generated by the CN (carbon-nitrogen) thermonuclear cycle, in which carbon acts as a nuclear catalyst. In a series of six steps, as illustrated below, four protons fuse into a helium nucleus, following the capture of a proton by the carbon nucleus C^{12}. At the end of this cycle, the carbon nucleus reappears.

THE CN CYCLE

C^{12} + H^1 → N^{13} + radiation.

N^{13} → C^{13} + e^+ (positron) + ν (neutrino).

C^{13} + H^1 → N^{14} + radiation.

N^{14} + H^1 → O^{15} + radiation.

O^{15} → N^{15} + e^+ (positron) + ν (neutrino).

N^{15} + H^1 → C^{12} + He^4.

The CN cycle is the energy-producing process in main-sequence stars of large mass. In less massive stars like our Sun, hydrogen is converted into helium during most of their lives via the proton-proton chain reaction. In both the latter and the CN cycle, energy is released, and this enables the star to maintain its equilibrium, as the forces of contraction (gravity) and expansion (gas pressure) exactly balance each other. Ultimately, after some 65 million years, the supply of hydrogen in the core of Betelgeuse was nearly used up and hydrogen was then burned in a shell around the core. At this stage, Betelgeuse was becoming chemically inhomogeneous (its chemical composition was no longer the same throughout its volume) and, graphically speaking, one might say moving day had arrived for the rapidly aging giant, as it took its departure from the main sequence.

In the course of 2.5 million years, a helium core was formed. This contracted for another ten thousand years. When the temperature at the surface of the core reached 40 million degrees K., high enough to ignite the CN cycle in the hydrogen-rich shell, the star reversed its evolutionary "track" and became cooler, redder, and somewhat less luminous. When the mass of the helium core reached 10 percent of the total mass

---

* *Convection: a natural mixing in which streams of hot gas flow out and cool flow in. In a homogeneous core, there is a thorough mixing of the stellar gases so that the mean molecular weight is constant throughout.*

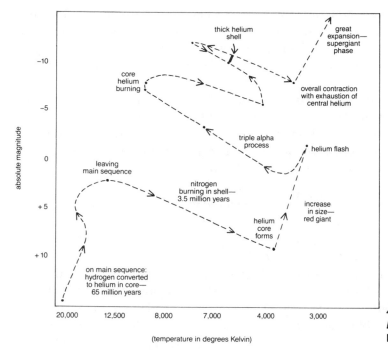

41. *Evolutionary track of Betelgeuse* (according to Iben).

of the star, the heavier core contracted under the increasing gravity, rapidly becoming much hotter and releasing tremendous amounts of energy. To accommodate the increasing output of energy, the outer envelope of the star had to expand rapidly. Betelgeuse became redder, very much larger, and therefore more luminous.

At the critical temperature of 100 million degrees K., the triple helium (or alpha) reaction started. In this process, two helium nuclei combine to form an unstable nucleus (beryllium 8), which combines very quickly with a single helium nucleus to form an excited carbon[12] nucleus, which then emits a gamma ray and becomes stable.

THE TRIPLE HELIUM REACTION

$$He^4 \; + \; He^4 \; \rightleftarrows \; Be^8.$$

$$Be^8 \; + \; He^4 \; \rightarrow \; C^{12} \text{ excited } \; \rightarrow \; C^{12} \; + \; \gamma \; \text{(radiation)}.$$

A runaway thermal process called the "helium flash" then occurred in the contracting core of Betelgeuse, accelerating the helium burning and cooling the core. From that time on, the triple helium process described above took over and stability was, for the time being, reestablished. As the helium-burning nuclear process, or triple helium reaction, progressed, the core gradually turned to carbon. Further contraction of the carbon core occurred, and production of the heavier elements proceeded as the temperature of the core rose. Ultimately, an iron core (gaseous in nature) surrounded by various shells such as calcium-burning, magnesium-burning, carbon-burning, and helium-burning was built up. These concentric shells were all surrounded by a thin hydrogen-burning shell (being transformed into helium), and all of this was wrapped in a radiating envelope of the original hydrogen-rich material of the star.

★
★   ★
★

In another few hundred thousand years or less, when Betelgeuse has used up all its remaining helium, a sudden buildup of large quantities of iron will occur; further contraction of the core and an increase in temperature will follow. As the core begins a rapid and final contraction, the outer layers of the dying supergiant will collapse violently onto the core and then bounce off in a vast explosion. At the awesome moment of its final eruption, the red supergiant will shine with an enormous brightness, perhaps greater than the combined starlight of the entire Galaxy, and Orion will appear to have caught fire under his armpit! Within a few weeks, however, as the outer layers of the star disperse, a new wonder will be revealed to the heavens: the exposed iron core, now a miniature star in its own right but one of a very singular nature, as we shall soon explain.

If Betelgeuse were a star of smaller mass (about the mass of the Sun), the scenario of its final years might read as follows: "Its atoms packed tightly together, the new star is scarcely larger than the Earth. Yet the gravitational pull on its surface is immense and its density is tremendous: one cubic centimeter of its greatly compressed carbon-like material has a mass of five tons. With a surface temperature of ten to twelve thousand degrees K., it radiates white light; but the diameter of this mini-Sun has become so small that its total luminosity is quite low. From a distance of 10 parsecs it is a dim, very faint object, if indeed visible at all. Whatever remains of the former star is contained in its dense matter, for it has become a 'white dwarf.' " But Betelgeuse, the huge "Armpit of the Central One," can find no peace, no refuge in this stage of its final collapse. It is too massive to live as a white dwarf. The mass of its core exceeds the allowable limit of 1.4 solar masses, which would enable it to become a white dwarf after the devastating explosion. The amount of supercondensed material in the surviving core will generate too great a gravitational inward pull for the forces of expansion (the gas pressure) in the star to compensate. Its outward gas pressure, exerted by free electrons alone, will be too weak, and it will no longer have the energy of thermonuclear fusion (which will have been quenched with the formation of iron nuclei and the cataclysmic final eruption) to support it against continued collapse.

In a white dwarf, the nuclei of its atoms are packed so closely together that the electrons ordinarily surrounding the nuclei are squeezed out of the atoms and move freely through the interwoven nuclei with very great speeds. The energy of this so-called "degenerate electron gas" is transported to the surface by conduction, and the outward pressure of this electron gas prevents the star from collapsing under its own gravitational pull. But because of Betelgeuse's large mass, these forces will not be in balance in its highly compressed core when the supergiant becomes a supernova! The core will continue to contract, and in the blinking of an eye, will pass right through the white-dwarf stage. The increasing force of its implosion will cause penetration of the atomic nuclei by the free electrons, which will then combine with the protons in the nuclei to produce neutrons. This stepped-up production of neutrons in the nuclei will continue until the neutrons greatly outnumber the protons in each nucleus, and neutrons will drip out of the nuclei, like water out of a wet sponge, until the star's interior is filled with neutrons. At this stage of fantastic density, a cubic centimeter of the star's material will have a mass of more than 100 million tons. When the contracting core's radius shrinks to a mere twelve miles, the outward pressure generated by the rapid motions of the neutrons and the inward force of gravity will about balance each other; the superdense core of Betelgeuse may then look forward to some millions of years' survival as a rapidly rotating *neutron star* or *pulsar,* and it will then continue to contract beyond this stage.

However, the great red star is so massive now that equilibrium will not be permanently established even at its extremely dense, neutron star stage, and the fearful implosion, the implacable contraction of Betelgeuse's core, will then continue until light itself can no longer escape from the intense gravitational field on its surface and it will have become a *black hole.* Within this totally collapsed star, space will be greatly curved and time will appear to grind to a halt. Any object happening to venture near it will be stretched out to a thin line and ripped apart by the immense tidal forces; its remnants will then fall instantly into the limitless depths of the black hole, never to return to the outer universe. For the earthly observer of these events, a great constellation will lose its symmetry. The undaunted Orion may even then take his stance against his enemies, but not without obvious personal damage: an irreparable black hole in that ill-fated armpit!

RIGEL (β) *Beta Orionis*
(RA 5 h 12 m   δ −08° 15′)   m 0.14

No less mortal than Betelgeuse are the other stars that suggest the constellation's outlines. For one, Orion sports a gorgeous, flashing diamond on his left knee. However, he had better not make the error of regression to his former habit of bragging— for which he was so severely punished—because that diamond is in such a rapid state of change that within a span of less than 20 million years it will follow the path of Betelgeuse: supergianthood, eventual supernova explosion, and death.

It bears an Arabic name, *Rigel,* meaning the "Left Leg" (of the Giant). Rigel has a visual companion, designated B, 9.1 seconds (at least 2,600 AU) distant from it and much fainter, with an apparent magnitude of only +6.8. Although put in the shade by the blazing light of the primary, Rigel B can be seen with a small telescope. But this is not a story of two stars! For here we are dealing with a triple system: Rigel's dim companion is a spectroscopic binary consisting of two components so near to each other that the most powerful telescope cannot "split" them. The duality of Rigel B can be detected only with special instruments that provide a detailed analysis of the light we receive from such double stars. Because B's components are so close together, they complete their cycle of revolution about their common center of mass in about ten days. The sum of their luminosities is approximately 150 L ⊙, and each of their diameters is about five times that of the Sun.

The Doppler Effect

A periodic displacement of the spectral lines due to orbital motion of these paired stars reveals their composite light source as a spectroscopic binary whose individual components are revolving about a common center of mass. This telltale shift of the

lines in Rigel B's spectral image is a phenomenon of invaluable usefulness to astronomers. Called the *Doppler effect,* it was discovered during experimental research on the
properties of light by the Austrian physicist Christian Doppler (1803–53). He noted
that the wavelength of light as received by the eye is affected by any motion that
changes the distance between source and observer. If this motion is recessive, thus
increasing the distance between them, the wavelengths of the spectral lines are increased and the lines are displaced toward the red. If the distance between the two
decreases, that is, if source and observer are approaching one another, these wavelengths are shortened and the spectral lines are displaced toward the blue.

Sound waves show the same effect: recessive motion of the source relative to the
listener results in an apparent lowering of the pitch of a given tone emanating from the
source; and a motion of approach results in an apparent raising of the pitch. This effect
is a part of everyday experience: the apparent rise in pitch of an automobile horn or
train whistle when it approaches, and lowering of the pitch after the vehicle passes by;
or the apparent rise and fall in pitch of warning bells at railroad crossings heard by
passengers in the moving train.

Strictly speaking, within a double star system two stars that are gravitationally bound
revolve around a common center of mass, as noted above. When a very massive
primary has a companion of small mass, this orbital center is situated deep within the
more massive star; for practical purposes it may then be said that the less massive star
revolves around the primary. If their masses are more or less equal, the center of mass
is midway between them and the components seem to circle one another; and if the
plane of their orbits is tilted toward our line of sight or is nearly edge on to it, the
situation is favorable for a pronounced and observable Doppler effect. In this case, as
one star approaches us and the other recedes (i.e., when their motion with respect to us
is radial), we observe a relative displacement of their spectral lines by an amount
proportional to their relative velocity. When the components are in the part of their
respective orbits that takes them across our line of sight (i.e., in transverse motion), the
Doppler effect disappears and the spectral lines are no longer doubled. Therefore the
Doppler effect is periodic, alternating with the eclipse of one star by the other.

Not infrequently in stellar society, a massive giant of high luminosity chooses to keep
company with a mini-star, much dimmer because of its tiny diameter. If one of these
unequal companions is more than one magnitude brighter than the other, only the
lines of the primary can be detected, and these oscillate back and forth about mean
positions. (In Rigel B, the spectroscopic components are about equal in size and
luminosity.) Analysis of the Doppler shift enables astronomers to determine the velocity curves of the components of binaries; these curves are related to their respective
masses and to the direction of the major axis of the system with respect to our line of
sight. From the velocity curves, the period of rotation of the binary under study can
then be determined.

Rigel A, the primary, is a variable bluish-white star of spectral class B8 Ia, which
means it is very young and has a high surface temperature (12,000° K.). The combined
spectral class of Rigel B, the companion, is B9. It has been estimated that perhaps 35
percent of the B-type stars are spectroscopic binaries. Most of these, like the ten-day
Rigel B system, have short periods, i.e., less than five years. The chemistry of youthful

Rigel A is still evolving, as hydrogen in its core is fused into helium by the CN cycle. Rigel's absolute magnitude, of -7.0, indicates that it is one of the most luminous stars known. Its density, however, is surprisingly low, only one hundredth of a gram per cubic centimeter in comparison to that of the Sun, 1.5 grams per cubic centimeter.

The Evolution of a Blue Giant

Rigel and Betelgeuse, sentinels of the great Orion quadrilateral, present an interesting color contrast as they maintain their unique posts at opposing corners. More remarkably, the dazzling blue of Rigel and the soft red of Betelgeuse afford us a significant picture of two important stages in the evolution of a massive star, for Rigel the "blue giant" is in every sense the young Betelgeuse of millions of years ago. Rigel's present is Betelgeuse's past. In its earlier years, the aging red supergiant was a vigorous blue giant, enjoying stellar "high life" to the hilt and burning up its fuel at a prodigious rate. Its younger colleague Rigel has not benefited from the elder spendthrift's past "errors" and is merrily following in its footsteps. Millions of years hence, it will become a member of that same red supergiants' club where Betelgeuse and its cronies lounge around, happily recalling their misspent youth.

Rigel A, the diamond of Orion, ignited its nuclear fuel and began its stellar life only a few million years ago (some authorities say a hundred thousand years ago), which indicates extreme youth in stellar terms. Rigel is about sixty-two thousand times as luminous and thirty times more massive than the Sun; it is consuming its fuel sixty-two thousand times faster than our own prudent star. The present blue giant will therefore enjoy only a short life span, probably remaining viable less than a few million more years. During this time, Rigel's position on the H-R diagram will move along a complex path, ultimately becoming a red supergiant as indicated, and when conditions are appropriate, it will explode as a supernova.

Although Rigel is receding from us in space at about thirteen miles per second, this causes insignificant change in its distance and will alter its apparent magnitude only slightly in centuries. The great star will therefore present an impressive spectacle to our distant descendants (or to future humanoid societies) when it suffers the cataclysmic explosion of a supernova. At a distance of only 270 parsecs, or 900 light years, it might be too close for comfort, which raises the question of whether the resultant radiation could be harmful; earthly inhabitants wouldn't experience this ultimate blinding flash, of course, until nine hundred years after it had actually happened. In any case, Orion would be wise to insure his flashing diamond; for after that future disaster, Rigel will be only a chip of its former self!

BELLATRIX (γ) *Gamma Orionis*
(RA 05 h 22 m   δ +06° 18′)   m 1.64

Not inappropriately, the Mighty Hunter carries a warrior—a "Female Warrior," in fact —on his left (northwest), or "distaff," shoulder. (According to the myth of ancient astrology, all women born under her influence are both lucky and loquacious!) Also called the "Amazon Star," Bellatrix has an apparent magnitude that places it among the more conspicuous stars. Like Rigel, it is blue-white, in fact slightly bluer than Rigel, having a color index of −0.22. Bellatrix's absolute magnitude of −3.3 indicates that it is one of the more luminous stars; its spectral class, B2 III, indicates that it is gigantic, although not so large as Rigel.

The "Female Warrior's" battles seem to be luring her away from us, for a Doppler shift toward the red in her spectral lines indicates that at a distance of 144.2 parsecs, or 470 light years (only half the distance of Rigel), Bellatrix is receding at a rate of 11 miles per second. We note here that Orion lies in the middle arm of the Galaxy, called the "Orion Arm," a structure 1,000 light years thick and 3,000 or 4,000 light years wide, the same arm in which our own Solar System lies. Most of the stars in the spiral arms of the Galaxy, including Bellatrix as well as Betelgeuse and Rigel, belong to population I. Like Rigel, Bellatrix probably began its life only a few million years ago at most, and therefore is a very young star: a maiden warrior who has not reached maturity, but who will achieve this metamorphosis with amazing rapidity!

A Conflict of Names

SAIPH (κ) *Kappa Orionis*
(RA 5 h 45 m   δ −09° 41′)   m 2.06

Saiph is an Arabic word for "sword." At present, it is the title given to κ, a prominent second-magnitude star that completes the Orion quadrilateral by indicating the right (southeast) foot of the Hunter (or his knee, depending upon how you draw him). Like so many stars of the Orion aggregate (p. 110) κ is blue-white (color index, −0.16) and belongs to spectral class BO Ia. Its absolute magnitude of −6.8 indicates that it is almost as intrinsically luminous as Rigel, but its apparent magnitude is 2 because of its distance, 700 parsecs, which is much greater than that of the other principal stars of Orion. Perhaps one of the hottest, most luminous supergiants visible to us (50,000 L ⊙), Saiph is unique; little wonder, then, that it stole the name of "Sword" from its former owner, a much fainter star.

SAIPH (η) *Eta Orionis*
(RA 15 h 22 m   δ −2° 26′)   m 3.32

Eta marks the handle of the sword, and was originally designated *Saif al Jabbār*, or "Sword of the Giant." It appears nearly as distant from the "blade" as κ Saiph. Of the third magnitude, bluish-white, η (spectral class B) can easily be observed with the naked eye. Strong telescopes reveal it as a close optical double, of special interest because its component A is an *eclipsing* spectroscopic binary similar to the famous Beta Lyrae (p. 222). The presence of a fourth star in this system is also suggested by variations in the radial velocity of η A.

THREE KINGS (*The Belt of Orion*)

And the Lord answered Job out of the storm and said:
". . . canst thou . . . unfasten the Sash of Orion?"
                                                    Job 38:31

Because of the impressive appearance and precise visual alignment of the Belt stars, there seems no limit to the number of colorful names bestowed upon them. It is fascinating how closely each designation reflects the culture that produced it. The early Hindus called these stars the "Three-jointed Arrow," while in Polynesia they were the "Three Canoe Paddlers." Both the Arabians and Chinese knew them as a "Scale," or "Weighing Beam"; while biblical influences in the Western world gave to the asterism the name of "Jacob's Staff," also the "Magi," or "Three Kings." French farmers called it "The Rake," and Germans, "The Three Mowers." To seamen it was the "Golden Yardarm," and to tradesmen, the "Ell and Yard," because the stars lie on a line just three degrees long (providing, incidentally, a measure for angular distances in this section of the skies). Fun-loving Australians knew the stars as dancing "Young Men" accompanied by the Pleiades (Taurus). Laplanders arrived at the most pleasurable image of all: a "Tavern"! In contrast, because of their dependency on the fishing industry, Greenlanders visualized a marine tragedy spelled out by the three great stars: "Seal Hunters," bewildered when lost at sea, and transferred together to the sky.

MINTAKA (δ) *Delta Orionis*
(RA 5 h 29 m   δ −0° 20')   m 2.2

The first "King" to rise, its name is derived from the Arabic *Al Minṭaka,* or "The Belt." A bluish-white variable of spectral class 09.5 II, δ is lettered according to its western-most position in the Belt rather than its magnitude in the order of Orion stars. Of these, it is one of the more distant, at 450 parsecs, or 1,500 light years. With an absolute magnitude of −6.0 and a color index of 0.21, δ is among the bluest and hottest of stars, a giant of great luminosity (20,000 $L \odot$). It is also of especial interest as a spectroscopic eclipsing binary, consisting of two similar stars that complete a revolution every 5.7 days around their common center of mass and exhibit a total light variation of less than 0.2 magnitude as they eclipse each other. A distant bluish companion 52.8 seconds to the north of this system, magnitude 6.7, is suitable for the small telescope. The triple system of Mintaka lies exactly on the celestial equator and is receding at the moderate rate of 12.5 miles per second.

Eclipsing Binaries

The ancients, it has been suggested, regarded stars whose brightness varied as possessed by demons. We have learned that the true cause of this fluctuation is not a devilish affair but, in the case of eclipsing binaries (nonintrinsic variables), purely mechanical: when the orbit of one component brings it in front of its companion in our

line of sight, and later, behind it, a nonintrinsic change occurs in the apparent brightness of the binary, because the eclipsing star cuts off some of the light reaching us from
its companion. Depending upon the spectral classifications of the eclipsing stars, the
spectral lines of the binary may also change during such an eclipse, because light from
the eclipsed star may be absorbed by the atmosphere of the eclipsing star; thus if a hot
B star eclipses a cool supergiant of type F to M, the lines of ionized metals, especially
calcium, are enhanced. If the amount of light reaching us per second is plotted against
time, we obtain the eclipsing binary's light curve, from which important features about
the binary can be deduced. The inclination of the orbit of an eclipsing binary to the
plane of the sky is nearly 90 degrees, and in many cases the actual tilt can be determined
accurately from the light curve. Combining the velocity curves (spectroscopically calculated) with the light curve, we can determine the masses of the binary's component
stars with great accuracy. If the brightnesses of the components differ by more than a
magnitude, a major eclipse, in which the binary undergoes a noticeable decline in
brightness when the fainter star passes in front of the brighter, is followed by a
secondary, minor eclipse, with slight dimming of the system, when the brighter star
passes in front of the fainter companion. Just before and after the primary eclipse, light
from the smaller star strikes the surface of the large star and is reflected to us, creating
additional brightness. If the stars are close to each other, tidal distortion also changes
the light curves, as do other factors, including the tilt of the orbit of the eclipsing star
with respect to our line of sight.

The Effect of Interstellar Matter on Mintaka's Spectrum

In 1904, J. Hartmann, of Göttingen, while studying the spectra of the more distant
galactic stars, turned his instruments toward Orion's Belt and detected sharp absorption lines of ionized calcium (now called K lines) in Mintaka's spectrum. Hartmann
found these lines remarkable, because they do not oscillate in accordance with the
Doppler effect associated with spectroscopic binaries, even though Mintaka has a
strongly variable radial velocity. However, he could not detect these K lines in the
spectra of the other Belt stars.

V. M. Slipher, of the Lowell Observatory, known for his work on the reflection of
stars' light by nebulosities, studied Hartmann's discovery and concluded that these
stationary lines, as they are called, originate in the atoms of gases lying in the voids of
interstellar space along our line of sight, rather than in the star from which the spectrum originates. These lines, he noted, are much narrower than those formed in stellar
atmospheres and sometimes have multiple Doppler-shifted components corresponding to absorption in clouds with differing radial velocities. These moving clouds of
interstellar gas therefore leave their own telltale "imprints" on the stellar spectrum
(fig. 42). The calcium K line can be detected only in the spectra of stars of classes O5 to
B3; and in the spectra of later types, from B4 onward, it is masked by the stronger and
broader K line produced by atoms in the atmospheres of these stars. With regard to
Alnilam and Alnitak, the other great Belt stars, Hartmann could not detect the stationary line of ionized calcium, because the radial velocities of these stars, unlike that of
Mintaka, vary little and do not differ from the velocity of the intervening calcium cloud.

This important discovery, like every milestone in the progress of scientific research, opened up new fields of study, as old notions of space, the "perfect vacuum," were revised. In 1930 Robert J. Trumpler, of the Lick Observatory, who had estimated distances for one hundred galactic clusters, rejected the inference, from their apparent brightness, that those farther away are larger than nearby clusters. He concluded that the interstellar granular matter, or "dust," tended to make the more distant clusters seem fainter, hence more remote and therefore larger. Trumpler's work led to a revision of the estimated size of the Galaxy, which had been too large, and the acceptance of a new concept, *interstellar extinction,* the dimming of starlight by intervening dust, in the ratio of one magnitude per kiloparsec. Owing to the interstellar dust, distant clusters appear redder than nearby ones; this is called the reddening effect.

Hartmann's and Trumpler's pioneering work revealed the presence in interstellar space of great gas/dust complexes, knowledge that was essential later to our understanding of the birth of stars. Other interstellar elements besides calcium were subsequently discovered: in particular, those already known from the stars' chemical abundances: sodium, potassium, titanium, iron, and hydrogen and oxygen in the form of water vapor and hydroxyl, as revealed by the stationary absorption lines in the spectra of such massive hot stars as Mintaka. Since the time of Hartmann, K lines have also been detected in the spectrum of Alnilam, Mintaka's neighbor and central Belt star, and in the spectra of several other stars as well, further indicating interstellar calcium clouds.

Since these discoveries, important new branches in astronomy closely related to the investigation of interstellar space have developed: *cosmology* deals with the origin, structure, and evolution of the universe; *astrophysics* is the study of the chemical and dynamical evolution of stars, from their origin as great globes of interstellar gas and dust to their ultimate death; *radio astronomy* studies radio galaxies and quasars and investigates the chemistry of interstellar space by studying radio signals originating in interstellar atoms and molecules; and *astrochemistry* studies the chemical processes that occur in the interstellar atoms and molecules (clouds of gas and dust). As we have already noted, the interstellar dust causes the apparent reddening of distant stars and accounts for interstellar extinction, which is dependent upon wavelength: the longer the wavelength of the light (for example, red light as compared to blue light), the smaller the extinction, because the dust particles are smaller than the longer wavelengths. They thus scatter the blue, or shorter, wavelengths more than they do the longer, or red, wavelengths, so that more of the red light reaches us than the blue. This same analysis applies to the light of the setting Sun, which glows with a deep red color because heavy layers of dust in the Earth's atmosphere, viewed edge on when looking toward the horizon, scatter the bluer wavelengths of sunlight and allow only the redder waves to reach our eyes. The rate of extinction varies inversely with the wavelength. This means that light with a wavelength one half that of red light is extinguished twice as effectively as red light. We determine the size of the dust grains by analyzing how they scatter various wavelengths. This law of extinction is effective in all directions of the heavens, and it applies to every type of star, which indicates that the dust is fairly uniform in character. Further investigations of the structure of these dust grains and their alignment indicate the presence of magnetic fields along the spiral arms of the Galaxy, so that, step by step, the study of interstellar dust has increased our knowledge of the universe.

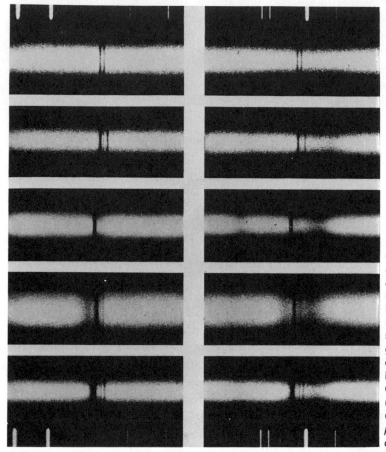

42. *Interstellar
absorption lines in
stellar spectra.* Multiple
interstellar H and K
lines of calcium II
(ionized calcium) in the
spectra of five stars
give evidence of
discrete clouds of
interstellar gas.
Calcium K (left) and H
(right). (*Yerkes
Observatory
photograph, University
of Chicago*)

In 1929, a year before Trumpler's discovery of interstellar extinction, Edwin Hubble began a systematic determination of the distances of nebulae,† utilizing Slipher's measurements of the Doppler shifts for nearby nebulae. In measuring the more remote galaxies, Hubble discovered that their velocities of recession are strictly proportional to their distances *(Hubble's Law of Recession)*, and from this he advanced the hypothesis that the linear increase with distance of the rate of recession of distant objects in the universe indicates that the universe is expanding. According to Hubble's law, the more distant a galaxy, the greater the speed of its recession as determined from the shift toward the red of its spectral lines (the Doppler effect). Milton L. Humason, who had investigated 546 galaxies for their radial velocities, confirming Slipher's results, took up Hubble's research and together they confirmed the accuracy of Hubble's law out to the most distant galaxies that could be measured in 1935. After that, many scientists investigated the red shifts of galaxies and deduced (for the rate of increase of the speed of recession with distance) a value of about 24 kilometers per second per one million LY, called the *"Hubble Constant."* However, spectroscopic evidence shows that Hubble's Constant

† *Following his work utilizing Cepheid variables to determine the distances of nebulae (galaxies).*

is decreasing with time, which indicates that the expansion of the universe may halt. If so, the universe in a very distant epoch may begin to contract. The question concerning various hypothetical models of the universe and its future development is a controversial issue and a focal point of scientific interest.

We thus see that our understanding of the nature of space, and the history, structure and evolution of the universe, traces a long path of discovery from the spectral lines of Mintaka and Hartmann's observations of interstellar lines down through the years of galactic research, to the high point of today's relativistic cosmology stemming from Einstein's discoveries.

ALNILAM (ε) *Epsilon Orionis*
(RA 5 h 34 m   δ − 1° 14′)   m 1.70

Its name derived from *Al Niṭhām,* or "The String of Pearls," ε, the middle Belt star, appears brighter than δ. One of the hottest stars known (about 45,000° K.), Alnilam's absolute magnitude is −6.8, about 40,000 $L \odot$, so that it is twice as luminous as δ. A blue-white supergiant, ε belongs to spectral class BO Ia, with a color index of −0.18. Unlike Mintaka and its other Belt neighbor, Alnitak, ε is a single star, having evolved without a companion. With a radial velocity of 26.1 kilometers per second, Alnilam, like Mintaka, is gradually receding from us in space, and at 500 parsecs, or 1,630 light-years, it is even more distant than Mintaka. Alnilam is 50 parsecs, or 163 light years, "behind" δ on our line of sight; taking into account the lateral distance between them as well, we deduce that these two stars are about 51 parsecs apart.

Our eyes can register the third dimension, or depth, only in our immediate terrestrial surroundings; thus all stars, and the planets as well, appear equally distant. In ancient days, it was assumed that these heavenly bodies lay at limited distances; the Greeks also postulated the existence of rotating concentric spheres (p. 4) which they believed contained the stars, planets, sun and moon. Because there was no comprehension of the immense distances of the stars, or of their basic nature, and certainly no understanding of the nature of our solar system, it was thought that the planets wandered among the stars; thus the astrological frame of reference‡ was developed wherein a planet might enter the "house" (sector of the sky) occupied by a zodiacal constellation. After the heliocentric concept of the Solar System was accepted, astronomy was free to progress and explore the concept of depth with regard to stellar distances, and ultimately to develop a technique for determining the parallaxes of stars (p. 13).

Light leaving Mintaka takes approximately 170 years to reach Alnilam, which means that despite their apparent closeness in the Belt of Orion, the two neighbors are as distant from each other as are our Sun and Schedar (Cass), or at least one third as distant from one another as the Sun and Bellatrix! A viewer on a hypothetical planet of Alnilam would see Mintaka and Alnitak, the lower Belt star, at "opposite ends of the sky." He would have no reason to suspect that along with his own blue-white sun they

---

‡ *Frequently used by astronomers who practiced astrology; astronomers often wrote horoscopes and were influenced by astrological belief well into Kepler's time.*

are regarded by us as part of the Great Hunter's glistening Belt: a situation similar to that of the "Megrezians" (p. 38), who cannot see the pattern of a "Big Dipper" in their heavens, even though their star, Megrez, is an important part of it in our eyes! Viewed from the vicinity of any of the great Belt stars, however, our own Sun would scarcely be visible to the naked eye. It is much too faint to make an impression at that distance; only the tremendous mass and resultant high luminosity of Alnilam and its regal colleagues of the Belt result in their striking prominence when we admire Orion.

ALNITAK (ζ) *Zeta Orionis*
(RA 5 h 38 m   δ −1° 58')   m 1.79

In 1807 the University of Leipzig named the Belt and Sword of Orion "Napoleon," while incensed Englishmen retaliated with the alternate title of "Nelson." Neither of these was recognized on star charts, but one name for Orion's Belt, "String of Pearls," did become quite popular in America, perhaps adopted by the jewelry-minded from the Arabic name for ε alone. The lowest-hanging pearl in this heavenly string is named *Al Niṭāk*, or "The Girdle." With an absolute magnitude of −6.2, this class-O9.5 Ib supergiant is a pronouncedly blue-white gem, its color index −0.21 and its luminosity 35,000 $L \odot$. Approximately 500 parsecs (1,600 light-years) distant, it lies more or less "side by side" with ε, but beyond δ in our line of sight. ζ is gravitationally bound to a much smaller B star: at 2.6 seconds (1,280 AU) apart, the components are too close to be resolved in smaller telescopes; in more powerful instruments, Zeta B appears slightly reddish in comparison to the blue-white radiance of Zeta A. About 22½ times farther out from the primary is a very faint (10th-magnitude) telescopic component, to which some observers assign the unlikely color "gray," this impression the result of its ghostly faintness. As no change has been noted in its position with relation to ζ A and B, it may not be gravitationally bound to them.

43. IC 434 (B 33), the Horsehead Nebula, photographed in red light with the 200-inch reflector. (Hale Observatories)

The Alnitak system is moving away from both Alnilam and Mintaka, which are also changing their positions relative to Bellatrix. Therefore the Lord will indeed, one distant day hundreds of thousands of years hence, "unfasten the Sash of Orion." He is in fact doing so now, albeit very gradually. The entire constellation will alter its shape, owing to the stars' changing positions; and an equal factor in Orion's altered appearance will be the evolutionary development of those stars. Old Betelgeuse will have burned out and exploded and left the Hunter with an expanding gaseous nebula as a right armpit; if some of the faint stars in the club remain, it will be poised by itself, with no arm to hold it aloft! Rigel will also have traveled the path of Betelgeuse and reached the end of a short life span, the penalty for burning its fuel too rapidly, leaving Orion at first with a red left knee, and later on, "without a leg to stand on." However, new stars born out of the gaseous matter in Orion's nebulae will appear, if not exactly in the proper places to conserve the Hunter's form, then perhaps aligned in such a fortunate manner as to create a new sky picture for our distant descendants.

Belt Vicinity

(σ) SIGMA *Orionis*
(RA 5 h 36 m   δ —2° 38')   m 3.73

Immediately southwest of Alnitak and lying 1,400 light years distant, we find σ, a multiple eclipsing binary of spectral class O9 V and luminosity 5,000 $L \odot$. The close primary pair is exceptionally massive, about 35 $\mathcal{M} \odot$. Astrometric variations indicate an additional unseen component. Three more components lie farther out; the most distant of these is notable for very strong helium lines in its spectrum; at the edge of naked-eye visibility (m 6), it also has an eighth-magnitude companion.

HORSEHEAD NEBULA (IC 434) B (Barnard) 33
(RA 5 h 39 m   δ —2° 32')

Between ζ and σ, lying in the bright nebulosity IC 434, we find the finest example of a so-called *dark nebula,* one of the most striking objects in the heavens when revealed to us through long-exposure photographs made with powerful reflectors. Bright nebula IC 434 was first detected photographically by Pickering in 1889; soon afterward the Horsehead itself began to show up on photos, but its true nature was not at first understood. Trumpler (p. 101), through his work with open star clusters, provided the first evidence of obscuration and absorption by interstellar gas in the plane of the galaxy. Visually, this dust and gaseous matter presents itself in the form of the various nebulae, some of them bright, hazy objects, others dense dark clouds that cut off our view of the stars lying behind them. These dark accumulations of interstellar matter are masses of gas not close enough to a star or group of stars to reflect much light or to be stimulated into emitting their own radiation. They appear as dark patches or rifts in the bright star fields.

B33 has the unmistakable form of a rearing horse, as it appears in dark contrast

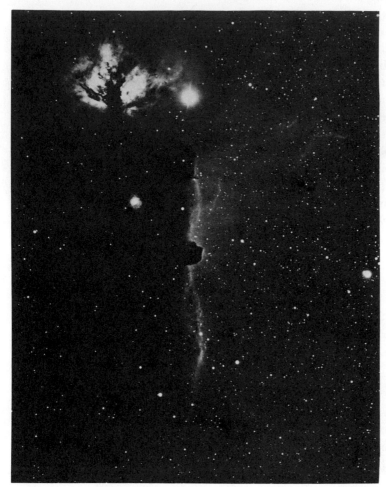

44. Field of the Horsehead Nebula. At the top of this remarkable photograph, immediately to the east of the Belt star Alnitak (Zeta Orionis), is the bright nebulosity NGC 2024, crossed by a dark, north-to-south absorption lane with multiple branches, giving it the appearance of a flaming tree. Photographed in red light with the 20-inch Astrograph. (*Lick Observatory*)

against the luminosity of IC 434 (fig. 43); with a total diameter of 70,000 astronomical units, or more than one light-year, it is a very great horse indeed! Its head is in profile; the ears slant forward, and with its mouth agape, one can fairly hear it snort! Because many astronomers believe that dark nebulae constitute the initial phases of star formation, it may well be that B33 is whinnying with joy and pride over the conception of some great star, to be born at a future date in its dark stables.

The Sword

Fully sheathed, this all-important yet neglected weapon hangs from a point below the Belt: it seems the Hunter prefers to use his upraised Club on Taurus instead! With its three naked-eye components vertically aligned and fainter and closer together than are the stars of the Belt, the Sword looks like a reflection of the Belt, but when subjected to scientific scrutiny it has new revelations for us. Its multitude of fascinating objects, most of them much too faint for the naked eye, provide us with a wealth of information about star formation. The Sword of Orion, uninvolved in the Great Hunter's forthcoming battle, is nevertheless of prime importance to astronomers and to all of us, for observation, study, and discovery.

NGC 1977
(RA 5 h 33 m δ −4° 52')

The upper starlike object of those marking the Sword is actually a bright nebula, which surrounds stars 42 and 45 (Flamsteed). Number 42 is a spectroscopic binary, type B3. Although visible with the naked eye, the nebula is much smaller and fainter than M42, its spectacular neighbor to the south.

The Great Nebula

M42/NGC 1976
(RA 5 h 32.9 m   δ −5° 25')

M43/NGC 1982
(RA 5 h 33.1 m   δ −5° 18')

One of the few gaseous nebulae that can be seen with the naked eye, M42 and its tiny adjacent companion M43 (fig. 45) appear together as a misty "star," the central of the three marking the Sword. Sometimes described as "greenish," when viewed through a small telescope it seems to be fan-shaped. One of the most impressive objects for viewing, it contains some very unusual stellar objects, many detectable only with a radio telescope. However, we commence our exploration of M42 with a famous group of stars, readily accessible to all observers.

THE TRAPEZIUM (Θ) *Theta-1 Orionis*
(RA 5 h 32.8 m   δ −5° 25')

Enveloped in the famous Great Nebula (M42), in the central portion of the Sword, is one of the most impressive multiple star systems suitable for the smaller telescope. At their approximate distance of 500 parsecs, or 1,630 light years, its four brightest stars form a distinctive quadrangle, called the Trapezium (fig. 46). Individual magnitudes range from 5 to 7; and spectral types O through A are represented by these hot young stars, which were formed about three hundred thousand years ago at most. The three brightest were first noted by Christian Huygens in 1656, and the fourth in 1684. Early in the nineteenth century, two more Trapezium stars were discovered, by Otto Struve and Sir John Herschel, and toward the end of that century two more very faint stars were observed. A group of more than three hundred stars are now known to surround the Trapezium. Nearby, and prominent within the same star field, lies *THETA-2,* a wide double, magnitudes 5.2 and 6.5 and spectral types O and B.

45. M42 (NGC 1976), the Great Nebula in Orion, together with NGC 1977 (the northern "star" of the Sword). Tiny companion nebula M43, with its multiple dark obscuration lanes, lies on the northeast rim of M42. All three bright nebular masses show a wealth of detail including dark gas dust clouds, luminous central areas, and intricate feathery structures with their faint outer parts. Photographed in blue-violet light with the 36-inch Crossley reflector. (*Lick Observatory*)

46. *The Trapezium,*
surrounded by the central
region of M42, showing
many faint stars and much
nebular detail.
Photographed in infrared
light with the Shane 120-
inch reflector. (*Lick
Observatory*)

Open Clusters

Together, Theta and these surrounding stars constitute another example of an "open" galactic cluster, a group of stars born at about the same time from the same chemical mixture, lying close to the plane of the Milky Way. These clusters thin out in number as we move away from the plane of the Galaxy. Their components are irregularly distributed population-I stars intermixed with the gas and dust of their origin. When the individual members of such a cluster are plotted on the H-R diagram or on the color-luminosity, or *color-magnitude (C-M)*, diagram, in which the color index (rather than the spectral class) is plotted against the absolute magnitude, we see that in the course of time these stars evolve away from the main sequence at differing rates owing to their differing masses, which predetermine the speed of their aging processes. The first stars to turn away from the main sequence are the massive, highly luminous blue-white giants, found high up on the diagram; the lower down on the main sequence this "peeling off" of the graph of the cluster is, the older the cluster under analysis is, because even its stars of smaller mass have begun to leave the main sequence, that is, to

become chemically inhomogeneous and to evolve into red giants. These less massive stars, such as the Sun, evolve much more slowly than stars of large mass and take much longer to leave the main sequence.

The Fish's Mouth Nebula

Near the Trapezium is another obscuration, or dark nebula, which resembles the mouth of a fish, perhaps gaping in awe at the beautiful young stars of M42. This dusty material is not in the *immediate* vicinity of any of these hot protostars, however, and therefore its atoms are not sufficiently stimulated to emit radiation. Therefore, like the Horsehead, it is seen as a dark patch in the otherwise bright area around it.

The Molecular Cloud (OMC 2)—the northerly cloud

Discovered in December 1973 in the vicinity of the Trapezium, and extending down through the Great Nebula, to its south, is a dense, opaque cloud of cool gas known as a "molecular cloud," because its relatively low temperature permits the clumping together of atoms to form molecules. The mass of this molecular gas ensemble from which the young Trapezium stars were originally formed is 2,000 $M \odot$. Hot stars of spectral classes O and B, known as the "Orion OB Association" (including the major Orion stars Alnilam and Sigma), extend 12 degrees northwest of the cloud, with the age of the stars increasing in that direction up to 10 million years. All these stars are in various phases of their early development. The molecular gas in their vicinity is ionized, owing to the newly formed stars' radiation, which has "burned away" the dust and revealed these stars to us. The cool molecular cloud seems to split M42 into its two parts, and the Fish's Mouth nebula marks this separation visually.

History of the Great Nebula

M42's discovery has been credited to various observers, but it was first described in 1610 by an unknown astronomer named Nicholas Peiresc. It was rediscovered telescopically in 1618 by the Swiss Jesuit astronomer Cysatus, who was comparing it with the comet of 1618, and Christian Huygens again discovered it in 1656, adding his observations on the multiple star Θ. The Great Nebula was the first object Herschel viewed after completing his great telescope. In 1786, cataloguing various nebular objects, he mentioned "nebulosity of a milky kind" near Θ, and in 1789 he described this nebulosity as "an unformed fiery mist; the chaotic material of future suns" (see fig. 47). Later he said that this "self luminous" material "seemed more fit to produce a star by its condensation, than to depend on the star for its existence." Far ahead of his time, Herschel had guessed that gaseous nebulae might be stellar birthplaces.

The first definite proof that those nebulae which cannot be resolved into stars by the telescope must be gaseous in nature was offered by the pioneer spectroscopist Sir William Huggins in 1864 (p. 64).

A tremendous gaseous nebula, M42 is about 500 parsecs, or 1,630 light years distant. It has a bright central region about 10 parsecs (30′) in diameter, which alone could

47. M42 (NGC 1976), the Great Nebula, a gigantic bright nebula some 30 parsecs in diameter, the birthplace of newly forming stars. Its fan-shaped structure displays many obscuration areas such as the Fish's Mouth Nebula, seen here as a dark bay overlying the bright central region, site of the Trapezium, a quadrangle of young stars. Photographed with the 100-inch reflector on Mount Wilson. *(Hale Observatories)*

more than cover the distance between our Sun and Vega, in Lyra, but M42's outer regions may extend an additional 20 parsecs in all directions. Telescopic measurements give its angular size as 20 by 15 minutes. With an absolute magnitude of −6, it is a very luminous object despite its diffuseness (its density is much lower than a laboratory vacuum). Nevertheless, at an average temperature of 70 degrees K., its dust causes stars within its boundaries to appear redder than they actually are, owing to the scattering of the shorter wavelengths of blue light by the dust particles. This dust also conceals faint stars deep within the nebula.

Bright Nebulae

In contrast to the dark Horsehead Nebula and the Fish's Mouth Nebula, M42 is a bright nebula, a gas-and-dust cloud strongly illuminated by hot stars in or near it. Bright nebulae exhibit bright-line emission spectra or continuous spectra that also show the presence of faint dark lines (p. 21). Nebulae illuminated by the hot stars of spectral class B1 or earlier, such as the Great Nebula, emit sharp bright spectral lines, owing to the excitation and ionization of their atoms by the stars' strong ultraviolet radiation. (Nebulae illuminated by the cooler stars, later than class B2, merely reflect the light of these stars, and thus show their characteristic spectra.)

Ionization of Oxygen

The principal chemical constituent of the Great Nebula is hydrogen but the nebula also contains other common chemical elements, whose unusual spectral lines, which give the nebula a characteristic greenish color, were erroneously thought to represent some completely new element which astronomers accordingly named "nebulium." However, these lines were eventually shown by I. S. Bowen to arise from doubly ionized oxygen and nitrogen atoms in unfamiliar (highly excited) states; that was the end of "nebulium."* We now know that ultraviolet photons from the hot stars in the nebula tear off the electrons from oxygen atoms, and these free electrons are then captured in excited, *metastable* states by atoms missing their required quota of electrons. Owing to the extreme thinness of the nebular material, however, collisions between such excited atoms, which would de-excite them before they could emit radiation and permit them to "jump down" from the metastable states in which they have landed into the *ground state*† in the usual interval of 10^{-8} second (as in the laboratory), are very rare. The captured electrons therefore are forced to remain in the so-called metastable state‡ for long periods of time (up to several hours), so that they have time enough to reach the ground state by emitting characteristic green photons, which results in the unusual spectral lines.

A Stellar Womb

Although the discrediting of "nebulium" seems to indicate that there is nothing *chemically* new "under the sun"—or in the Sword of Orion—some new objects are certainly appearing in that Sword! The most interesting aspect of the Great nebula is its function as a "baby-star cradle." In its mists, concealed from us by shrouds of dust, new stars are forming; this exciting process, a necessary forerunner of life itself, merits a detailed description.

Before the earliest stages of the formation of a future star, interstellar clouds of gas must be brought together by the force of gravity and molded into massive globes. Gravity accomplishes this wonder with the aid of a weak silent partner, the galactic magnetic field. The galactic magnetic lines of force exhibit a characteristic bending, which, under appropriate conditions, gives rise to an instability, an initial compression, the *Rayleigh-Taylor instability,* which permits the gravitational field to carry on its work of molding the cloud into a protostar. This important phenomenon can be described in simple terms: hollows or depressions develop in the systems of the magnetic field lines, and great quantities of dust and gas are guided into these hollows by their mutual

---

It was soon followed into the Great Beyond of obsolescence by two more chemical miscalculations: "mysterium" and "coronium."

† The smallest permissible orbit of an electron around its atomic nucleus, according to the laws of quantum mechanics.

‡ A sort of purgatory (or no-man's-land) for electrons!

gravitational attraction. The more this gaseous matter accumulates, the deeper become the hollow wells and the more effectively does gravity compress the gas so that still more material builds up at an ever accelerating pace, eventually resulting in the formation of great prestellar gas/dust globes.

Supernova explosions are an additional factor in the process of very early star formation. As they explode, dying stars send powerful shock waves, traveling about six miles per second, through the interstellar gas and dust, which also contribute to the formation of the great globes. Thus, a globular gas/dust complex begins to contract because of its own gravitational pull, the effects of the Rayleigh-Taylor instability, and the supernova compression waves. The globe then breaks up into smaller globes, which continue to contract, eventually forming clusters of associated protostars. At first, the radius of such a smaller globe may, nevertheless, be as large as the orbit of our outermost planet, Pluto, but as the globe condenses, it begins to shrink, and the surface falls inward. This is known as the *free-fall* stage. With amazing rapidity, after only ten years, it contracts to the size of Mercury's orbit. Within such a contracting sphere, the temperature of the central region increases, so that energy is continuously radiated away until a starlike nucleus is formed. This hot nucleus, or embryonic star, has a complex destiny: it undergoes a complicated metamorphosis, a dynamic series of constantly changing phases in its prenatal life. In a sense, this baby star might remind us of an insect in its pupal stage, for a dense shroud of dust, impenetrable to light, now forms about it. This region is known as the stellar "cocoon." Furthermore, if the newly developing star is massive enough, it goes the insect world one better and acquires both an inner and an outer cocoon!

These star cocoons are obviously not spun of fine silk; rather, they are composed of materials aesthetically less pleasing but nevertheless vital to the future destiny of all organic life: namely, great quantities of the inorganic gas and dust that originated in the cool molecular cloud. The formation of protostars out of great gas/dust complexes may be observed in other parts of the Galaxy and is undoubtedly a universal process, one that resulted in the birth of our own Sun and its planetary system.

Astronomers differentiate between two principal types of gas/dust complexes: *HI zones,* which are neutral hydrogen regions, and *HII zones,* regions in which the hydrogen is *ionized;* that is, the hydrogen atoms have lost their electrons as a result of bombardment by strong ultraviolet radiation from hot stars already formed.

Boiling Stars

Deep within its cocoons, as if to announce its own birth, the stellar nucleus, or protostar, sends up a great infrared flare. This results, temporarily, in a considerable increase in the object's luminosity.

At this stage of its development, all energy released from the nucleus is transported outward by the boiling, "mixing" process known as convection (p. 92, footnote). The Japanese astrophysicist Chushiro Hayashi was the first to analyze this convective process in contracting protostars, and so this phase of their metamorphosis is called the *Hayashi stage.*

In the Great Nebula many groups of protostars are in the Hayashi convective stage. The material in their outer layers is churning vigorously, like water in a teakettle brought to a full boil; and their atmospheres are in a state of violent convective turbulence.

The first star of this type to be discovered, an irregular short-term variable called T Tauri, was observed more than thirty years ago in the domain of Taurus, Orion's opponent. All protostars of this category are now referred to as *T Tauri variables*. They tend to herd together in groups known as T-associations, and they run the gamut in temperature, although often very hot and massive. They are charted above the main sequence of the H-R diagram, and according to the general law of nature already discussed, those of greater mass are evolving much more rapidly than the less massive stars and will reach the main sequence millions of years sooner. While on the main sequence, these stars may lose much of their original mass, owing to turbulence in their interiors.

Just as the Hayashi stage gets underway, the protostar sends up a second flare. After this event, the temperature deep in its interior steadily rises, until at several million degrees K., the first thermonuclear reactions begin. When the central portions of the protostar reach a temperature of at least 10 million degrees, the proton-proton chain of nuclear reactions begins, and a state of balance is achieved between the outward gas pressure and the inward pull of gravity. The birth process is now complete.

We have drawn quite a complex picture, which tells us that nature chooses to wrap her new goods very carefully: after some hundred thousand years of development, the protostar might be compared to a fragile object that has been enclosed in its own tiny gift box (the inner cocoon), placed in a much bigger box (the extended, HII region, or ionized gas/dust complex), and all of that, finally, surrounded by the outer package (the original cool molecular cloud).*

Figuratively speaking, of course, this starlike object, well protected and superbly packaged, will be delivered in due course to its future address on the main sequence.

Infrared Sources

We have already mentioned the two major types of gas/dust complexes: HI, the neutral zones; and HII, those consisting of ionized hydrogen. The Great nebula is a good example of a young HII zone. This zone is, in its turn, situated within the dense, opaque molecular clouds near Trapezium already described.

Within these HII regions we find areas that emit *infrared* radiation. This means they emit light in the longer wavelengths just outside the range of visibility at the cooler end of the spectrum. Again, inside these "extended" infrared sources are much smaller, hotter, and more concentrated areas of radiation, the so-called "point" sources, which probably immediately surround the gas/dust cocoons of newly forming stars.†

The detection of infrared light waves requires the use of a special infrared photometer attached to the telescope and mounted on a metal block that is in thermal contact with a container of liquid nitrogen coolant.‡

---

* This packaging is in the inverse order, more or less, starting with the outer wrapping.

† One of the brightest infrared clouds is the "Ney-Allen" source (discovered in 1969), surrounding Theta, the Trapezium, but centered on Theta 1 C Orionis, one of its stars.

‡ For a detailed explanation of the process of detection of infrared light, the reader is referred to Chapter 2, p. 21, Infrared, the New Astronomy, by David A. Allen.

The Kleinmann-Low Nebula

In the year 1967 the astrophysicists D. E. Kleinmann and F. J. Low published an important paper, *Discovery of an Infrared Nebula in Orion,* which has become one of the earliest landmarks in the development of a new branch in the dynamic and ever changing science of modern astronomy. The *Kleinmann-Low Nebula,* or *KL Nebula,* as it came to be known, is an interesting addition to man's knowledge of the heavens and an early and important achievement of the new science called infrared astronomy.

The KL Nebula is a new type of celestial object, detectable only in the infrared part of the electromagnetic spectrum, with the aid of the aforementioned process. This sub-nebula is the more compact central region of a much larger infrared cloud in the Great Nebula. KL is in the upper part of M42, quite near the Trapezium (1' of arc south of it) and deep in the densest part of the cool molecular cloud.

Becklin's Star

In the year of the discovery of the KL Nebula, Eric Becklin, then a young research student at the California Institute of Technology, and G. Neugebauer, a professor at the Institute, jointly discovered a bright point source of infrared radiation inside KL. Professor Neugebauer, a pioneer in the field of infrared astronomy, was already carry-ing out an infrared survey of the entire northern sky, along with Professor Leighton, also of CalTech.

Becklin's Star, variously called Becklin's object, the Becklin-Neugebauer source, and just plain BN, is an unusual item with a color temperature of about 550 degrees K. and is the brightest star-in-formation to be detected in the infrared wavelengths. BN is gigantic, with a diameter of fifty astronomical units as computed from its radiation: in other words, fifty times the radius of Earth's orbit around the Sun! This infrared nebula has a total luminosity thousands of times that of the Sun. In the course of its future development, its radius will contract to a "normal" size, if it really is a protostar now in its earlier stages.

The correctness of this assumption is not assured. For there is a rival theory concern-ing Becklin's Star which became the center of a scientific controversy a few years after its discovery. Through their research into the nature of this fascinating object, the astrophysicists M. V. Penston, D. A. Allen, and A. R. Hyland concluded that BN is not a protostar at all, but is much farther along on its evolutionary path, its infrared radiation being *intrinsic,* rather than stemming from the dust and gas that surround it. According to this theory, the star has already arrived on the main sequence, but its extremely high luminosity is obscured by surrounding envelopes of gas and dust.

The Future Star of Orion

More recent investigations have revealed the hydrogen content of a very small and compact HII region, thought to be the star's inner cocoon. Becklin's Star cannot yet be photographed with the usual blue-sensitive or red-sensitive (visual) plates, nor can it be directly viewed even with the aid of the most powerful telescope, for it is embedded in these inner and outer cocoons (p. 113) and they are so compact that they block all light

from this secret star except the infrared wavelengths. However, if the newer theory is correct, BN has already passed through its earlier evolutionary stages to settle down on the upper left-hand segment of the main sequence, and these gas/dust envelopes can be expected gradually to dissipate.

We may anticipate, therefore, that after about one hundred thousand years, the Great Hunter, already honored by his mythological ascent into the heavens, will receive one additional meritorious award: a brilliant new jewel in his Sword!

HATSYA (ι) *Iota Orionis*
5 h 33 m   δ −5° 56')   m 2.76

Hanging below the Theta star cluster and the Great Nebula, third of the three and in a position that marks the deadly tip of the sheathed sword, is a bright star sometimes called "Hatsya" and also designated as the Arabic *Na'ir al Saif,* or "Bright One in the Sword." However, noting its position between the Sword and Orion's feet (and perhaps mindful of the advantages of the swordless resolution of conflicts), the practical Chinese named the star *Fa,* or "Middleman."

Iota is 2,000 light years distant, blue-white, and of spectral class Oe V, which indicates emission lines that characterize a star just approaching the main sequence. It is thus another young star of the type so frequently found in Orion, one of very high temperature, great luminosity (20,000 $L \odot$), and gigantic size. Iota is a triple, suitable for small telescopes, whose companions are a blue-white B9 star and a very faint (m-11) reddish star much farther out from the AB pair. Iota A is also a spectroscopic binary, with a period of about 30 days.

THABIT (υ) *Upsilon Orionis*

To the south of ι is a fourth-magnitude star, *Fl. #36,* lettered υ in the old Heis *Atlas Coelestis Novis,* of 1872; the star's proper name was apparently derived from the Arabic *Al Thabit,* or "The Endurer," perhaps in patient tolerance of its obscurity.

The Head

Noticeably smaller than his more famous shoulders, waist, and feet, the Great Hunter's cranium seems rather stunted. We wonder if this light-headedness might have been responsible for the boastfulness that led to his ultimate downfall. The three stars of the head, which form a small triangle, were important to the Babylonians, known as the "Constellation of the King" and rising at dawn at the summer solstice. They were "Twins" to the Euphrateans and others; and a "Coronet" to the Persians. In India, they were the third lunar station, or "Head of the Stag," and sacred, perhaps because they marked the vernal equinox in 4500 B.C. To the Chinese, however, despite their faintness they were "The Head of the Tiger."

MEISSA or HEKA (λ) *Lambda Orionis*
(RA 5 h 32 m   δ +9° 54')   m 3.40

The star at the apex of the triangle, the brightest of the three, is much fainter than ι, at the tip of Orion's Sword, visual proof that the Hunter's Sword is stronger than his head! "Meissa" originated as a misnomer, but "Heka" is derived from *Al Haḳah* or "The White Spot." At 1,800 light years, it has a luminosity of about 9,000 L ☉. A double star of class O8, its components are suitable for amateur viewing.

(φ) *PHI 1 and 2 Orionis*   m 4 and m 5

The faint stars forming the base of the "cranial" triangle are barely visible to the naked eye on a clear night. Through a telescope, φ is revealed as two stars (not gravitationally bound, however); the fainter component is of the sixth magnitude.

The Lambda Orionis Nebula

Surrounding λ is a dense shell of absorbing clouds, expanding at a rate of 8 kilometers per second. Possibly another "baby-star cradle," it is now believed to be the birthplace of λ and others, which were formed as a result of compression in the cold gas of this dark shell. The shell's diameter is 180 light years, and its age is estimated at 2 million years.

Region Near the Head

PLANETARY NEBULA NGC 2022 (H 34[4])
(RA 5 h 39 m   δ +9° 3')

Our telescope should be fairly powerful. If we shift it about 7 minutes to the southeast of Heka, we encounter a miniature but excellent example of a *planetary nebula,* an expanding gaseous shell surrounding an exploded star. Its green spectral line, so characteristic of bright gaseous nebulae, which is especially apparent in planetary nebulae, was discovered by Huggins, and it immediately disclosed to him the gaseous nature of the so-called "planetaries." Of tremendous importance, his discovery indicated that space need not be empty, but enormous volumes of it could be filled with luminous gas. This concept led to such investigations as the studies on Mintaka by Hartmann.

According to recent theories, a "young" planetary may be the ejected product of an old star, one late in its evolution but only in the earlier stages of its dissolution, in which it begins to throw off material. However, we see that NGC 2022 displays a brighter elliptical "ring," apparent magnitude +12, its dimensions 22 by 17 seconds, enclosed within an extremely tenuous outer ring (28″ × 27″), which is barely visible through the telescope. The double shell formation may indicate that the fainter (m-14) central star has already undergone two outbursts. The inner ring would therefore be the product of the second, more recent ejection, and the outer, very thin ring, already in a state of dissolution, would indicate a primary ejection of the star's outer layers at a more remote time in the past. This nebula may therefore be regarded as a "mature" planetary, whose unstable central star is getting rid of its excess mass and settling down to the proper life of a stable white dwarf. The dense gases of this central star generate a continuous spectrum, without emission or absorption lines. Therefore it is probably a very compact, hot object in such a high state of ionization that no neutral atoms can exist near its surface. Such conditions are to be expected in a star approaching the white-dwarf stage.

FU ORIONIS, a "peculiar" star
(RA 5 h 43 m   δ +9° 3')

If we again shift our line of sight just .08 degree to the east of planetary nebula NGC 2022, we observe one of the most intriguing mysteries in Orion. Like a bright topaz resting upon its own black velvet cushion, the brand-new star FU gleams at us from the central portion of dark nebula B35. Classed as spectral type F, G, or "F peculiar," this object challenges our scientists, for its history presents a puzzling picture. In 1937, as a very faint star in an almost starless black void not far from the Hunter's head, it became 250 times brighter than before, increasing from the fifteenth magnitude to the ninth in about three months' time. It remained at magnitude +9.7 for two years but then diminished in brightness very gradually so that in 1960, twenty-three years after its sudden flare, it was still at magnitude +11. Now at +12, it is still fading very slowly.

Astronomers have differed about its true nature. At first, many considered FU to be an extremely slow nova, but newer theories of stellar formation point to the probability of a brand-new "baby" star, whether in the Hayashi stage or further along in its development. It is thought to have entered the flare stage at the time of its sudden increase in brightness; however, its surface temperature is about 5,000 degrees K., much higher than usual for a protostar at that stage of its formation. Other young stars of similar type tend to decline in brightness after the flare. Although FU's behavior resembles that of novae whose expanding gases emit rich emission-line spectra, only a few weak emission lines have been observed during FU's maximum as a type-F5 star and during its subsequent gradual change to classes G0 and G3.

It has also been classified as a T Tauri irregular variable (p. 146). FU Orionis has in fact given its own name to this type of protostar, often called a *fuor*. However, some authorities suggest that it may not be a protostar in the flare stage, but a newly arrived main-sequence star. If so, it has provided observers the unique experience of witnessing the transition of a protostar from its final natal stage to its earliest beginning as a full-fledged "normal" star—a moment of greatest interest, whose imminence cannot be anticipated with certainty in any given object. Astronomers have justly dubbed FU's stellar classification "peculiar"; not only is its temperature too high for a protostar in the flare stage, but its increase in brightness has been far greater than present theory allows for a star entering the main sequence. Nevertheless, the star V 1057, in Cygnus (the Swan), has followed FU's example, exhibiting similar behavior; therefore, some long-cherished notions may have to be scrapped.

THE CLUB (Right, Betelgeuse Arm)

This weapon is marked by five rather faint stars; we therefore wonder how reliable it would prove in the heat of battle! Its batting end is delineated by *(χ) CHI 1 and 2,* both fifth-magnitude stars. The base of the Club—and Orion's presumed hand—is more or less indicated by *(ξ) XI* and *(ν) NU,* each magnitude 4. Orion's uplifted elbow, to their south, is marked by (μ) *MU,* also of the fourth magnitude a class-A2 spectroscopic binary with a period of 17½ years.

Immediately to the east of $χ_1$, (Chi₁) Fl. #54, at the tip of the club, is the long-period red variable *U Orionis.* It has an amazing magnitude range but is visible only near its maximum of +5.2, and "fades out" all the way down to +12.9 with a period of 372 days. $χ_1$ and Fl. #57, magnitude 5.9, to which U Orionis may be compared, lie in its field. About 800 light years distant, U is considered similar in type to "Mira the Wonderful" (p. 304).

More Open Clusters

In the lower-arm or wrist region, south-southeast of ν and west-southwest of ξ, is *NGC 2169,* a loosely associated group of nineteen stars; although this object appears to

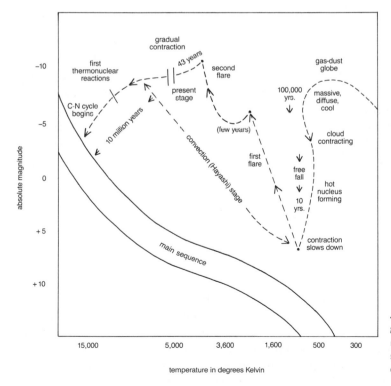

48. *Evolutionary track of the protostar FU Orionis* descending to the main sequence in the H-R diagram.

the naked eye as a sixth-magnitude star, it is really a loose and poorly formed cluster 2,570 light years distant. *NGC 2194,* a very remote (16,300 LY), rich cluster of the ninth magnitude, consisting of a hundred stars, lies in the same region.

THE SHIELD (Left, Bellatrix Arm)

Legend has it that Orion slew a lion to demonstrate his prowess; he then fashioned the unfortunate beast's skin into a shield, which he now holds in his upraised left arm as he awaits the Bull's charge. The Shield consists primarily of six third- to fifth-magnitude stars in a curved vertical row, all designated π and numbered 1 to 6 from top to bottom. Seen variously by different cultures, they were a "Crown" to the Persians, "Orion's Sleeve" to the Arabians, and "Three Flags" to the Chinese.

(π) PI$_3$ (RA 4 h 47 m δ +6° 53'), referred to as "Tabit" in some charts (but not to be confused with υ), is somewhat brighter than its fellows in the Shield, because it is only 26 light years distant. Of spectral class F6 V, it is therefore not the usual white-hot gigantic star so often seen in Orion. *(π) PI$_5$* (RA 4 h 52 m δ +2° 22'), marking the lower part of the Shield, is of interest as a variable visual double, type B2, consisting of a fourth-magnitude primary and a faint companion to its northwest.

To locate some minor stars of Orion, where imagination allows, we differentiate the Shield from the arm that holds it and indicate the anatomical parts of that arm: *(ρ) RHO,* in the "upper arm" about halfway between Bellatrix and π^6, of the "elbow," is a double, magnitude 4.46; its components, orange and blue, are 7 seconds apart. Immediately south of ρ is Σ *627,* a faint (m 6) double; the type-AO components, a beautiful pair for viewing through the low-power telescope, are 21 seconds apart.

The Hunter also sports a faint white star lying approximately midway between Betelgeuse and Bellatrix and marking the pit of his neck or his Adam's apple. Despite its vital location, this unfortunate is unnamed and carries no Bayer designation, the Greek letters having been used up on brighter stars, but it is *HD 37320* in the Henry Draper Catalogue, and also *SAO 112979* in the Smithsonian Astrophysical Observatory Star Catalog '66 (RA 5 h 38 m δ +7° 32')   m 5.88, spectral class B8.

(ω) OMEGA lies almost due south (and slightly to the east) of HD 37320, directly in the center of the chest. Although bearing the last Greek letter, this variable, magnitude 4.4 to 4.6, is no insignificant object, but a class-B3 giant 310 parsecs distant, similar to famed γ Cassiopeiae, a highly irregular shell star showing bright hydrogen spectral emission lines. On either side of ω are two symmetrically aligned faint stars; somewhat over 3 degrees west-southwest of ω is B2 subgiant *(ψ) PSI,* a 4.5-magnitude double. All these form a curving row that outlines Orion's rib cage.

Completing the Picture

To assure the Hunter a good connection between his famed blue-white left knee and his torso, *(τ) TAU,* a faint but vital star, lies a quarter of the way up his left thigh, not far to the north-northeast of Rigel. Obscure τ, like its famous and far more prominent neighbor, is a blue-white binary (m +3.6).

We have dallied longer than usual within Orion's boundaries, because it offers so many attractions. In taking our leave, we note that the entire package is appropriately tied up with a great celestial "ribbon" called **BARNARD'S LOOP,** a faint ring of gas that surrounds Orion. Covering a visual area 14 by 10 degrees, but 19 by 14 degrees in ultraviolet light, this nebula encompasses the Belt and Sword areas like a great cosmic bubble. Believed to be remnants of a supernova, this cosmic bubble generated shock waves that may have contributed to the early formation of protostars; a "bulldozing effect" is now visible on material outside the optical loop that has become superdense owing to radiative pressure from the hot stars born in the nebula.

As more of these young stars are formed, the Hunter's future destiny will be increasingly dependent upon them, for the great Orion stars familiar to us will be gone from the heavens, "burned out" from their headlong evolutionary pace. The short life span of stars like Rigel and Saiph (κ) are nevertheless adequate for the birth and death of countless civilizations on Earth who may observe such stars. These societies of the future will plot their own sky charts, according to the folklore and stellar images of their remotely distant ages. Through these events, the benign Sun, our own dwarf star, virtually unchanged as it evolves at a snail's pace, will continue to sustain life on its third planet, Earth, the Sun's chemical evolution practically unaffected by the passage of a mere few million years!

ERIDANUS

The Hunter stands on the banks of this great, sprawling river, his left foot almost touching its waters. The Northern Stream of Eridanus extends from that point westward to the vicinity of some faint stars in Cetus, the Whale. The Southern Stream circles back and forth around the boundaries of Cetus and Fornax (the Chemical Furnace) and, continuing its long flow far to the southwest, arrives near the junction of Phoenix, Tucana, Hydrus, and Horologium, at the borders of the south circumpolar zone, a point whose distance from Rigel would cover a path more than three and a half times the full length of Orion!

Although Eridanus' winding pattern has been imaged from earliest times as a river or as water, authorities have never agreed concerning its correct name, or the correct body of water with which it should be identified. Hesiod associated it with the mythological Phasis, in Asia, and Homer with the "Ocean Stream" flowing around the Earth, an early concept of the world. Eridanus was said to be the River Po, of Italy; the Ebro, of Spain; the Granicus, of Alexander the Great; and also the early Rhine and Rhône. It may relate to the story about Apollo's son Phaëthon, who lost control of the sun chariot and was hurled into a river, which was later transferred to the heavens. Eratosthenes said that Eridanus represented the Nile, while others suggested the Tigris-Euphrates, which, like the Nile, is a long and winding stream with two great branches. Even the River Jordan and the Red Sea have been suggested as its counterparts.

Although Eridanus' boundaries enclose a larger area of sky than any other constellation, most of its stars are of third magnitude and fainter. An exception is **ACHERNAR (α) ALPHA Eridani,** its southernmost star. Derived from *Al Āhir al Nahr,* "The End of the River," its name has many variations. Lying 32.5° from the south pole (RA 1 h 36 m δ −57° 30′), Achernar cannot be viewed from farther north than southern Florida and Texas. A blue-white giant of the Orion type, it is the ninth-brightest star (m 0.53), with a luminosity 650 L ⊙.

At the opposite (northern) end of the River lies *CURSA (β) BETA Eridani* (RA 5 h 5m δ −5° 09′) m 2.79. Sometimes included with the stars of Orion, Cursa is about 3° northwest of Rigel. If counted as an Orion star, Cursa would mark the knee of the Hunter's bent left leg; but the Arabians called β *Al Kursiyy al Jauzah,* or "The Footstool of the Central One." Cursa, 78 light years distant, is a class-A3 star, which denotes pure white, but it has been described as "topaz yellow," and the Chinese named it *Yuh Tsing,* the "Golden Well"! This color perception may be caused by various factors: β's contrast with nearby blue-white Rigel; atmospheric conditions near the horizon, where stars of Eridanus are seen from the north at certain times; and, for the modern viewer, the chromatic aberration in the telescope objective; which result in many differences of opinion regarding the colors of stars. Look for the wide double 66 Eridani (Σ 642), of similar spectral type, nearby.

Moving along the stream some distance to the west, we find *ZAURAK (γ) GAMMA Eridani* (RA 3 h 56 m δ −13° 39′), m 2.98. A reddish (M0 III) star, γ has a small companion, first discovered by Herschel. To Zaurak's northwest, of note because it is one of the nearest bright stars (10.8 LY), lies *(ε) EPSILON Eridani* (RA 3 h 31 m δ −9° 38′), m 3.73. An intriguing unseen companion accompanies this star, its mass less than 0.05 $\mathcal{M}$ ⊙, which makes us wonder whether it might be the largest and most massive (Jupiter-like) member of a great planetary system; or one of the least massive stars known!

From *(η) ETA Eridani,* which is farther to the west, we now journey along the bend of the river, past a long series of (τ) *TAU* stars, numbered 1 to 9. τ 4 is of interest as a spectroscopic binary, an M3 red giant. Winding along just south of the constellation Fornax, we reach *ACAMAR (θ) THETA Eridani,* its proper name similar to that of α, whose name, Achernar, it bore in earlier days, when θ marked the end of the long River. Look for this third-magnitude white wide double quite far to the south on early-winter evenings.

Wonders of the Ostrich's Nest

Cursa (β) and some nearby stars were known as the "Chair" to Arabians; but nomads long before them saw it, in combination with *(o) OMICRON 1 and 2 Eridani,* as an "Ostrich's Nest." Early Arabia called o 1 *Al Baid,* or "The Egg" (now written as *BEID),* owing to its white color. Immediately to Beid's southeast lies o 2, or *KEID,* from *Al Kaid,* or "The Eggshells"! Also *#40* (Flamsteed) (RA 4 h 13 m δ −7° 44′) m 4.48, Omicron 2, is a triple system of great interest, only 16 light years distant (the eighth-nearest naked-eye star), and known for its very large proper motion. The primary is an orange main-sequence star, somewhat cooler than the Sun. Its faint companion was discovered in 1783 by Sir William Herschel. Separated by 400 astronomical units, the components (A and B) complete an orbit around their common center of mass only once in several thousand years. In 1851, Otto Struve detected duplicity on the part of the companion

(B), and an amazing phenomenon was disclosed: B, a close visual binary, consists of an extremely dense white dwarf and a larger aged, underluminous red dwarf of exceptionally small mass (0.2 $\mathcal{M}\odot$), separated by about 34 astronomical units. Suitable for low-power viewing, the white dwarf is only twice the size of the Earth, but, owing to its much larger mass, its surface gravity is about thirty-seven thousand times greater! With a density approximately sixty-five thousand times the Sun's, a cubic inch of its highly compressed matter has a mass of nearly 2 tons. Yet the interpenetration of its atoms is in harmony with natural law: this "degenerate" star has already traversed the evolutionary path of stars like Betelgeuse and has met the explosive destiny the red supergiant will face in a few hundred thousand years.

The dwarfs complete one orbit in about 248 years. We note that the center of their mass lies much closer to (or even within) the white dwarf, because of its much greater mass. Therefore, despite its smaller size, the white dwarf is closer to the central point of the dwarfs' orbit, and the red dwarf, although the larger of the two, would appear to revolve about the white dwarf. Red dwarfs may be among the most ancient objects in the universe; owing to their small masses, which they have had from their inception (and that in some cases also prevented them from achieving nuclear ignition), these stars evolve at a very slow pace and therefore have life spans correspondingly longer than stars like the Sun.

LEPUS

The Hare lies below Orion's feet, westward from the bright star Sirius. The Greeks of Sicily were happy to see it placed near the Hunter, because hares were exceedingly destructive to their crops. Early Arabs, however, called it "Chair of the Giant" or "Throne of the Central One"; but its four brightest stars were sometimes seen as camels drinking from the Milky Way. Lepus was also the "Boat of Osiris" to early Egyptians, who identified their god with Orion.

Double Stars

ARNEB (α): a very luminous (5,700 L $\odot$) but distant (900 LY) second-magnitude class-F star, with very faint (m-11) optical companion (separation of 35.5″). About 10 degrees south-southeast of Rigel (RA 5 h 31 m   δ −17° 51′).

NIHAL (β): a yellow giant with eleventh-magnitude class-K dwarf companion, suitable only for large telescopes. Look for β a few degrees south-southwest of α (RA 5 h 26 m   δ −20° 48′)   m 2.85.

GAMMA (γ): a yellow main-sequence third-magnitude star with sixth-magnitude orange companion; 29 light years distant. With separation of 900 astronomical units, wide and easy double for small telescopes. East-southeast of β (RA 5 h 42 m   δ −22° 28′).

A Symbiotic Story

We have discussed the recurrent nova-like eruptions caused by the exchange of material between the components of a red-giant/white-dwarf binary (p. 80); certain types of so-called *"shell stars"* with rapidly expanding atmospheres (up to 200 km./sec.) and high rotational velocities seem to be related to the ordinary recurrent novae. These class-B stars with broadened emission lines in their spectra resemble the final stage in the outbursts of such novae. When two close, unequal companions produce these eruptive effects through a mutual exchange of material, they are called "symbiotic" stars. *No. 17 Leporis* (RA 6 h 3 m δ −16° 29′) is a rather exceptional example of a symbiotic binary, because its magnitude variations are relatively small (m 0.07). Its components are a bluish (late-B or type-A "peculiar") primary with a gaseous shell, and a red-giant companion close to the primary. As the pair complete one revolution about every 260 days around their center of mass, material from the red star is drawn to the more massive blue primary, forming part of the latter's gaseous shell and causing outbursts that correspond approximately to the orbital period of the binary.

A Celestial Ruby

Lying at the border of Lepus and Eridanus (about 7° SSW of Rigel) is the red variable **R Leporis**, popularly known as **HIND'S CRIMSON STAR**, discovered by J. R. Hind in 1845. (RA 4 h 57 m δ −14° 53′). Its remarkable brightness range is about five magnitudes in its 432-day period. At its smallest magnitude, +5.5, it may be seen with the naked eye when viewing conditions are optimum; it is suitable for small telescopes at all times, and until it reaches maximum brightness, its intense crimson color is quite startling; at maximum, however, when its magnitude is smallest, the reddishness tends to pale. R Leporis is classed as an N, or carbon, star, because its molecular-band spectrum is dominated by carbon molecules (see p. 75, S Cephei). The pulsation period of R Leporis is quite irregular, alternating between fainter and brighter maxima, similar to Mira, its famous counterpart in the Whale.

MONOCEROS

The Unicorn crouches near Orion, eager to witness the combat between Bull and Hunter. Three 4.5- to 5-magnitude stars, Flamsteed numbers 15, 13, and 8, almost perfectly aligned from northwest to southeast and covering a distance of about 6½ degrees, form the distinctive single horn of the beast. **S Monocerotis** (#15), is a slightly variable fifth-magnitude class-O giant, about 8,500 L ⊙, with an eighth-magnitude companion 3 seconds distant from it. Fl. #8 is a white class-A giant (m 4.5) accompanied by a small, class-F companion (m 6.5). The three horn stars lie in an area roughly 9 degrees east of Betelgeuse, deep in the bright nebulosity of the Milky Way. Monoceros, first appearing in Bartsch's catalogue as "Unicornu," is relatively modern, although some astronomers and scholars believe it was preceded by earlier figures, including a horse. Since none of its stars is brighter than the fourth magnitude, the constellation is not favorable for naked-eye viewing except under optimum conditions. (β) **BETA** (Fl. #11), approximately 10 degrees south of our three stars, is worth mentioning as a fine triple star (RA 6 h 27 m   δ − 7° 0′); its beautiful class-B white stars create a slim triangle for the small (6-in.) telescope.

Contrasting Binaries

Among the most remarkable features of this faint constellation are two binaries that represent the extremes—large and small—of stellar mass. North of β (about 2° north of cluster NGC 2232, or Herschel 25[8]) and east of RU Orionis, is the very faint (m-11) binary **ROSS 614,** first detected at Yerkes Observatory in 1927. This is a pair of red dwarfs, with the mass of the primary only .14 $\mathcal{M}\odot$ and that of the m-15 companion only .08 $\mathcal{M}\odot$—the smallest known stellar masses with the exception of UV Ceti, in the Whale. First observed with the 200-inch Palomar telescope, in 1955, the B component has a radius only about one-eleventh the Sun's, its mass is about ninety times that of Jupiter, its surface temperature is 2,700° K., and its luminosity is only .000016 $L\odot$. Red, cool stars such as Ross 614B, although of very low mass, are nevertheless very dense: Ross 614B, according to its luminosity and mass, has a mean density about a hundred times the Sun's. (White dwarfs, however, are far denser.) Although a mere 13.1 light years distant, the component stars of Ross 614 can be perceived separately only with large telescopes like the Palomar at the time of their greatest separation (1.2″).

In marked contrast to the miniature Ross 614 binary is famed **PLASKETT'S STAR** (RA 6 h 35 m   δ +6° 11′)   m 6.06, lying about 1½° southeast of the horn star #13. One of the most massive binaries known, it consists of two class-O giants only 50 million miles apart, or a little more than half the Earth's distance from the Sun. At such close range, the components are probably exchanging material like the components of ordinary recurrent novae described on p. 78, although not so advanced in their evolution as the novae. The total mass of the system is now believed to be about 100 $\mathcal{M}\odot$; the secondary, less luminous component has been reported as the more massive, in violation of the mass-luminosity relation; it may be that the primary, less massive, will evolve rapidly toward the white-dwarf stage by reason of its loss of mass to the secondary star; and the latter, because of this accretion of mass, will evolve until it outshines the primary. In a still more distant epoch, it may then return this hydrogen-rich material to its companion, which will have evolved into a white dwarf, and the previously mentioned nova-like effects will follow.

The present luminosity of Plaskett's Star is about 3,000 $L\odot$, its absolute magnitude almost −4.0. Even the largest supergiants known, the *infrared giants*, are much less massive than the components of this unusual binary.

Clusters and Nebulae

Monoceros is notable for a variety of interesting and picturesque deep-sky objects. Of these, the so-called **ROSETTE NEBULA, NGC 2237** (fig. 49), is the best known. More suitable for fairly large telescopes, long-exposure photographs bring out its beauty. As its name implies, this glowing cloud of gas creates the image of a huge celestial bouquet as it surrounds bright galactic star cluster NGC 2244 (RA 6 h 30 m δ4° 54′), about 2 degrees southwest of Plaskett's Star, or directly south of #13). At a probable distance of 5,200 light years and with a diameter of about 93 light years, it is one of the most massive nebulae known (11,000 $\mathcal{M}\odot$). Of especial scientific interest are the many dark spots, or "globules," seen in the Rosette, which, it is theorized, may be identical to the great gas/dust globes which are the forerunners of protostars. In this, the Rosette Nebula resembles the Great Nebula of Orion as a "baby-star cradle." The gaseous material of this so-called *emission nebula* (i.e., one that shines by its own light, rather than the reflected light of stars) has apparently been "cleared away" from the nebular center by the winds of one or more hot, young stars of NGC 2244, which ionize the surrounding "bouquet" or "wreath" of bright nebulosity with ultraviolet radiation. The displaced gas forms a thick shell around the hollowed-out and relatively darker central region of the nebula; for this reason nebulae of this type are now called "interstellar bubbles."

Radio astronomer T. K. Menon initially investigated the Rosette, using the 85-foot radio telescope of the National Radio Astronomy Observatory at Green Bank, West Virginia, in 1961; his studies indicate a deficiency of gas in the nebular center. Subse-

49. NGC 2237, the Rosette Nebula, called an "interstellar bubble," is an exceptionally beautiful emission-type nebula. In the dark center, note star cluster NGC 2244; its hot stellar winds and radiation have apparently influenced the structure and appearance of the Rosette. Dark globules that may develop into protostars are visible in this photograph, taken in red light with the 48-inch Schmidt telescope on Palomar Mountain. (Hale Observatories)

quent findings with a more advanced instrument show a central cavity almost 70 light years in diameter surrounded by a gas shell 23 light years thick, in which most of the nebular mass is contained. William T. Mathews, of Mount Wilson and Palomar Observatories, proposed three theories in an attempt to explain the formation of the Rosette nebula. His first theory suggests that the missing gas condensed to form NGC 2244, the cluster in its center. His second theory ascribes the hole in the nebula to the action of outward-flowing hot stellar winds; the third theory proposes that microscopic dust particles in the nebula are electrified by ultraviolet light from the central stars and pushed outward by light pressure from newly formed stars in the cluster, this electrically charged dust dragging adjacent gas outward with it, clearing the central hole.

Recent observations support the stellar-wind theory, indicating a gas flow of 6 miles per second, well above the speed of sound, and showing that the two bright stars in the cluster emit strong stellar winds. The mass of the most luminous star is fifty times that of the sun, and its luminosity is 1 million $L \odot$, with a wind probably strong enough to blow out the hole in the Rosette. Tests with the International Ultraviolet Explorer (IUE) satellite relative to the temperature of the thin gas in the cavity will be conclusive in determining the birth process of the Rosette.

50. NGC 2237, the Rosette Nebula, enlarged section. Prominent lanes of dark obscuration matter crisscross the bright nebular mass. The dark globules are clearly visible. The nebular center, lower left, shows the young stars of NGC 2244, but the brighter stars in the picture may be foreground objects. (Hale Observatories)

1908 Yerkes   1913 Lick   1916 Yerkes

51. NGC 2261, Hubble's Variable Nebula, photographed over a period of eight years, showing changes in its appearance; note the nucleus, R Monocerotis, at its southern tip. (*Yerkes Observatory Photograph, University of Chicago*)

A Variable Nebula

In 1916, eight years before he began measuring the distances of galaxies by applying the period-luminosity relation to their Cepheid variables, Hubble discovered that a comet-shaped gaseous nebula associated with R Monocerotis (a variable star of uncertain type lying about 1 degree south of S, or #15) is itself variable, its nebular details constantly changing. Originally discovered by Herschel in 1783, "*HUBBLE'S VARIABLE NEBULA*" (NGC 2261) has been photographed over a period of thirty years, and its puzzling changes, which seem completely irregular and unrelated to R's variations, have inspired a number of hypotheses, ranging from "changing light conditions and moving shadows . . . cast by dark masses" (R. Burnham, Jr.) to ". . . the flow of light-pulses over the nebula [from a pulsating nucleus]" (Slipher). It is not certain that R, the nucleus, is a true star; Low and Smith, in 1966, first suggested that R may be embedded in dust grains and is thus a "protoplanetary system," but it has recently been classified either as an RW Aurigae or a T Tauri type of protostar (pre-main-sequence object); hence, unlike the Cepheids, it cannot be used to determine the distance of the nebula. Observations made in 1983 with the Mayall Telescope, at Kitt Peak, and with the Infrared Telescope Facility and the U.K. Infrared Telescope, at Mauna Kea, tend to confirm the thesis that R Monocerotis is the central sun of a protoplanetary system existing at the present time in the form of a core-halo structure. The halo probably consists of relatively slow-moving matter bound gravitationally to R and may represent a very early stage of a planet-forming disk around the young star. The total mass of the small particles is ten times that of the Earth, and owing to the "scattering" of the shorter (blue) light waves by these particles, the halo light appears bluer than the stellar light. The mass of the gas is comparable to the mass of our great planets Jupiter and Saturn. Inasmuch as mass loss from stars is almost certainly insufficient to produce the halo matter, the protoplanetary thesis is probably correct; furthermore, mass loss by the star and radiation pressure on the dust grains would tend to disrupt the halo as R evolves, but its halo is apparently stable. Similar observations were made at Kitt Peak and Mauna Kea of a Solar System-sized halo around the young star HL Tauri.

Two important highlights of the faint but content-rich Unicorn merit our attention: *THE CHRISTMAS TREE, NGC 2264* (RA 6 h 38 m   δ 9° 56′) is an open galactic cluster, about 2,600 light years distant, associated with S Monocerotis (#15), containing many very young pre-main-sequence stars and surrounded by a bright nebulosity visible only in very large telescopes. At the southern extremity of this bright area is a dark nebulosity, the famed *CONE NEBULA.* Like the Horsehead Nebula (B33), in Orion, the Cone is caused by turbulence in an underilluminated gas/dust cloud where protostars are forming. Long-exposure photographs reveal its spectacular beauty.

FIVE

Completing the Orion Scene

CANIS MAJOR

Apparently leaping after the Hare, Orion's faithful Dog probably has his eye on much bigger game, ready to assist his master in combat with Taurus. According to an early Greek myth, Canis was a gift of the dawn goddess Aurora to Cephalus, hunter and Sun symbol; in honor of its great speed, Zeus later placed the Dog in the heavens. Its chief star, owing to its great brilliance, was recognized long in advance of the entire constellation, and references to the "Dog" often meant Sirius. Arabians called Canis Major *Al Kalb al Akbar,* or "The Greater Dog," and also *Al Kalb al Jabbār,* meaning "The Dog of the Giant." This concept was emphasized in Euphratean astronomy and was familiar to the Etruscans. In India, our Greater Dog was known as *Mrigavyadha,* "The Deer Slayer," and also enjoyed the title we presently give to Orion, "The Hunter" (*Lubdhaka*). Prehistoric dwellers in India called it *Sarama,* one of the twin watchdogs of the Milky Way.

SIRIUS (α) *Alpha Canis Majoris*
(RA 6 h 43 m   δ −16° 39′)   m −1.42

The brightest he, but sign to mortal man
Of evil augury.

                    Homer

Popularly called the "Dog Star," its present name is apparently derived from the Greek *seirios,* or "scorching," which at first was applied to any bright star. Various authorities, however, have differed as to its origin, because in many early languages similar-sounding names appear, such as the Egyptian *Cahen Sihor,* the Greek *Osiris,* and the Celtic *Syr.* In fact, the name has untold variations, differing in every star catalogue of centuries past.

Sirius was an object of veneration in many ancient lands. The earliest Hindus knew it as *Sukra,* the rain god. In northern Egypt, it was extensively worshiped and many temples were oriented to its rising at the summer solstice, which heralded the flooding of the Nile, an event of supreme importance to Egyptian farmers. Numerous influences, good and evil, have been attributed to the star, and in this regard it has been the object of many superstitions. Roman farmers sacrificed fawn-colored dogs to it at festival times and tried to complete their farm work before its ascent. Chinese astrologers warned of attacks from thieves when Sirius was unusually bright. Owing to medical (as well as astronomical) ignorance, peoples of centuries past attributed famine and plagues to its "baleful influence" on the weather, and even the great Hippocrates (460–377 B.C.), Greek physician known as the "Father of Medicine," fell prey to this delusion! These notions of Sirius' unhealthy association with the heat of midsummer have persisted to the present time in the form of the expression "dog days." As a winter star, however, Sirius was also blamed for uncomfortable extremes of cold! On such a winter's evening we observe it now as it rises, its white radiance mirrored by the gleaming snow covering the countryside. We say that Sirius is *in opposition* at this time, for the Sun has just set and the two bodies are in opposite parts of the sky. In the summer and early fall, when Sirius' rise heralds the sunrise and denotes "curtain time" for the night's stargazing, we say it is *in conjunction.* These terms refer to the star's *elongation,* that is the angle formed by the intersection of the line from Earth to Sirius, and the line from Earth to Sun. When the elongation is 0, the star is in conjunction; when the elongation is 90 degrees, in quadrature; and when 180 degrees, in opposition. (These terms are frequently applied also to the Moon and the planets.)

Although Sirius has the greatest apparent brightness of any star (and therefore the smallest apparent magnitude), it is not the most luminous. A class-A1 white hydrogen star on the main sequence, its mass is $2.35 \, \mathcal{M}\odot$, its radius only 1.8 times the Sun's, and its luminosity about $23 \, L\odot$. However, at a distance of only 8.7 light years, Sirius, in our eyes, outshines every other star.

Observations of Aldebaran (Taurus), Arcturus (Boötes), and Sirius in the eighteenth century initiated a new era in astronomy with Dr. Edmund Halley's discovery, in 1718, of the proper motion of stars (p. 41). More than one hundred years earlier, in the sixteenth century, Giordano Bruno, later burned at the stake in Rome for heresy, correctly guessed that the stars are not fixed, but move about in space. Halley, by comparing his observed positions of Sirius, Aldebaran, and Arcturus with Hipparchus' data, found that the three stars had altered their positions with respect to background stars in the same fields; thus Bruno's early hypothesis gained credence, fundamentally changing prevailing concepts and hinting at forces in the universe that act on the stars.

The Companion of Sirius

Like all scientific milestones, Halley's discovery opened the doors to further revelations. In 1834 Friedrich W. Bessel, who first measured the parallax of a star (61 Cygni, p. 280), detected an oscillation, or wavy curve, in Sirius' path through the heavens and ultimately concluded that the bright star must have a faint, massive companion that produces the observed orbital motion of the brilliant primary about their common center of mass. At Bessell's death, in 1846, no telescope was large enough to reveal the companion, and many astronomers dismissed his theory. In 1862, however, the American telescope builder Alvan Clark directed his new 18½-inch telescope at Sirius and saw a faint point of light in the correct position predicted for Bessel's companion. Other hypotheses of Bessel, e.g., the existence of an unseen companion to Procyon (the bright star of Canis Minor) and an eighth planet (Neptune) in the Solar System, were also ultimately confirmed, and Bessel is now known as the founder of "the astronomy of the invisible."

Once the companion of Sirius was telescopically viewed, it became an object for subsequent research, and its measurable properties were either observed or calculated. The brilliant primary and the eighth-magnitude Sirius B (now called "The Pup") complete one revolution about their common center of mass in about fifty years; their mean separation is 24 astronomical units. As the Pup's nature was gradually revealed, the astronomical world experienced a series of surprises! Because the luminosity of a normal star depends essentially upon the temperature of its photosphere as well as on its surface area, the faintness of the companion—its luminosity less than one four-hundredth that of the Sun—suggested a very low surface temperature to astronomers of that time, even though a high temperature was already suspected owing to its white color. In 1915, however, Dr. W. S. Adams, of Mount Wilson Observatory, discovered that the companion's surface temperature is about 8,000 degrees (now revised upward to 8,500° or 9,000°), or more than 1.5 times as high as the Sun's. This implied that Sirius B radiates more than 5 times as much energy per square foot as the Sun! From this figure and the Sun's luminosity, 360 times the companion's, Adams deduced that the Sun's surface area is 1,370 times that of the companion: a tiny Pup indeed, its radius only 21,000 miles (currently revised to 19,000), or less than 5 times the Earth's radius. The astonishing implication of these figures was that if Sirius B's mass is nearly equal to the Sun's but its radius is only about 5 percent of the sun's radius (432,000 miles), then, gram for gram, the matter of the companion is unbelievably compressed; its density (mass per cubic centimeter) is 65,000 times that of water (this figure has been revised upward to 125,000), a condition scarcely conceived of in the earlier part of the century. Such a compression of matter means that Sirius B's constituent atoms do not exist in their normal states but are squeezed so close together that many atomic nuclei are crowded into a space previously occupied by a single "normal" atom.* Under such conditions, the electrons of these atoms are squeezed out of their orbits and move about freely like electrons in a metal (a degenerate state). A cubic foot of this star's material has a mass of nearly 2,000 tons; a matchbox full of such material would contain 1¼ tons of this so-called *degenerate matter*.

Several years later, Adams, continuing his investigation of this strange body and seeking confirmation of his amazing conclusions, obtained spectroscopic evidence for the companion's great density. Here Einstein's relativity theory entered the drama. According to the laws of relativity, when a *quantum* of radiation (a photon) of a given wavelength leaves a star's surface, it yields up part of its radiant energy in escaping from the star against the pull of gravity and its wavelength increases accordingly, since the energy of a quantum of radiation varies inversely as its wavelength. A spectroscopic analysis of the escaping radiation must therefore show a displacement of its spectral

---

* *The companion's high surface temperature would result from the inability of the "shattered" atoms in its photosphere to absorb radiation from its hotter, central region; thus the temperature of the entire star is about equalized.*

lines toward the red, the degree of displacement for a star of a given mass varying inversely as the star's radius: the greater the radius the smaller the displacement. This "Einstein red shift" is barely detectable in the stellar spectra of normal main-sequence stars, because their radii are large. But the companion of Sirius is so small that the displacement of any of its spectral lines is thirty-six times larger than for any solar spectral line, or a shift of .35 angstrom for a 5,000-angstrom line. From the total displacement (Einstein plus Doppler) of the companion's spectral absorption lines toward the red, Adams obtained an excess radial velocity (from the excess displacement) for Sirius B of 12½ miles per second, after subtracting the conventional Doppler displacement arising from the radial velocity of the entire system from the total Doppler displacement for the companion. Since the theoretically deduced relativistic shift of the companion's spectral lines caused by the foregoing gravitational energy loss is 12 miles per second, almost exactly in agreement with the observations, Adams had proof of the immense density and tiny radius of the companion, for its mass had already been calculated from the binary's orbit.

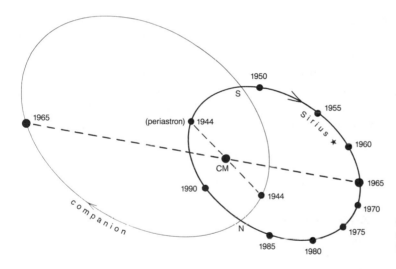

52. The orbits of Sirius and its companion relative to their center of mass.

Other stars sharing the Pup's strange characteristics were ultimately discovered, and they became known as "white dwarfs," the tiny, superdense stars occupying the lower left-hand corner of the H-R diagram (p. 87). With increasing knowledge about the evolution of stars, astronomers ultimately realized that the white dwarfs represent a final stage in the life of average main-sequence stars. Sirius' companion has already traversed a long evolutionary path, which first led to the loss of its outer layers in a series of nova explosions many aeons ago and then to its present stable, although seemingly strange, conditions of life as a white dwarf.

This account may arouse the reader's desire to see such an extraordinary star, which can be done even with a 12-inch telescope, provided the instrument is of high quality and has a sharp focus. Seeing the Pup for the first time is an exciting experience, but patience is the most important ingredient here, as well as good vision for distant

53. Three exposures of Sirius; shows the companion as a tiny point of light to its right; taken with the Shane 120-inch reflector. (*Lick Observatory*)

viewing. At first, the dazzling brightness of Sirius A is quite disconcerting; from its center, it seems to emit long, sharp rays (actually produced by the observer's eye), and these have a rather blinding effect. If you avoid looking directly at Sirius A's image and retain an attitude of "relaxed concentration," the following phenomena may result: at first, a number of tiny points seem to dance all about the circumference of the primary, but, the product of an optical illusion, they quickly disappear. As your eyes grow accustomed to the situation—this may take several minutes—a ghostly, almost color-less point of light suddenly appears near the primary; it tends to fade out, but always reappears, so that there is no doubt of its reality. You have seen the most famous white dwarf, the companion of Sirius.

MURZIM (β) *Beta Canis Majoris*
(RA 62 h 5 m   δ −17° 56′)   m 1.98

Sometimes called "Mirzam," its name is derived from *Al Murzim,* or "The Announcer," for when β rises, Sirius soon appears. Similar in type to β Cephei, Murzim is a class-B "bright giant," varying by about 0.03 magnitude in two different superimposed six-hour periods; all such variables are known either as Beta Canis Majoris stars, or, currently, as Beta Cephei stars. Sometimes called "quasi-Cepheids," these variables obey a period-luminosity law similar to that governing the Cepheid variables. They are believed to be in a state of overall expansion, with their pulsation periods lengthening as they begin to evolve away from the main sequence.

Other Stars in the Greater Dog

(γ) **GAMMA,** a class-B giant, called "Muliphen" by the eighteenth-century American astronomer Elijah Burritt, "Mirza" in the old *Century Atlas,* and "Isis" by Bayer and Ideler, is not only confusing as to its proper name, but also in regard to its magnitude: presently 4.10, but reportedly much brighter in previous centuries, supposedly having "faded out" since its early detection.

WEZEN (δ) DELTA: from *Al Wazn,* or "The Weight," presumably because its rather low position in the constellation and consequent late rising led desert dwellers to imagine that the star had difficulty in appearing above the horizon! On a summer morning, the dawn would also add to this problem. Wezen is a bright (m 1.82) but very distant (about 2,100 LY), late-class-F supergiant, shining with the luminosity of sixty thousand Suns.

ADHARA (ε) EPSILON, or Adara, from *Al Adhārā,* "The Virgins," a group including δ, η, and ο. Adhara is a wide double, with its m-8 companion at a separation of about 1,600 astronomical units.

FURUD (ζ) ZETA, of uncertain derivation, possibly *Al Kurud,* "The Apes," referring to surrounding faint stars; a blue-white main-sequence star and spectroscopic binary, 750 times the Sun's luminosity.

ALUDRA (η) ETA: one of the Virgins; another supergiant, class B, fifty-five thousand times the luminosity of the Sun.

(ο) OMICRON 1 and 2: centrally located at the beginning of the Dog's foreleg: ο 1 (Fl. #16) is a fourth-magnitude reddish star; ο 2 is a third-magnitude supergiant, very distant, at 3,400 light years, and comparable to Rigel in luminosity. Together with *(τ) PI,* a fourth-magnitude optical double 3 degrees to the northwest, they form an elongated triangle, which, along with some fainter stars, the Chinese saw as *Ya Ke,* the "Wild Cock"! To the northeast of Sirius are *(μ) MU,* a faint yellow-and-white double, and *(θ) THETA,* a red star at the Dog's northern border.

Clusters

Blue-white *(τ)* TAU (Fl. #30), lying just at the border of the Milky Way, in the territory of the Virgins and about 2 degrees northeast from δ, is surrounded by an attractive cluster of very young, hot giants *(NGC 2362).* Immediately north of τ is *UW* (#29), an exceedingly close binary that rivals Pierce's Star (AO Cas) because of its great mass and luminosity. UW's period is about 4 days, the separation of its egg-shaped components less than 17 million miles. As expected, at such close range they are exchanging material and are surrounded by a gaseous shell.

A particularly noteworthy cluster, suitable for low-power viewing and even visible to the naked eye, is *M41 (NGC 2287),* lying in the Wild Cock's territory, about 4 degrees south of Sirius. Aristotle, in 325 B.C., called M41 a "cloudy spot," perhaps the earliest reference to a deep-sky object. It contains about a hundred to a hundred fifty stars of varying spectral types. Inhabitants of an imaginary planet of any of them would enjoy a night sky filled with stars of greater apparent brightness than Sirius!

CANIS MINOR

Even as the star Murzim is Sirius' "Announcer," Orion's second dog heralds the rising of Canis Major. The Lesser Dog, or *Al Kalb al Asghar,* as the Arabians knew it, had various mythological identities, and with cataloguers centuries later it suffered a confusing array of titles. For today's astronomy, however, it bears especial interest as the setting of a remarkable discovery, echoing the story of Sirius and its companion. *PROCYON (α) ALPHA Canis Minoris* (RA 7 h 37 m   δ +5° 21')   m 0.35, sometimes called "the northern Sirius," is its focal point. A yellow-white class-F star with approximately twice the Sun's radius, α is only 11.3 light years distant, closer than the standard 10 parsecs, so that its absolute magnitude is +2.6. Its fairly large proper motion reveals certain irregularities, from which A. Auwers, in 1861, concluded that a massive companion was orbiting Procyon in a forty-year period (its existence had been predicted many years earlier by Bessell, who correctly deduced the existence of Sirius' companion). In 1896, J. M. Schaeberle, at Lick Observatory, using a 36-inch refractor, was the first to see the very faint stellar point of Procyon's companion. Because it is three to five magnitudes fainter than Sirius' companion and much closer to Procyon than the Pup to Sirius, the mean separation being only 15 astronomical units (and the system somewhat farther from us than Sirius), Procyon's companion, which can be viewed only in large telescopes, took much longer to be detected. At least fifteen thousand times fainter than Procyon, its mass is 0.65 $\mathcal{M}$ ⊙ and its diameter, a little more than twice the Earth's, is about 17,000 miles. Thus, at more than two tons per cubic inch, its density is even greater than that of Sirius' companion, and it is unquestionably a white dwarf.

Several optical companions share Procyon's field, but none is gravitationally bound to it. The brightest is only of twelfth magnitude. Four degrees to the northwest of α is third-magnitude *GOMEISA (β),* and nearby is 4.5-magnitude *(γ) GAMMA.*

TAURUS

". . . a snow-white bull with great dewlaps and small, gem-like horns, between which ran a single black streak." Thus the disguise with which Jupiter, wisest of the divinities and most glorious, concealed his identity in order to possess the fair Europa. In this legend, the most famous of numerous myths about the Bull, Jupiter fell in love with a princess in Canaan. While she was playing with her companions in a meadow near the seashore, he appeared to her in the form of a beautiful white bull and behaved so gently that Europa and her maidens decked its horns with garlands of flowers. The bull lowered its head, inviting Europa to sit on its back; but as soon as she complied, it sprang up and leaped into the sea, bearing her over the waters with fantastic speed to the land of Crete. Jupiter then revealed himself in his divine identity and became Europa's lover. In this context, the constellation of the Bull came to be called *Portitor Europae.*

The Bull appears in many ancient stories: Another is the legend of Pasiphaë, daughter of the sun god Helios. Pasiphaë's husband, Minos, a son of Zeus (Jupiter, to the Romans), offended the sea god Poseidon by failing to sacrifice a white bull miraculously sent to him by the gods in support of his claim to the Cretan throne. In revenge, Poseidon caused Pasiphaë to fall in love with the white bull, which now belonged to Minos' herds. Daedalus, an Athenian doll maker, granted her plea for help in this matter of the heart by building a hollow wooden "cow" upholstered with a cow's hide, and complete with folding doors and concealed wheels. Pasiphaë hid inside this contraption, which was left standing near the grazing bull while she waited for the deception to affect the beast. It did, and she later gave birth to the famed Minotaur ("Minos' bull"), a fire-breathing creature of human frame and bull-like head. Greeks found the story of the Minotaur's birth distasteful; in their properly altered version, Pasiphaë had her affair with King Minos' cruel and arrogant general, who was also named Taurus. These myths gave the title *Amasius Pasiphaës,* or "The Lover of Pasiphaë," to the constellation.

Taurus is also associated with the Argonaut Jason, either as a brazen-footed bull of fiery breath tamed by him, or as the divine ram that carried Phrixius to safety and whose pelt, after its sacrifice, became the Golden Fleece recovered by Jason. The Bull was everywhere an object of worship: the Druids held their "Tauric" festival when the Sun "entered" its boundaries, and white bulls were considered sacred to the Moon and figured in a number of ancient fertility rites. The Romans dedicated a temple to Serapis, the "incarnate bull," lit great bonfires, and sacrificed burnt offerings to the gods of Earth and fertility. In ceremonies of the "sky-bull" cult at Cnossos, circling dancers represented the annual courses of the heavenly bodies. Mythologists say that the constellation is also related to the Egyptian bull god Osiris. The Bull has been imaged from earliest times and in diverse lands on coins and on gems. The pattern of the Hyades also suggested the jaw of an ox to South American tribes. Our individualistic Chinese, however, saw Taurus as part of a "White Tiger," and also as *Ta Leang,* or "Great Bridge."

ALDEBARAN (α) *Alpha Tauri*
(RA 4 h 33 m δ +16° 25') m 0.86

I saw on a minaret's tip
Aldebaran like a ruby aflame, then leisurely slip
Into the black horizon's bowl.
                    William Roscoe Thayer

Its name is derived from *Al Dabarān,* or "The Follower," because it rises after the Pleiades and pursues them across the skies, but it was designated *Parilicium* in all the old catalogues, including Flamsteed, from *Parilia,* the feast of Pales, marking the birthday of Rome, on April 21, when the star sets at twilight. It has appeared in contrasting guises: with desert dwellers as a "camel"; in early England as the familiar "Bull's Eye"; and owing to its red color, as *Rohini,* or "A Red Deer," long ago in India. For us, however, enjoying the revelations of modern astronomy, Aldebaran's ruddy hue assumes a vastly different significance. We see no woodland creature but, rather, a scientific key—one that promises to unlock the riddle of our own Sun's future, as well as that of all other average main-sequence stars.

The Evolution of an Orange Giant

Aldebaran's reddishness (somewhat paler than Betelgeuse) relates to its spectral classification (K5) as an "orange giant," or a member of that important group of stars charted to the upper right of the main sequence on the H-R diagram (p. 87), below the supergiants Betelgeuse and Antares. This position on the graph indicates an evolutionary path that, although it has elements in common with that traced by the supergiants, those stars of very great mass, nevertheless differs from them in many respects. Aldebaran's present diameter is forty times that of the Sun, but this was not so in its youth. With a mass not in excess of 3 $\mathcal{M}\odot$, it spent the greater part of its life span as a normal main-sequence star like our Sun. For billions of years, the hydrogen burning in the core of the star was gradually converted to helium by the proton-proton chain reaction (p. 91); Aldebaran thus remained stable as its size and luminosity increased very gradually. Like the supergiant Betelgeuse, it exhausted its central store of hydrogen and thus transformed its central region into a helium core surrounded by a hydrogen-burning shell, but this process took billions of years, whereas the much more massive Betelgeuse accomplished this change in only 2.5 million years. Aldebaran then suffered a slow increase in luminosity and size, with its diameter doubling in a billion more years. As the star began to leave the main sequence, it entered the "subgiant" phase of its evolution, expanding rapidly in size but with its luminosity remaining constant and its surface temperature decreasing rapidly. Owing to a great increase in its rate of hydrogen burning, after another 100 million years its luminosity increased again, reaching 125 times the Sun's luminosity, and it expanded quickly to about its present size, its expansion causing a further drop in its surface temperature. These last, accelerated stages in its evolution quickly led Aldebaran along a vertical track on the H-R graph, qualifying it for membership in the club of the aging orange giants, where we presently find it.

The helium core of an orange giant is very dense, containing about half the star's mass, and is surrounded by a thin hydrogen shell a few thousand miles thick. Around this, there is a tenuous envelope of hydrogen gas, about 100 million miles thick. When the temperature in the contracting core reaches 100 million degrees K., the triple helium (alpha) reaction begins in a flash (the helium flash) and helium nuclei are fused in groups of three to produce carbon (p. 93). The helium flash will occur deep within the core of Aldebaran, and this explosion, causing an expansion in the core, will drop the temperature of the core and shell as the rate of hydrogen-burning in the shell decreases. As a consequence, the star's luminosity and size will diminish again, and it will move back toward the main sequence for another ten thousand years. This will be but a brief interlude in the star's life, for in a short time the gravitational compression will cause the helium in the core to burn more rapidly and release increasing amounts of nuclear energy, while the thickness of the hydrogen shell decreases. After that, with the helium in the center fusing rapidly to form carbon, a carbon core will form, surrounded by a helium-burning shell. In a few million years, Aldebaran will make a rapid return to its orange-giant phase. It may then possibly experience further helium flashes in the helium shell and move back and forth on the diagram; however, its ultimate fate will differ considerably from that of a supergiant.

A normal orange or red giant is expected to get rid of its outer regions gradually, forming a gaseous shell akin to that observed in the so-called planetary nebulae. Aldebaran will therefore not undergo the vast, cataclysmic explosion (supernova) of a supergiant, but after it sheds its outer layers, through which its dense core will be visible, it will settle down to its final stage of existence as a "white dwarf." Were Aldebaran a close binary, recurrent novic explosions might result because of exchange of matter between the components; however, a 13th magnitude red dwarf companion, 0.001 L $\odot$, lying about 650 AU from the primary, seems much too distant from it for this type of process to occur. An 11th magnitude dwarf binary also lies in Aldebaran's field, but is not gravitationally bound to it.

THE HYADES

A famous V-shaped open cluster outlines Taurus' "head," of which Aldebaran on its eastern border is optically, but not intrinsically, a member. These stars figure in many ancient myths, as Atlas' daughters and half-sisters of the Pleiades, and as the nymphs of Dodona and those of Mt. Nysa, who cared for the infant Bacchus. Hesiod named five: Kleea, Eudora, Koronis, Phaeo and Phaesula, but their various names were not related to specific stars. From early times, they were believed responsible for great rains, possibly because of the Hyades' copious tears when their brother Hyas was killed by a wild boar. Roman country-people, however, called these stars "Suculae" or "Little Pigs," while the Chinese saw them as a "Rabbit-net."

As Orion awaits Taurus, we now have some cosmic evidence that the Bull really is charging the Hunter! Together with a large surrounding group of faint stars called the Taurus moving cluster, the Hyades are moving through space at 26 miles per second toward Betelgeuse, and are also receding from us, so that millions of years from now they will no longer be visible to the naked eye. At present, however, they are only 130 light-years distant, one of the nearest galactic clusters.

(γ) GAMMA Tauri marks the Bull's nose, the apex of the V, and was long known as Hyadum I, or "The Leading One of the Hyades Cluster," but at magnitude 3.9 it is no longer regarded as the brightest Hyades star. Northwest of α, at the top of the head or face, is (ϵ) EPSILON, also called Ain, or "Eye," by the Arabs, and Oculus Boreus, or "Northern Eye," by Flamsteed. At magnitude 3.6, it is somewhat brighter than γ, and like it, a yellow giant. Between γ and ϵ lies δ, or Hyadum II.

(θ) THETA 1 and 2 lie between Aldebaran and Gamma; θ 1, as well as δ, are also yellow giants. Theta 2, at magnitude 3.34, is the brightest cluster member, a white star (A7 III) with the luminosity of fifty Suns. The two Thetas form an easy pair for binoculars.

The Bull's Horns

EL NATH (β) BETA Tauri, marking the tip of the northern horn, formerly belonged to the Charioteer as γ Aurigae. Its name originated with the Arabs as Al Nāṭiḥ, or "The Butting One." Taurus is the first ecliptic constellation we are visiting, and six thousand years ago, at the time of the vernal equinox (when the Sun in its apparent annual motion crosses the celestial equator), the Sun and El Nath were nearly on the same line of sight. Astrologers promised eminence and fortune to those who could claim it as their natal star; however, we make no such guarantee and soberly note that β Tauri, about 300 light years distant, is a hot, class-B giant with a luminosity of 1,700 $L \odot$.

(ζ) *ZETA*, lying at the tip of the southern horn, south-southeast of β, is the Babylonian *Shur-narkabti-sha-shutu*, the "Star in the Bull Toward the South," a third-magnitude class-B giant, rather remote at 940 light years. A notable example of a shell star, Zeta's spectrum shows tremendous turbulence of its outer layers, in which material is both ejected and sucked in by the star at speeds up to 100 kilometers per second. Astrometric variations reveal a close companion of small mass; in close binaries where a shell or envelope is present, an exchange of mass between the components may occur, which will cause them to evolve much more rapidly.

The spectral emission lines of such a shell star are broadened, owing to superimposition of differing Doppler shifts arising from differing radial motions of various parts of the shell; the star's atmosphere is therefore expanding in a complex way, and the various factors causing this may be related to the mechanism that ultimately produces a nova. Class-B emission stars like ζ also have rapid rotational velocities, which induce instabilities beneath the stars' surfaces because of equatorial centrifugal acceleration as large as or larger than the gravitational acceleration. This causes material to stream out of the equatorial planes and form the expanding atmospheres indicated by their broadened emission lines.

THE CRAB NEBULA (M1) (NGC 1952)
(RA 5 h 31 m   δ +21° 59')

To Zeta's northwest lies the object that started Messier on his career as a cataloger of "nebulae" while he was observing ζ and a nearby comet (the latter his primary interest). However, M1 was first discovered, in 1731, by J. Bevis, an amateur astronomer, and it was rediscovered by Messier in 1758 and observed sixteen years later by J. E. Bode and William Herschel; they differed as to its nature, with Bode correctly guessing that it is a nebula and Herschel incorrectly assuming it to be a star cluster. Lord Rosse, at Birr Castle, discovered its expanding filaments and likened them to a "crab's legs" (see fig. 54), which gave the nebula its name; both he and Lassell imagined it as a nebula containing a number of stars. We now know it, however, as the rapidly expanding gaseous remnants of the supernova of 1054, a great star that exploded with catastrophic force and was recorded as a "guest star" by Chinese astronomers.

The Crab Nebula is 3,500 light years distant, which means the star actually blew up about 4,434 years ago, or 2,447 B.C., when the most ancient constellations were being named in Egypt and China; light departing from the "disaster site" traveled for 3,500 years through deep space before reaching the Earth, in A.D. 1054, conveying to observers the image of a supernova whose apparent magnitude rivaled that of Venus at its brightest. The "Crab" extends for about 2 parsecs, so that light takes 6.52 years to traverse it. An intense radio source, it consists of two interpenetrating structures: an external system of relatively slender filaments, forming at the surface an almost elliptical envelope that is expanding at the rate of 1,300 kilometers per second (over 600 mps), and a somewhat S-shaped mass completely filling the interior region and contributing most of the luminosity. The filamentary structure exhibits an emission spectrum consisting primarily of the lines of helium and hydrogen; the S-shaped mass emits a continuous spectrum, with no bright or dark lines. A sixteenth-magnitude late-B-type dwarf telescopically visible near the center of the nebula was tentatively identified as the core of the exploded star. At first, the faintness of the star presented a puzzle, because it could not account for the luminosity of the gaseous envelope. As a solution, I. S. Shklovsky proposed that relativistic, or very rapidly moving, electrons ejected by the star are decelerated in local magnetic fields in the nebula and emit what is now called *synchrotron radiation.* In this mechanism, the electrons spiral around the magnetic lines of force and radiate energy of a frequency determined by the strength of the magnetic field and the velocity of the electrons. It takes only a small number of these high-velocity electrons to generate very intense radio waves and thus account for the observed optical radiation. The light of the nebula is *polarized,* † which demonstrates the presence of local chaotic magnetic fields; if we study the nebula through a rotating polaroid filter, we can trace out the magnetic fields and determine their directions in the gaseous matter. Synchroton radiation is polarized in a definite way, because the decelerated electrons in the large synchrotons in laboratories on Earth emit radiation in the direction of their motions. The origin of the large number of relativistic electrons probably lies in collisions between small local masses of gas moving at high relative speeds which produce very high temperatures at the interfaces of the collisions and thus impart high kinetic energies to the interface electrons; this energy increases in a "snowballing effect" as the electrons repeatedly undergo random collisions with masses of gas carrying magnetic fields.

The Radio Source in the Crab Nebula

In 1967, a new large radio telescope picked up unidentified signals of a periodically recurring nature, a series of pulses of fixed duration (⅓ sec.) occurring at precise intervals of 1.3372795 seconds, emanating from a fixed point in space (in the constellation of Vulpecula). Similar discoveries of such pulsating signals followed; the sources of these signals, named *pulsars,* (they are actually "rotators") have been identified as the rapidly spinning remnant cores of supernovae. Some of the pulsars have very short periods and are believed to be very young, because they are still emitting energy in the visible part of the spectrum as well as in the radio part. The first of these, *NP 0532,* rotating at the incredible speed of thirty rotations per second, was identified as the hot central star in the Crab Nebula when in 1969 it was discovered (by using a special technique, the equivalent of a stroboscope) that this central star blinks in its visible radiation at the same rate as the radio pulses of NP 0532. Since then, NP 0532 was found to emit X-ray pulses; remarkably, it emits a hundred times as much energy in the X-ray region as in the optical, and a hundred times as much in the optical as in the radio spectrum.

Initially, many astronomers suggested that pulsars were pulsating white dwarfs, because the size of the emitting area of a pulsar cannot exceed the duration (which is small) of a single pulse multiplied by the speed of light. White dwarfs, however, were

---

† *I.e., the electric fields of the light beams vibrate in one direction, as opposed to nonpolarized light, in which they vibrate in all directions.*

54. M1 (NGC 1952), the Crab Nebula, in Taurus. The remains of Supernova A.D. 1054, taken in red light with the 200-inch reflector on Palomar Mountain. The filamentary system is clearly delineated, overlying and surrounding the shaped gaseous mass. (Hale Observatories)

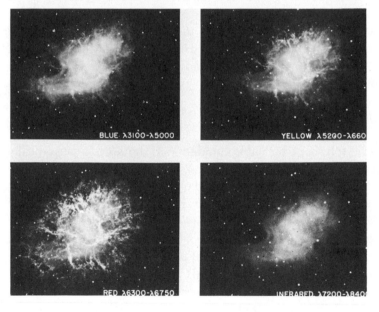

BLUE λ3100-λ5000

YELLOW λ5200-λ660

RED λ6300-λ6750

INFRARED λ7200-λ840

55. The Crab Nebula photographed in blue, yellow, red, and infrared light with the 100-inch reflector. The detailed structure of the filaments in red light indicates hydrogen as their primary source of visible radiation. Blue and infrared light show only the central gaseous system, which exhibits a continuous spectrum, indicating the presence of a dense, incandescent gas but giving us no clue to its chemical nature. (Hale Observatories)

they pulsating, could not give the observed periods of most pulsars, nor can they (owing to their size) rotate fast enough to account for these observations on the basis of rotation. Conversely, the very much tinier and still more condensed neutron stars would pulsate much too fast for the observed period. Their *rotation*, however, which would be fast enough, was first proposed by T. Gold as the mechanism accountable for the phenomenon, as follows: The initial collapse of the supernova compresses its core so greatly that immediately after the supernova explosion, the hot residual core ("neutron star") has a vast amount of rotational energy, owing to the principle of the *conservation of angular momentum,‡* which states that as the core of fixed mass contracts

‡ *The unchanging vector quantity (mvr) of a particle of mass m moving with constant speed v in a circular orbit of radius r.*

and its radius gets smaller, its rotational speed and hence rotational energy must increase. For a neutron star like the pulsar of the Crab Nebula, rotating thirty times per second, this energy is enormous, enough to account for the luminosity of the nebula $(30,000\,L\odot)$, and its total radiation in the X-ray, ultraviolet, visible, infrared, and radio parts of the spectrum.

A Lighthouse in Space

CM Tauri, as the Crab pulsar is now named, does not, to our knowledge, direct extraterrestrial spaceships away from dangerous shoals in their interstellar voyages, but this amazing stellar object may indeed be likened to a cosmic lighthouse, its searchlight beam signaling new scientific information to astronomers on Earth. Seeking reasonable explanations for the nebula's high luminosity and the mechanism by which CM Tauri radiates away its enormous rotational energy, astronomers turned their attention to the powerful magnetic field on the surface of the pulsar, resulting from the intense collapse of the supernova's core. The magnetic axis of CM Tauri does not coincide with its axis of rotation, so that the intensity of its magnetic field, as seen from the Earth, varies as the star rotates, giving rise to electromagnetic pulses of the observed period. Astronomers reasoned that this rotating magnetic field greatly accelerates the charged particles (electrons and protons) emitted by the neutron star, and these particles then spiral around the magnetic lines of force and emit synchroton radiation.

This cone of radiation, corotating with the star, blinks out the message of CM Tauri's dynamical secrets as it sweeps across our line of sight in regular intervals. (Occasional changes in CM Tauri's period occur, which are believed to be caused by cracks developing in the exceedingly dense but very thin iron skin constituting the surface of the star, fissures resulting from the severe dynamical stress of its rapid rotation. Following these so-called "starquakes" CM Tauri readjusts itself and, when equilibrium is reestab-

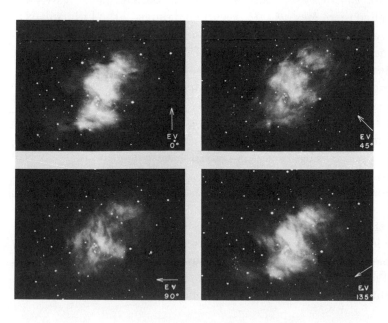

56. *The Crab Nebula* photographed in polarized light with the 200-inch Palomar reflector. The direction of the electric vector is indicated. The arrow gives the direction (as seen by an observer on Earth) in which the electric vector of the light reaching the observer is vibrating. The direction of polarization determines the direction of magnetic fields in gaseous matter. The hot central star is visible as the lower of the two dots.

lished, the period returns to its initial value and life goes on as usual in the bizarre environment of the pulsating neutron star.)

THE PLEIADES (M45)

"Canst thou tie up the bonds of the Seven Stars?"
—Book of Job

The misadventures of the Great Hunter also include an unrelenting attachment to the seven daughters of the Titan Atlas and the Oceanid Pleione; Alcyone, the brightest Pleiad; Maia, the most beautiful; Electra, the saddest; Taygete, Celaeno, Merope, and Sterope (Asterope). For seven years, Orion chased Pleione's daughters with ill intent, until Zeus turned them into a cluster of seven stars to save them from his attentions. Orion also was transferred to the heavens, where he continues his hot but hopeless pursuit, rising after the Pleiades in the eastern skies and following them across the heavens until they sink below the western horizon.

According to a second myth, however, the seven sisters were so distressed at the death of the Hyades that they all killed themselves and were placed among the stars. In any case, whatever the cause of their heavenly transfer, it seems that Orion was not the sole hot-blooded pursuer of the fair sisters: in yet another of these ancient yarns, Zeus himself was guilty of the same unworthy behavior toward Taygete, so that Artemis was obliged to turn her into a hind to help her avoid the impetuous god.

The unfortunate sisters' title is apparently derived from *pleō*, "sail," for their rising in May began the season ancients reserved for navigation; or from *pleios*, "full" or "many," also the approximate meaning of the Hebraic *kimah*, "heap" or "cluster" (a reference to the Pleiades in the Book of Job), and of their Arabic name, *Al Thurayya*. The list of the "Seven Sisters' " names is as endless as the many varying myths concerning them. Hesiod called them the "Seven Virgins," the Anglo-Saxons *Sifunsterri* ("Seven Stars"), and they were *Schiffahrts Gestirn,* "Sailors' Stars," to the Germans of Ideler's times; the Hungarians called them *Fiastyuk* ("Hen with Fledglings"); the Lithuanians a "Sieve," and French peasants saw them as a "Mosquito Net." Diverse tribes and nations in far-flung corners of the Earth at various times in history all regarded these stars as a "Hen with Her Chickens." Australians, however, saw them as young girls playing for three dancing young men: Orion's Belt stars!

From earliest days, the Pleiades' universally significant appearance was celebrated in sacred ceremony and festival, their exact rising or setting marking the seasons and the commencement of sowing and harvesting. Nevertheless, Mohammed saw these stars as harmful, and the great Hippocrates believed they caused disease in autumn! The word "Pleiad" became synonymous with "seven," and as such, was applied to literary groups of noted historical importance, among them seven poets of our own Revolutionary War.

Imaged as the Bull's shoulder, the Pleiades (fig. 57), when viewed naked-eye (on a very clear night) or with low-power glasses, resemble a stubby dipper, often ignorantly called "The Little Dipper." At the point where the "dipper's" short handle joins its bowl lies the brightest Pleiad, *ALCYONE (η) ETA Tauri (Fl. #25)* (RA 3 h 45 m δ 24° 06'), a third-magnitude blue-white (late-class-B) giant of about 1,000 $L \odot$, its emission spectrum indicating a fairly hot stellar atmosphere; three faint companions to its northwest form a small triangle suitable for low-power viewing. Appropriately named "Light of the Pleiades" (but *Al Jauz*, or "The Walnut," by early Arabs), η represents the Atlantid nymph who became the mother of Hyrieus by the sea god Poseidon. Her sister Maia, the first-born and most beautiful, after whom our month of May is named, is represented by *Fl. #20*, lying in the upper right of the dipper's bowl, a fourth-magnitude late-B-type giant. Representing Taygete, the mother of Lacedaemon by Zeus, *Fl. #19* is another fourth-magnitude class-B giant (with a faint companion), mentioned here because of its location just to the northwest of #20.

57. *NCG 1432, the Pleiades,* an open star cluster in Taurus, photographed with the 100-inch Mount Wilson telescope. The feathery effects around each star are diffuse bright blue nebulae reflecting the light of these hot, B-type giants. Counterclockwise, the stars shown are Alcyone (left), Merope, Electra, Celaeno, Taygete, and Maia. (*Hale Observatories*)

The "Lost" Pleiad

The Sister Stars that once were seven
Mourn for their missing mate in Heaven.
                    Alfred Austin

According to legend, one of the Sisters disappeared, accounting for the difficulty in seeing more than six stars with the naked eye. Her exact identity seems as mysterious as the circumstance of her disappearance, with four Pleiades vying for the role of the lost one. *ELECTRA (Fl. #17),* at the lower right end of the bowl, has been visible as a magnitude-4 star in recent centuries, but in ancient times it was said that she covered her face in mourning on witnessing the destruction of Troy (according to Hyginus, she wandered off as a comet to watch it fall); or, with a sudden spurt of "mythological proper motion," joined Mizar in the Big Dipper as its fainter companion, Alcor. An Australian tribe believed that Canopus (α Carinae), their "Heavenly Crow," carried Electra away as his bride. Similar in spectral type to η, #17 is a B giant with an emission spectrum.

MEROPE (Fl. #23), fourth magnitude, is a late-class-B subgiant, marking the lower left corner of the bowl and southernmost of the major M45 stars. The legendary Merope hid her face in shame, because she was the only sister to marry a mortal: Sisyphus, king of Corinth. Her name means "Mortal" and, as an astronomical symbol of her disgrace, #23 is enveloped in a faint gaseous reflection nebula: *NGC 1435,* the brighter part of a diffuse, wispy nebulosity that covers the entire cluster.

CELAENO (#16), a subgiant of Merope's spectral type, lying north of Electra and at the very limit of naked-eye visibility (m 6.5), has also been called the "Lost Pleiad," supposedly because she was struck by lightning! Another candidate for that title is *ASTEROPE (#21 and 22)*, more correctly called *STEROPE I and II,* a very wide double consisting of sixth- and seventh-magnitude main-sequence stars lying to the north of Maia.

Marking the "dipper's" short handle (east of bright Aleyone) are Atlas and Pleione. *PLEIONE (#28), BU Tauri,* mother of the brood, is the most interesting of all these "lost" stars by far. Its bright hydrogen spectral lines, discovered in 1888, led Pickering to suggest that this "peculiar" late-class-B main-sequence star may have enjoyed a temporary high luminosity in ages past and then faded, thus qualifying it as the true Lost Pleiad. The bright hydrogen lines have disappeared and recurred several times during the half century following their discovery; spectroscopic evidence relates these outbursts to atmospheric turbulence caused by Pleione's fast rotation (100 times the Sun's), as the result of which it ejects a series of expanding gaseous shells from its equatorial regions, owing to centrifugal acceleration (see p. 140). In recent years, however, Pleione's variability has been slight.

ATLAS (#27), a 3.5-magnitude class-B giant, lies just south of Pleione, the two forming an easy optical double. One myth tells us that his daughter's stellar transfer resulted from their grief over his labors as Bearer of the World; their setting supposedly relieves him of some of this burden. Atlas has an extremely close faint companion, discovered by Struve in 1827.

Variable Stars

(λ) LAMBDA Tauri (RA 3 h 58 m   δ 12° 21'), lying beyond the muzzle and southwest of γ, is a famous nonintrinsic variable, an eclipsing binary of the Beta Lyrae type (p. 222), known to astronomers since 1848; its light changes, which range between magnitudes 3.4 and 4.1, may be compared to those of γ, and also to *(ξ) XI,* to Lambda's west-southwest. (The three stars form the arm of a slightly bent cross-bow, oriented north to west-southwest, with λ in the middle.)

T TAURI (RA 4 h 19 m   δ 19° 25'), a very important irregular variable, located about seven minutes west and slightly to the north of ε Tauri, the northern eye, and associated with variable nebula *NGC 1555,* was discovered in 1852 by J. R. Hind. We have already discussed protostars in the Hayashi convective stage (p. 114), now called "T Tauri variables," very young turbulent stars still evolving toward the main sequence, many of them lying in nebulosity, the gas and dust from which they were created. T itself, a dwarflike class-G emission-line star, is not dissimilar to our Sun as it was billions of years ago, in its formative stages. T's magnitude range is from 9 to 13; its changes may occur in a few weeks to several months!

A recent discovery made by research teams at several observatories, using telescopes coupled with extremely sensitive instruments, adds an exciting dimension to the story of T Tauri. These observations indicate that T may possess a planet; if so, the first ever detected outside our Solar System. Named "TIRC" (T Tauri Infrared Companion), this dim, gaseous object, some five to twenty times Jupiter's size, lies 7.5 billion miles south of T Tauri. Three University of California astronomers (Manson, Jones, and Lin) have redetermined the optical position of T, indicating that it coincides with a weak radio source and that its apparent satellite, TIRC, is coincident with a strong radio source lying 0.6 second to T's south, a projected separation of 80 astronomical units.

Ruling out other possibilities, they conclude that TIRC is embedded in a gaseous disk surrounding T Tauri and powered by accretion from the disk. According to their theory, TIRC's increasing mass will soon cause tidal truncation of the disk, ending the current accretion phase of "protoplanet" TIRC. Nevertheless, other astronomers disagree with their conclusions: T. Simon of the University of Hawaii, one of the original discoverers of TIRC, cautions that the enigmatic object might be a very small star, rather than a planet. Within ten years, however, it will be possible to see if TIRC is orbiting T Tauri; such an orbit, according to Mr. Lin, would be another strong indication that TIRC is indeed a planet.

Previous investigations carried out by Philip Morrison* indicate that T Tauri stars are sources of cosmic rays, the high-energy particles that move about in the galactic magnetic fields, accumulating for millions of years in interstellar space. T Tauri itself shows emission lines of hydrogen, neutral oxygen, singly ionized silicon and calcium, and in particular, an as yet unexplained overabundance of lithium. Approximately 9 degrees to the north of T lies a second T Tauri variable, *BP Tauri,* cooler and rosier, its later spectral class (dK5e) indicating an orange dwarf with an emission-line spectrum.

HL Tauri (RA 4 h 29 m   δ 18° 07′), lying southeast of ε and northeast of the binary Fl. #68 (both of which delineate the Bull's face), is a very faint T Tauri-type star of about one solar mass and apparently no more than one hundred thousand years old. HL Tauri is 160 parsecs distant (about 521.6 LY), and has a bolometric luminosity (taking into account the invisible wavelengths, i.e., ultraviolet, infrared, and radio waves, as well as visible light) of 7.2 *L* ⊙. In February 1982, new photometric measurements were made in the far infrared wavelengths from the Kuiper Airborne Observatory (KAO) at an altitude of 41,000 feet which support the view that this protostar is seen through an edge-on circumstellar disk of small particles, surrounded by a more extensive halo of larger grains, thus indicating that a planetary system may be forming. Astronomer Martin Cohen of NASA Ames Research Center and the Radio Astronomy Laboratory, University of California, who is studying the star, previously established the presence of ice and silicate absorption lines in its spectrum, features not present in the spectra of other T Tauri stars. Of the latter, HL also has the largest optical linear polarization (13%)—another indication of the edge-on ring of ice and dust grains.

Combining all the data that have been gathered so far, Cohen compares his model of the proposed protoplanetary disk of HL Tauri to the nebula which (according to current theory) surrounded our own Sun in the early stages of its planet formation. The

---

* *Theoretical physicist at MIT who researched cosmic-ray particles observed in our part of the Galaxy.*

new measurements indicate that HL Tauri is now in the convective stage of a contracting protostar, having recently completed its accretion phase; Cohen suggests that most of the circumstellar particles may already have agglomerated into larger bodies, contributing to an early process of planet formation. Thus our Sun with its retinue of planets may well be a model for the baby star HL Tauri in its future state, some billions of years from now.

RV Tauri, lying 8 degrees west-southwest of β, belongs to a class of variables that form a bridge between the periodic Cepheids and semiregular variables, also closely related to the W Virginis stars (p. 190), which are very old population-II Cepheids. RV is a class-G emission star that changes to type K as it fades, showing spectral characteristics of a red giant when its apparent magnitude is *largest* (and the star therefore *dimmest* —did we catch you?). Its unstable light curve exhibits short erratic fluctuations superimposed on a cycle of thirteen hundred days.

RW Tauri, 6 degrees north-northeast of the Pleiades, and about 1 degree southeast of the double star **(Ψ) PSI,** in the northernmost part of the Bull's territory, is an exceedingly close eclipsing binary, consisting of a B and a K star separated by only 5 million miles. Spectroscopic evidence shows a glowing hydrogen ring rotating about the B star. The close components are undoubtedly exchanging material, thus producing a stream of gas from which the ring is formed.

GEMINI

> . . . Fair omen of the voyage; from toil and dread,
> The sailors rest rejoicing in the sight,
> And plough the quiet sea in safe delight. . . .
>
> Homer

Leda, the princess of Aetolia, was given in marriage to Tyndareos, king of Sparta. But the god Zeus was smitten with her beauty, and on her wedding night he heedlessly came to the young bride, disguised as a swan. As a result of this "bigamous" visitation and the consummation of her marriage to Tyndareos that same night, Leda bore two pairs of twins, each pair enclosed in a single huge egg: one contained Polydeuces and Clytemnestra, who were Zeus's children and immortal; and the other, Castor and Helen, Tyndareos' offspring and therefore mortal. Castor and Polydeuces ("Pollux," in Latin), who came to be known as the "Dioscuri" (from *Dios kouroi,* or "sons of Zeus"), were extremely devoted to each other; Castor was a famous horse tamer and soldier, and Pollux was his country's leading boxer. Among their many exploits, they took part in the expedition of the Argonauts and were instrumental in saving the fleet of the *Argo* from a violent storm. Because of this deed, Castor and Pollux became the patrons of sailors, who carved their images on the bows of ships and called upon them when in distress. The Greeks and Romans also invoked their aid in war and, according to legend, the Twins participated in a battle at Lake Regillus, helping the Romans, and immediately afterward appeared at the Forum, miles away, to announce the victory!

58. *The Abduction of Leucippus' Daughters.* Cupid and angelic assistant hold the rearing horses as Castor and Pollux carry off Phoebe and Hilaira. Painting by Rubens. Pinakothek, Munich. (*The Bettmann Archive*)

Out of this magical event arose the cult of the Dioscuri, and a temple of Castor and Pollux was erected at the Forum. Their images also appeared on coins of ancient Greece, in southern Italy from 300 B.C., and on early silver coins of the Roman Republic.

Heroic though they were, the brothers were not free of the weaknesses of ordinary men. They had taken a strong fancy to Phoebe and Hilaira, daughters of their uncle Leucippus of Messenia, but the young women became engaged to Idas and Lynceus, sons of Tyndareos' other brother, Aphareus, king of Messenia. Properly invited to the wedding, Castor and Pollux behaved quite boorishly (at least according to present-day mores) in seizing the young brides, who apparently had little to say regarding their marital choices, and carrying them away by force to Sparta.

At a later date, the four male cousins were feasting together after a cattle raid in Arcadia. When Idas and Lynceus had already eaten most of their portions of meat, Idas suddenly announced a "meat-gobbling" contest, in which the man who finished his portion first would have half the cattle, and the runner-up, the rest. Enraged at this unfairness, Castor and Pollux appropriated the cattle in Messenia and drove them back to Sparta, angrily pursued by Idas and Lynceus. Overtaking the Twins, Idas slew Castor with a spear, and Pollux then pursued Lynceus, fatally spearing him, while Zeus joined in the melee, killing Idas with a thunderbolt. The god then offered eternal residency in Olympus to his grief-stricken son Pollux, who refused to accept it if his beloved Castor was to remain in the underworld. Touched by Pollux's devotion, Zeus allowed him to share his immortality with Castor under a special arrangement whereby each of the brothers would spend alternating days in Hades and in Olympus. Under the even more merciful mythological pen of the late Greek writers, Zeus then transferred the Dioscuri to the starry abode where we presently view them.

CASTOR (α) *Alpha Geminorum*
(RA 7 h 31 m   δ +32° 00')   m 1.59

Ovid named it *Eques*, or "Horseman," but from classical times until relatively recent years astronomers called this star Apollo, also the ancient name for our planet Mercury, which, like Castor, so often rises at dawn. With the Babylonians, α was *Mashmashu-Mahrū*, the "Western One of the Twins"; more exactly, it is situated 4½ degrees north-northwest of Pollux. Castor is a remarkable example of a star that "leads a double life"; under observation between 1719 and 1759 by James Bradley, Astronomer Royal of England, who noted the 30-degree change in position angle of the components during that time, it was, in fact, the first binary recognized. In 1802, Sir William Herschel announced that the components are gravitationally bound (they are now known to have a period of 380 years, with a mean separation of about 90 AU), which led him to coin the word "binary." His son Sir John called Castor "the largest and finest of all the double stars in our hemisphere."

In 1895, the Russian astrophysicist Aristarch A. Belopolsky announced that the primary component (Castor A) is itself a spectroscopic binary, with a period of less than three days. The two A stars, virtually in contact at only 4 million miles' separation (their duality thus not resolvable in any telescope), are both class-A main-sequence stars, and the diameter of each is about twice that of the Sun; each has a luminosity of 12 $L \odot$.

Castor B, the visual companion observed by the Herschels and others nearly two centuries ago, is now known to be a spectroscopic binary also, whose spectral-class-A components are even closer to each other than those of Castor A, the separation of the B stars being only 3 million miles, and their period of revolution, in a nearly circular orbit, less than three days. A very distant ninth-magnitude companion, now called *YY Geminorum*, lying 100 billion miles, or more than 1,000 astronomical units from the A-B pair, enjoys a similar history, having been detected as a spectroscopic double and also as an eclipsing binary; but its extremely close components, only 1.67 million miles apart, are red dwarfs of class K, completing one orbit about their common center of mass every 19½ hours. All in all, the amazing Castor system consists, therefore, of six stars, four of them decidedly larger than the Sun and two of them smaller. The multiple star Castor shows a combined spectrum of A1 and A5, the white, "Sirian" type.

POLLUX (β) *Beta Geminorum*
(07 h 42 m   δ +28° 09')   m 1.2

He was Ovid's *Pugil* ("The Pugilist"), the Babylonians' "Eastern Twin," and the Arabs' "Head of the [Hindmost] Twin," *Al Rās al Jauzā'*, from which was derived the Alfonsine "Rasalgeuse." At 35 light years, or 10.7 parsecs, Pollux's apparent and absolute magnitudes differ by only 0.2 (see p. 19). Its color has been variously described, from

Ptolemy's "yellowish" to Agnes Clerke's "fiery red," but its spectral class, KO III, and color index, +1.01, indicate that Pollux is orange in tint. The Pugilist is also the closest giant to our Solar System, its diameter twelve to twenty times the Sun's. Several optical companions are in the field but not gravitationally bound to β.

Other Stars of Gemini

ALHENA (γ) GAMMA, from *Al Han'ah,* "The Brand" (on a camel's neck), also called *Almeisam,* "The Proudly Marching One," lies within the Milky Way and marks the Twin Pollux's ankle or foot. An Arabian astronomer saw γ and other stars in the Twins' feet as the "bow" with which Orion took aim at an earlier star pattern of a lion (not our present Leo). The Babylonians, however, saw γ and η as the "Little Twins." Gamma is a second-magnitude white (Sirian) subgiant.

WASAT (δ) DELTA, from *Al Wasat,* "The Middle," lies about midway between Pollux and Alhena, and bears the intriguing Chinese name of *Ta Tsun,* the "Great Wine Jar." A class-F subgiant, δ has a faint, cool dwarf companion of class K at a separation of 95 astronomical units, so that it takes many centuries for the completion of an orbit. Southwest of α, *MEBSUTA (ε) EPSILON* (RA 6 h 41 m   δ +25° 11′)   m 3, from *Al Mabsuṭāt,* or "The Outstretched" (i.e., the paw of that Arabian lion), marks the hem of Castor's tunic in modern charts. Very distant at 1,100 light years, ε, a star of our Sun's spectral class (G), is, in contrast to our dwarfish luminary, a supergiant of 5,700 L ⊙.

MEKBUDA (ζ) ZETA (southeast of ε) is from *Al Makbūḍah,* "The Contracted," the Arabic lion's drawn-in paw. To modern astronomy it is of interest as a giant Cepheid variable, ranging between magnitudes 3.7 and 4.1, with a period of about ten days, its variations having been discovered by J. Schmidt at Athens in 1847; ζ alternates between classes F and G in its light changes.

PROPUS (η) ETA, or "Praepes" (RA 6 h 12 m   δ +22° 31′), is a red giant 200 light years distant, lying in the region of the Milky Way, named by Hipparchus and Ptolemy to indicate its position in front of the Twin Castor's foot, but known to the Chinese as *Yuö,* "Battle-ax." Schmidt also discovered η's slight variability (in 1865), and in 1881 S. W. Burnham detected a 6.5-magnitude subgiant companion of solar type. The third-magnitude primary, about 160 L ⊙, is also a spectroscopic binary, its secondary component believed to be another red giant.

Third-magnitude *(μ) MU,* an irregular variable of very small range, lying immediately east of η and also in the Milky Way, is sometimes called *"Nuḥātai,"* a title derived from the collective name for all the stars of the Twins' feet, seen by the Arabs in earlier times as *Al Nuḥāt,* "The Camel's Hump," because they form a curved line. A few degrees southeast of γ lies fourth-magnitude *(ξ) XI,* marking Pollux's toe and called *Al Zirr,* "The Button," by the eleventh-century astronomical writer Al-Bīrūnī. Xi is a class-F subgiant, 65 light years distant.

U Geminorum: A Rare Miniature Nova (RA 7 h 52 m   δ +22° 08′)

Although very faint, its apparent magnitude ranging from 8.9 to 14, this is an important variable, first discovered by J. R. Hind in 1855. Lying near the border of Cancer and just a degree north of the ecliptic, about 2 degrees north of fifth-magnitude

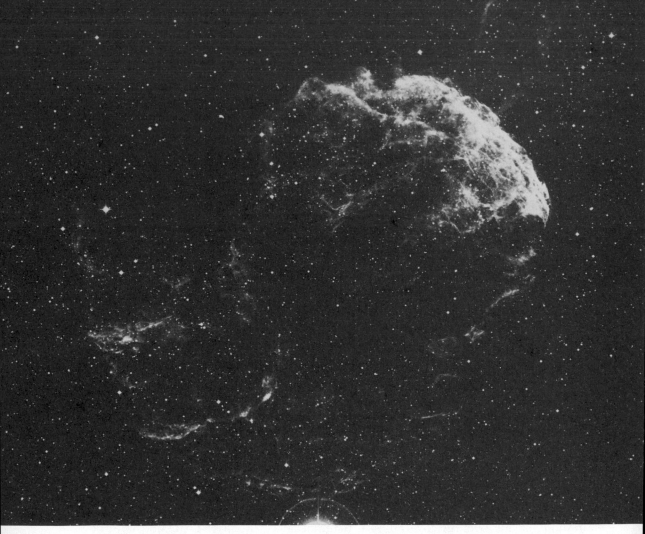

59. *IC 443, gaseous nebula in Gemini*, photographed in red light with the Schmidt 48-inch telescope on Palomar. As with the Crab Nebula, the complex filamentary structure indicates that most of the visible radiation arises from the Balmer lines of hydrogen. (*Hale Observatories*)

#85 (and immediately east of seventh-magnitude #84), U can be seen with a small telescope. The classical example of its type, U's nova-like light curve rises to maximum with great rapidity, taking anywhere from twenty-four hours to two days and exhibiting a large increase in brightness; it remains at maxima that alternate between seventeen and nine days' duration with each successive outburst. The intervals between these eruptions are quite irregular, with a mean of about one hundred days. Its brightness declines very slowly, but more rapidly than that of classical novae. At minimum, U's spectrum is characterized by wide emission lines, which weaken and merge into the continuous background as the star brightens, until at maximum they reappear as absorption lines.

For all stars of this type, which includes SS Cygni (p. 282) and SU Ursae Majoris (p. 44), there is a correlation between their magnitude ranges and the average intervals between outbursts, i.e., the greater the period the greater the magnitude range; for stars with fifteen-day periods, the range is almost three magnitudes; for sixty-day periods, four magnitudes; and for periods of two hundred days, five magnitudes. According to this relationship of magnitude range and period, the bright galactic novae should have outbursts approximately every thousand years, in accordance with present belief; therefore dwarf novae like U Geminorum are probably similar in some way to ordinary recurrent novae and to the "classical" galactic novae of great luminosity.

We previously described the processes in the evolution of ordinary recurrent novae, which are close binaries exchanging material. Since spectroscopic studies show that U Geminorum is also a binary, its behavior is probably caused by gravitational interactions on a miniature scale; its interacting components are a hot, dense blue subdwarf surrounded by a radiative gas cloud, and a much cooler, orange dwarf of later spectral type. They are separated by only a few hundred thousand miles, and their orbital period, one of the shortest known, is about four and one half hours. Adding to the complications of U's light curve, these tiny components also eclipse one another, reducing its brightness for fifteen minutes and increasing its magnitude by about 0.9, so that U is both intrinsically and nonintrinsically variable! As with ordinary novae, for many years its eruptions were believed to be caused by material from the extended, cooler star falling onto the surface of the hotter dwarf; in 1965, however, this theory was challenged by W. Krzeminski, who discovered that the light changes caused by the eclipse of the hot dwarf cannot be detected at the times of U's eruptions. He therefore concluded that the smaller, bluish dwarf, completely concealed by the more extended red companion during a major eclipse, cannot account for the great changes in magni-

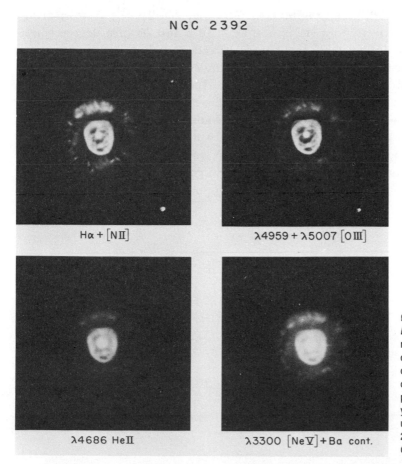

NGC 2392

$H\alpha + [NII]$

$\lambda4959 + \lambda5007 [OIII]$

$\lambda4686$ HeII

$\lambda3300 [NeV] + Ba$ cont.

60. NGC 2392, the Eskimo Nebula, planetary nebula in Gemini showing dark condensations on the central disk and a diffuse outer ring, as photographed in red, yellow, violet, and ultraviolet light with the 200-inch reflector. (Hale Observatories)

tude caused by the nova-like outburst. In other words, "covering up" the blue dwarf makes little difference in the light curve when U is erupting; therefore, the source of the explosion must be the red component! According to this theory, a strong tidal effect of the dense bluish star, causing the distended red star to increase in size and temperature, contributes to the greatly increased luminosity of the system as it ejects material which is then captured by the ring of gas surrounding the blue star. Questions concerning the U Geminorum stars and other "cataclysmic novae" are still unanswered, however, and the door of scientific inquiry is therefore wide open for new hypotheses.

Clusters and Nebulae

M35 (NGC 2168) (RA 6 h 06 m   δ +24° 20′), lying 2½ degrees northwest of η and Castor's foot, is a beautiful open galactic cluster approximately 2,200 light years distant, containing stars of varied spectral types and suitable for the small telescope. *OΣ 134,* a spectroscopic binary of solar spectral class, lies in its field.

IC 443 (RA 6 h 14 m   δ +22° 48′), about 2 degrees southeast of M35 and ½ degree northeast of η, is a large, wispy nebula showing filamentary structure; a source of radio energy, of interest as a probable supernova remnant (fig. 59).

NGC 2392 (Herschel 45⁴), THE ESKIMO (RA 7 h 26 m   δ +21° 01′): A clown brings our tour of the Orion scene to a close! Southeast of δ and ⅔ of a degree south-southeast of the wide double #63, this popular planetary nebula is visible in small telescopes, but its quaint, whimsical "face," surrounded by a "fur parka hood," becomes discernible only in larger telescopes (fig. 60). This pictorial illusion is produced by a luminous inner ring, containing bright condensations, surrounded by a much larger, more diffuse, outer ring about ½ light year in diameter. The tenth-magnitude central star, one of the most luminous in any planetary nebula, is a very hot (40,000° K.) class-O8 dwarf. Its radiation is particularly intense, and therefore the usual doubly ionized oxygen is strongly evident in the blue-green color of the nebula.

Spring

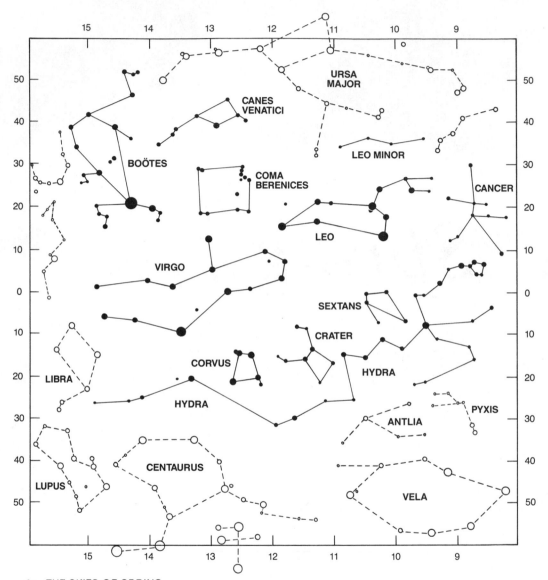

61. *THE SKIES OF SPRING*

SIX

More Animals

CANCER

Never was there a stranger alliance than that between crab and swamp monster, for which Hera, Queen of the Heavens and proudest of all goddesses, was responsible. King Eurystheus had ordered Hercules to perform twelve penitential labors; the second of these was to slay the Hydra, a monstrous water serpent with a hound's body and a hundred heads, which dwelled in the swamps of Lerna, Greece. Although the goddess Athena was on Hercules' side and pointed out to him the monster's lair, Hera was opposed to Hercules' assigned task. As Hercules began clubbing off the Hydra's many heads, Hera sent a giant crab, named Cancer, to assist the Hydra by creating a diversion; Cancer grabbed Hercules by the toe with his powerful pincers, but the mighty hero smashed the crab with a blow of his foot and continued his battle against the Hydra, which he ultimately slew. Hera, however, pitying the unfortunate crab, which had lost its life obeying her wishes, rewarded it with a place in the heavens as a zodiacal constellation.

Aratos, the poet author of *Phainomena,* called the constellation *Karkinos,* Latinized to *Carcinus* in the Alphonsine Tables. Ovid and Propertius, the Roman poets, called it *Octipes,* and Ovid, together with Manilius, the Latin astronomical poet of the first century A.D., also gave it the appropriate title *Litoreus,* "Shore-inhabiting." The Roman writer Columella, who connected stars to the weather, called it *Lernaeus,* after its habitat, the swamp of Lerna. When highest above the horizon, it is near the zenith for observers in most northern latitudes; therefore, astrologers named Cancer "Horoscope of the World!" The ancient Egyptians dubbed it "Sacred" (*Scarabaeus* [Scarab] in Latin), an emblem of immortality in the form of a beetle that rolls the Sun through the sky with its front claws. Names like "Crayfish" and "Lobster," invented by catalogers of centuries past, read like a seafood menu (and a tiny, shrimplike object to its west, "Cancer Minor," was also added). For naked-eye viewing of this tempting dish, however, we need clear skies, for no star of Cancer is brighter than about the fourth magnitude.

Stars in the Crab

Lying to the north-northeast of ζ Hydrae (a star in the head of Cancer's erstwhile ally), *ACUBENS (α) ALPHA Cancri* (RA 8 h 56 m δ +12° 3′) m 4.3, marks the Crab's southern pincer; its name is a distorted derivation of the Arabic *Al Zubanāh*, "The Claws." A double star, its spectral class-A components, of magnitudes 4.5 and 11, are relatively fixed at a separation of 10.9 seconds. The northern claw is represented by (ι) *IOTA*, (RA 8 h 44 m δ +28° 57′), also a double star; its primary is a magnitude 4.5 yellow (class-G8) star, and its magnitude-6.5 component, at a separation of 30.5 seconds, is a white (A3) "Sirian" star, which may appear slightly blue in contrast.

AL TARF (β) BETA, meaning "The End" (of the southern foot), lying west-southwest of α (RA 8 h 14 m δ +9° 21′), is an orange (K4) star, and it, too, is a double, the components being magnitudes 3.5 and 14 (RA 8 h 15 m δ +9°30′). The "Donkeys," *ASELLUS BOREALIS (γ) GAMMA Cancri,* and *ASELLUS AUSTRALIS (δ) DELTA,* are the northern and southern "Ass Colts," which the Romans called *Aselli* or Asini, and the Arabians, *Al Ḥimārain,* "The Two Asses." Delta is a class-K optical double; the separation between the magnitude-4 and -12 components, which are not gravitationally bound, is presently declining. The Romans used γ and δ to forecast the weather: if these stars were obscured by fog, rain was predicted; and astrologers warned that those under their influence could expect violent death!

TEGMINE (ζ) ZETA Cancri, Σ 1196 (RA 8 h 8 m δ +19° 11′) m 5.10, lies on the rear edge of the Crab's shell, as its name, "In the Covering," implies. This triple star is worth a special look, but the A-B pair is difficult to "split" with a small telescope. Zeta A and B are main-sequence stars of types F8 and G, their masses close to the Sun's; they complete one revolution around their common center of mass every 59.6 years. The distant companion, an ordinary yellow dwarf of magnitude 5.5, lies 175 astronomical units from the A-B pair, and its computed period of revolution about the close pair is 1,150 years. The ternary, ζ, was discerned as a double star in 1756, but Sir William Herschel first noted the close component of the primary with his powerful telescope in 1781.

Important Star Clusters

THE BEEHIVE, or PRAESEPE (NGC 2632, M44), lies on Cancer's head (RA 8 h 37.5 m δ 19° 52′); although it appears as a misty spot to the naked eye, when viewed through a low-power telescope its many bright stars resemble a swarm of bees (fig. 62), whence its popular name. "Praesepe" is of Latin derivation; the Copts called it *Ermelia,* "Nurturing." With the Arabians it was *Al Ma'laf,* "The Stall," from which followed many derivations by catalogers; its popular names, from Aratos' time down to the seventeenth century were "Manger" and "Crib." As usual, the Chinese imagery differed, but most unpleasantly: M44's misty look reminded them of *Tseih She Ke,* an "Exhalation of Piled-up Corpses!" As with γ and δ, Greek and Roman authors used this cluster to predict the weather; its obscuration by fog signified the approach of a violent storm.

About 200 parsecs, or 652 light years, distant and containing over 350 stars, Praesepe was the only universally recognized "nebula" before the invention of the

62. *M44 (NGC 2632), the Beehive Cluster, or "Praesepe," in Cancer,* an open cluster containing over 350 bright stars; photographed by Barnard with the 10-inch Bruce lens. (*Yerkes Observatory Photograph, University of Chicago*)

telescope, its stellar content unsuspected. Galileo was thus the first to resolve it as a cluster, estimating that it contains more than forty stars and publishing his findings in his *Nuncius Siderius,* in Venice in 1610. Astronomers subsequently learned that M44 is a vast object, its bright, central portion about 13 light years in diameter; but if its farthest stellar members are included, the diameter extends to about 40 light years, more than four times the distance between our Solar System and Sirius!

The varied stars of Praesepe, mostly normal main-sequence objects of classes A2 to K6, range from apparent magnitudes 6.3 to 14; if our Sun were among them, it would be invisible to the naked eye, with its apparent magnitude at that distance only 10.9. The brightest star in M44 is *(ε) EPSILON,* a type-A giant, 70 $L \odot$, apparent magnitude 6.3. The cluster also contains four orange giants (class K III) and at least five white dwarfs; also of note is ***TX Cancri,*** a variable dwarf eclipsing binary of class dF8; similar to W Ursae Majoris (p. 44), this celestial whirligig consists of two extremely close, tiny components, and its period is only .38 day!

The Parallax of Clusters

Since cluster members are gravitationally bound and move through space as a unit, with a common motion, their proper motions, as represented by directed arrows on the sky, appear to converge to (or diverge from) a point on the celestial sphere. This *convergent point* enables us to find all the stars belonging to a given cluster, regardless of their present locations. Relative to the Sun, the Praesepe stars are moving toward a convergent point west-southwest of the cluster's present position; Praesepe's proper motion in that direction seems almost parallel to that of the Hyades cluster in Taurus, at a speed of 25 miles per second as compared to 26 miles per second for the much closer Hyades. Their ages are also rather similar, at least 250 million years for Praesepe and 400 million years for the Hyades; these facts suggest the possibility of a common origin for the two clusters.

The convergent point is also useful in determining the parallax, and thus the distance, of a moving cluster, provided we also know the radial velocity of at least one of its stars. The line of sight from our Sun to the present position of any of the cluster's stars forms an angle (θ) with the direction of motion to the cluster's convergent point; by utilizing a formula involving this angle and the star's radial and transverse velocity relative to the Sun, astronomers can determine the distance of any star in the cluster with great accuracy.

From this procedure, we know that the center of the Hyades cluster is 130.4 light years, or 40 parsecs, distant; Praesepe, at roughly 525 light years, or 161 parsecs, is also one of the nearer clusters, but it is nevertheless too distant for trigonometric parallactic methods to be accurate.

The foregoing method of obtaining stellar distances, an extension of the parallactic tool, is of tremendous value in refining fundamental data about stars.

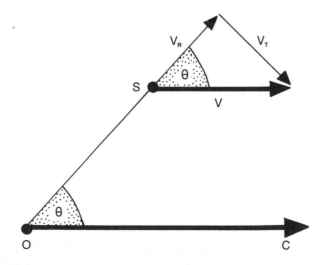

63. Determining the parallax of a moving cluster when the convergent is known.

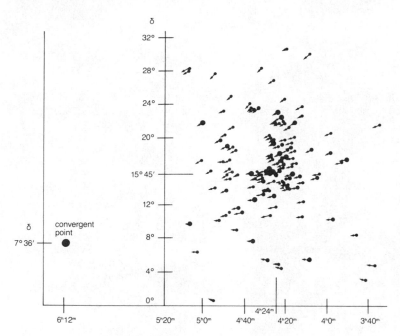

64. The convergent point of the Hyades cluster as determined from the proper motions of its members. Adapted from a figure by Van Beuren, Leiden University.

M67 Cancri, NGC 2682 (RA 8 h 48 m δ + 12° 00′), about 1.8 degrees due west of α and 9 degrees south-southeast of the Beehive, is one of the oldest clusters known—next to NGC 188 in Cepheus. Estimated to be 10 billion years old, about 2,500 light years distant, and very compact, M67 contains five hundred stars ranging from class B9 to K4 and covers an area with a diameter of about 15 minutes, or 12 light years. In addition to its most luminous member, a tenth-magnitude class-B9 star, it contains about eleven cool, type-K giants, their absolute magnitudes ranging from 0.5 to 1.5. Nearly all of its original bright blue stars have completed their evolution and died; those remaining (and even the stars of smaller masses) have already turned away from the main sequence, with the exception of a few so-called "blue stragglers," whose longevity presents a mystery.

Determining the Ages of Open Clusters

In 1954, Allan R. Sandage undertook a detailed analysis of the evolution of the stars in M67, from the initial main-sequence distribution to their positions on the presently observed C-M (color-magnitude) diagram (fig. 66), using a graph in which absolute magnitude is plotted against effective temperature, rather than color. Sandage had to know the *luminosity function* (the number of stars of a given absolute magnitude) for the original main-sequence distribution, so that he could determine how many stars lying in a given small stretch in the present C-M distribution originated from a definite segment on the *zero time line.* * To do this, he altered the present observed luminosity function by taking into account the rapidly moving stars that have escaped from the

* *The line in the H-R diagram representing the original main sequence, on which all the stars began their lives.*

65. *M67 (NGC 2682), star cluster in Cancer*, a very old, rich open cluster. Most of its stars have left the main sequence. Miniature replicas of various constellations seem to appear within the cluster; which demonstrates how frequently these accidental configurations occur throughout the heavens.

cluster during its lifetime. He then plotted seventeen evolutionary tracks for M67 and gave detailed mapping data for the seventeenth track, one that reverses its direction after reaching the top of the giant region and then runs back toward the original main sequence.

The age of any open cluster can be computed by considering how long it takes a main-sequence star of bolometric magnitude 4.4 to use up 12 percent of its hydrogen and begin to enter the subgiant phase of its evolution, its M_{bol} now 3.46. This time interval is obtained by dividing the average luminosity over the entire evolutionary track of the star into the total energy released when 12 percent of the hydrogen is transformed into helium. Using this procedure, we find that the higher up along the original main sequence the turn-off point from the observed main sequence occurs (i.e., the higher the percentage of more massive stars still on the main sequence), the younger the cluster is (see p. 109). Thus we utilize the natural law already discussed:

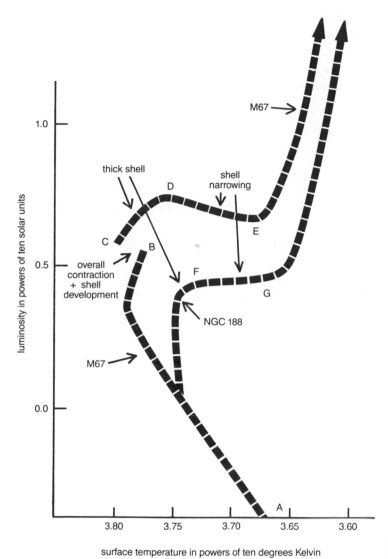

66. Evolutionary tracks of M67.

the more massive a star is, the more quickly it runs through a given segment of the observed C-M diagram during its evolution, i.e., the more rapidly it evolves. From Sandage's data, we can give the time any star spends along successive segments of its evolutionary track.

Plotted on the color-magnitude chart, M67 seems similar to the aged globular clusters that surround our Galaxy. Like them, it lies about 1,500 light years above the galactic plane, instead of along the central plane where we find most open clusters. A particular feature of M67 that is similar to the globulars is its *horizontal branch,* on the C-M diagram, Sandage's seventeenth track, which represents a number of older stars that were red giants but are now moving back toward the main sequence, following the onset of the triple-alpha thermonuclear process (p. 93), which releases new energy in the stars' cores. There is, however, an important difference between these evolved population-I M67 stars and those of the globular clusters, consisting almost entirely of population-II stars: owing to their deficiency in the heavier metals as compared to population-I stars, the members of globular clusters are twice as luminous as their counterparts in M67 and other open clusters. The accompanying diagram (fig. 67) shows how the Pleiades, Praesepe, and M67 represent youth, middle age, and old age in the life spans of galactic clusters!

HYDRA

It was no easy task for Hercules to destroy the great, multiheaded water serpent of Lerna. Its breath was poisonous, and ordinary men could not tolerate even a glimpse of this monster without dying of fright. Bravely, Hercules attacked it with arrows and a club, but each time he struck off one of its numerous mortal heads, three new heads grew in its place! Hercules overcame this difficulty by means of an ingenious plan: his charioteer Iolus followed up each decapitation by immediately cauterizing the stump with burning bushes, so that no new heads could grow. The two men labored ardently, until at last only a single snake head remained; this one was immortal and encased in gold. Valiant Hercules severed it with a single blow of his sword and buried it under a rock.

The Hydra was dead, and the people of ancient Argos were thankful to be free of its menace; nevertheless, we must concede that life is as precious to a hundred-headed

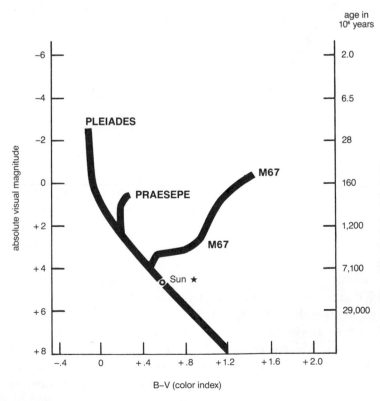

67. *Pleiades, Praesepe, and M67* plotted on the color-magnitude diagram. The numbers on the right-hand vertical axis refer to the turnoff point of a cluster from the main sequence. These numbers do not apply to the ages of individual stars on the main sequence like the Sun, not in a cluster. The Sun's position on the main sequence is given as a reference point for absolute magnitude and color.

68. Hercules' slaying of
the Hydra. Engraving.
(The Bettmann Archive)

water serpent as to any other creature: indeed, a hundred times more so! Apparently recognizing this fact, the gods transferred Hydra's image to the skies as an immense constellation, nearly 95 degrees long, twisting all the way from the borders of Monoceros and Canis Major in the west and the Serpent's former ally, Cancer, in the northwest, to Libra, in the east, and Centaurus, in the south (i.e., from RA 8 h 20 m δ about $+7°$, to RA nearly 15 h δ about $-35°$), but placed at a safe distance from Hercules!

Besides the unfortunate Lernian monster, Hydra has many other identities: its head was the seventh Indian lunar station, *Aclesha*, the "Embracer"; and also the seventh Chinese lunar house, a "Willow Branch" or a "Circular Garland." From α, to δ, in the head, the Chinese imaged the neck and beak of a "Red Bird," worshiped as an emblem of immortality at the summer solstice. A Euphratean stone of 1200 B.C. depicting the heavens shows a snake identified with Hydra as the source of oceanic fountains, and, like Draco, Hydra is a symbol of the Euphratean dragon Tiāmat, overcome by an ancient sun god who was later identified as the formidable Hercules.

In its heavenly form, however, the Water Serpent is not so menacing as it was in the swamps of Lerna, for chartists usually show it as quite decently possessing one head only (undoubtedly the immortal one), a ring-shaped asterism, just south of Cancer, comprising six third- and fourth-magnitude stars. The brightest of these, *(ζ) ZETA Hydrae* (RA 8 h 53 m   δ +6° 8′)   m 3.12, is an orange giant about 230 $L \odot$ and some 220 light years distant. Second-brightest is *(ε) EPSILON,* to its west-northwest (RA 8 h 44 m   δ +6° 36′)   m 3.36, a complex system of five stars. The close, difficult-to-observe A-B pair, magnitudes 3.7 and 4.8, was discerned as a binary by Schiaparelli in 1888; its period is 15.3 years and the mean separation of the components is about 8.5 astronomical units; their most recent close approach to each other was in 1977. The mass of the primary, a yellow giant (type G0 III) is 1.75 solar masses, and that of its companion is 1.60 $\mathcal{M} \odot$, or nearly the same. In 1830, F. G. W. von Struve discovered the eighth-magnitude companion of this binary, a normal class-F dwarf that revolves around the A-B pair about once every 650 years at a distance, of 130 astronomical units, that is steadily increasing. A cool, class-K dwarf of 2.5 $L \odot$ lies at the perimeter of the system, some 825 astronomical units from the close binary. A fifth star belonging to this group is also believed to exist.

West-southwest of ε is fourth-magnitude *(δ) DELTA,* a white class-A main-sequence star; if you have a small telescope, look for double star *Σ1255 SAO #117000, ADS #6913* (class G5), immediately to its east (RA 8 h 37 m   δ +5° 57′)   m 7 and 8. South of δ and aptly described by Ulug Beg as the "Snake's Nose" is *(σ) SIGMA,* a magnitude-4.5 orange giant. Southeast of ζ, we find *(θ) THETA* (RA 9 h 12 m   δ +2° 32′), a bluish-white optical double, magnitudes 4 and 10, uppermost star of the Serpent's neck. To

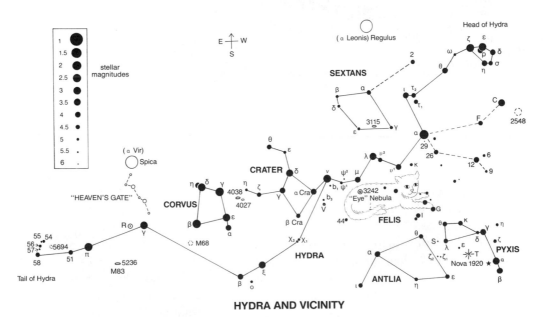

HYDRA AND VICINITY

69. *Hydra and vicinity.*

θ's southeast, at the bend of the neck, lies fourth-magnitude *(ι) IOTA,* which the Chinese called *Ping Sing,* a "Tranquil Star," perhaps because of its soft reddish hue. Immediately to its southwest and west, respectively, lie (τ) TAU 1 and TAU 2, magnitudes 4.8 and 4.5, pale cream and bluish. Tau 1 is a wide double, spectral type A3 III (giant), 48.9 light years distant, its seventh-magnitude companion separated from it by 66 seconds of arc.

Where the snake's neck joins its body (south of the τ's) we find *ALPHARD (α) ALPHA Hydrae,* from *Al Fard al Shujā,* "The Solitary One in the Serpent," intended as a reference to its position in Hydra, but appropriately named, it so happens, because it is a single star. Tycho Brahe called α "Cor Hydrae," the "Hydra's Heart," and for the Chinese, its position (RA 9 h 25 m δ −8° 26′) was strategic, combining seven lunar divisions of the southern quarter of the heavens. A slightly variable orange giant, about 95 light years distant, Alphard has a luminosity of approximately 110 $L \odot$, but it is only of the 2nd magnitude; the immense Hydra contains very few bright stars. Immediately to the south of α lies #29, class A0, a binary consisting of two close magnitude-7 components circled at 10.8 seconds of arc by a magnitude-11.5 companion.

Third-magnitude *(ν) NU* lies at the western border of the constellation Crater; an orange giant of 100 $L \odot$, it is part of a zigzag line of stars that includes fifth-magnitude Kappa, Upsilon 1 and 2, Lambda, Mu, Psi 1 and 2, Chi 1 and 2, as well as β Crateris, and ξ and β Hydrae—all of them delineating *Al Sharasif,* "The Ribs," to the sixteenth-century Arabian astronomer Al Tizini. Farther to the east along the Serpent's twisting body, near the northern border of Centaurus, lies the southernmost "major" star of Hydra: *(β) BETA,* (RA 11 h 50 m δ −33° 38′), strangely, at magnitude 4.3, fainter than *(ξ)* XI, magnitude 3.5, to its northwest. Beta, which is a close double of magnitudes 4.5 and 5, spectral class A0, can be seen only in the south. With ξ, it was a "Green Hill" to the Chinese. Much farther east along the Serpent's tail, south-southwest of brilliant Spica in Virgo, lies third-magnitude *(γ) GAMMA Hydrae* (RA 13 h 16 m δ −22° 54′), a giant with a "late" G-type spectrum, about 115 light years distant, and 65 $L \odot$. An irregular line formed by fainter stars between γ and Spica resembles a gate, invitingly ajar, which the Chinese called *Tien Mun,* or "Heaven's Gate."

An important variable star, *R Hydrae,* lies 2.6 degrees east of γ; a "Mira-type" red giant, R's magnitude ranges from 3.5 to 11, but like many others with similar chemistry, its red color fades slightly when its luminosity approaches maximum. Its period, of about four hundred days, has decreased markedly in the past two centuries. Long-period variables are too distant for reliable parallactic measurements except by statistical methods, which indicate that R is about 325 light years distant. A hallmark of all such variables is their bright emission lines of hydrogen at maximum luminosity, which probably originate deep below the photospheres and vary in intensity because of the stars' pulsations. At minimum, when the stars are most compressed, the relatively dense atmospheres obscure these emission lines. At maximum luminosity, however, the stars' atmospheres are sufficiently attenuated to allow the emission lines to stand out strongly.

A very interesting variable of fainter magnitude than R is the rare carbon star *V Hydrae* (RA 10 h 49 m δ −20° 59′), about 5 degrees south/southeast of ν. It ranges from about magnitude 6 to 12 in a period of approximately 530 days, but this cycle shows an eighteen-year fluctuation; such a star is called a *semiregular variable* (see

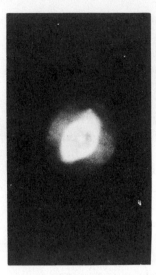

70. *NGC 3242, the Eye Nebula,* a planetary nebula in Hydra photographed in red light with the 200-inch reflector. (*Hale Observatories*)

"Hind's Crimson Star," p. 124, and S Cephei, p. 75). V's distance, based on its approximate absolute magnitude at its maximum, −1.5, is estimated to be 1,300 light years.

Nebulae, Clusters, and Galaxies in the Serpent

NGC 3242, H 27⁴ (RA 10 h 22 m   δ −18° 23′)   m 8.9, partway along the Serpent's body, about 1.8 degrees south of (μ) Mu, is a planetary nebula whose component parts —a hot blue central dwarf (m 11, 25 *L* ⊙), a bright elliptical inner ring, and a diffuse spherical outer shell, resemble a human eye in larger telescopes (at least 10 in.) (fig. 70). Estimates on its distance vary from 1,900 light years to 3,300 light years, and its diameter may extend for 36,000 astronomical units, or 0.6 light years.

M48, one of Messier's so-called "nebulae," which he incorrectly located in an empty spot and which astronomers long regarded as a "missing" Messier number, is now identified with the galactic cluster *NGC 2548* (RA 8 h 11 m   δ −5° 38′)   m 5.5, 4 degrees to the south of Messier's erroneous position. A large triangular group of about fifty stars, 1,700 light years distant, but easily viewed with a small instrument, it contains mostly class-A main-sequence stars, and also three yellow giants.

M68, NGC 4590 (RA 12 h 36 m   δ −26° 29′) south-southeast of β Corvi, is a rich globular cluster resolvable in telescopes of 6 inches and over; it contains more than a hundred thousand stars. The total number of stars in a globular cluster can be estimated only from star counts on photographic plates, and the results thus obtained are somewhat unreliable, but they nevertheless indicate that globulars must contain hundreds of thousands or even millions of stars. Light from a star in the outer regions of M68 takes over a century to traverse the cluster and reach the opposite side.

NGC 5694 (RA 14 h 36 m   δ −26° 19′) in Hydra's eastern tip near the stars numbered 54 to 57, is one of the most distant globulars, at about 100,000 light years. Although limited by the great distances involved, measurements have been taken of the proper motions for eight or nine globulars, indicating that such clusters have large space velocities. This is confirmed by their radial velocities, easily measured by the

Doppler shift, which shows that they range from about 300 kilometers per second of approach to 400 kilometers per second of recession. The motion of the Solar System around the galactic center and that of the globulars about this center account for the high peculiar velocities of these clusters. The radial velocity of NGC 5694 is 116 miles per second of approach; but its true space velocity relative to the galactic center, 273 kilometers per second (about 170 miles per second), is unusually large. Lying about 85,000 light years from the galactic center and traveling at this speed, NGC 5694 may eventually escape from our Galaxy altogether and wander off into intergalactic space.

 M83, NGC 5236 (RA 13 h 34 m δ −29° 37′) m 8, lying on the Hydra-Centaurus

71. M83 (NGC 5236), an outstanding spiral galaxy in Hydra. Photographed with the 60-inch reflector at Harvard's Boyden Station, South Africa. (*Yerkes Observatory photograph, University of Chicago*)

border, about 7½° south-southeast of γ, is an outstanding spiral galaxy at a distance of 10 million light years, its beautiful swirling form easily viewed in small telescopes by our readers far to the south (fig. 71). With a luminosity of 5 billion $L_{\odot}$ and its visible diameter 30,000 light years, M83 is known for its unusually high number of supernovae, four having been observed between 1923 and 1968. The last of these, centered in the nucleus, had a luminosity of nearly 100 million $L_{\odot}$ at its maximum.

A Forgotten Cat

The eighteenth-century French astronomer Joseph Jérôme Le Français de Lalande included eighty-eight constellations in the third edition of his *Astronomie*. Among these, for the first time, nestling beneath the great coil of Hydra, there appeared a familiar animal named "FELIS." Lalande explained: "I am very fond of cats. I will let this figure scratch on the chart. The starry sky has worried me quite enough in my life, so that now I can have my joke with it."

Sharing his cat-loving sentiments and regretting Felis' subsequent disappearance from star charts, the authors have undertaken, at least pictorially, to restore Lalande's kitty to its rightful place among the constellations! Look for Felis from about the tenth to the eleventh "meridians" of right ascension, lying between Hydra and Antlia (the "Air Pump"), its head to the south-southeast of α Hydrae.

SEXTANS

Another of the minor figures that hug the form of Hydra is Sextans Uraniae, the "Heavenly Sextant," created by Hevelius in honor of the instrument he used for stellar measurements. Hevelius placed Sextans south of Leo and just above the Serpent's back because of the "fiery nature" of these creatures, as a commemoration of the destruction of his instruments when his house in Danzig burned in 1679. The brightest star of this faint constellation is *(α) ALPHA*, about 12 degrees due south of Regulus; a blue-white class-B5 main-sequence star, α Sextantis is only of magnitude 4.5.

In the southwest corner of the frame of the instrument, about 9 degrees southwest of α and a generous 6 degrees east-northeast of α Hydrae (Alphard), is the triple star *(γ) GAMMA Sextantis* (RA 9 h 50 m   δ −7° 52′)   m 5, consisting of an A-type close binary with a period of seventy-five years and a distant, very faint companion at 35.8 seconds. Other stars of interest are *#35*, in the northeast part of Sextans, a binary consisting of two orange giants of magnitudes 6.5 and 7.5, and *#41*, in the southeast, an A-type binary of nearly the sixth magnitude.

A "Hybrid" Galaxy

Midway between ε and γ, which mark the southern corners of Sextans' frame, some 7.5 degrees south of α Sextantis and about 20 degrees south of Regulus (α Leonis), lies the nearly edge-on lenticular galaxy *NGC 3115*, the *SPINDLE* (RA 10 h 2 m   δ −7° 28′)   m 9.8, its apparent dimensions about 1 by 4 minutes. Some 25 million light years distant, it was formerly classed as an elliptical galaxy but is now regarded as a possible intermediate, or SO, type because of its bright oval nucleus and its flattened equatorial plane, which combine the characteristics of an elliptical and a spiral galaxy.

CRATER

Interminable Hydra was sensibly divided by Flamsteed and others into four parts, including "Hydra et Crater." Now charted separately, Crater, "The Cup," is a group of very faint stars lying above the Serpent's back. When discernible, they form a beautifully shaped wineglass; its universal identification as a cup is therefore no surprise. For the Greeks, it represented the "Goblet of Apollo" and also a "Water Bucket"; to the Euphrateans, it was the "Mixing Bowl" of divine mythology; other identities are the Arabians' *Al Batiyah,* an earthen wine vessel, and the English "Two-Handled Pot." (To us, it resembles a sundae dish, thus revealing our confectionary weakness.) Another Arabic title, *Al Kās,* a shallow basin, and the Hebrew Cōs, or "Cup," both suggest the star name *ALKES* with its variants, "Alker" and "Alhes": *(α) ALPHA Crateris* (RA 10 h 55 m δ −18° 20′) m 4.2, marking the base of the stem of Crater, or *Fundus Vasis,* "The Base of the Cup"—its Roman name. A reddish star of late spectral type, α has several optical companions, not gravitationally bound to it. *(β) BETA* to its south-southeast (m 4.5), a white subgiant, was the first of Tizini's *Sharāsīf,* "Ribs" (of Hydra), which we have mentioned. *(δ) DELTA,* one of two stars at the bottom of the bowl, is now the brightest star in Crater, at magnitude 3.8; the other, *(γ) GAMMA,* south-southeast of it, magnitude 4.1, is a class-A7 close double, the components magnitudes 4 and 9.

CORVUS

Another of Hydra's earlier divisions was "Hydra et Corvus," the latter a Crow resting on the Serpent's back and formed of third- and fourth-magnitude stars to the east of Crater. The precession of the Earth's axis has "moved" Corvus to an area crossed by the twentieth degree of declination, but two thousand years ago it lay on either side of the celestial equator. The Greeks called it "Raven"; according to mythology, Apollo, the sun god, assumed its shape during the gods' war with the giants. A legend linking the three constellations tells how Apollo sent Corvus for water with a cup (Crater), but the crow, distracted by the ripening fruit of a fig tree, returned late with a water serpent (Hydra) in his claws, which he falsely blamed for this delay. As a punishment, Corvus was placed in the heavens together with the Cup and the Snake, the latter now guarding the Cup and its contents. Thirsty Corvus rests midway on Hydra's back and pecks away at one of the serpent's folds, presumably to remind it that he desires a drink from the forbidden Cup.

Although Corvus and Cancer do not set simultaneously, a coin of Mindaon depicts these constellations as a raven and an ass together (the latter a reference to the *Asini,* γ and δ Cancri). In early Arabia, four stars of Corvus were a throne for Spica, the

"Unarmed One." Euphratean myth identified Corvus as a descendant of Tiāmat (represented by Hydra); the Akkadians regarded it as *Kurra,* the "Horse," and it was also the Hebraic *Ōrev,* or "Raven." Writers drawn to biblical legend linked Corvus to the Flood as Noah's raven, alighting on Hydra's back to keep dry!

Desert dwellers called Corvus *Al Hibā,* "The Tent," which its shape strongly suggests; this is also the title of *(α) ALPHA Corvi* (RA 12 h 8 m   δ −24°), now a magnitude-4 orange star supposed to have diminished in apparent brightness. (β) BETA, marking the southeast corner of the tent and called *Tso Hea,* the "Left-Hand Linchpin," by the Chinese, is a magnitude-2.65 yellow giant. *GIENA (γ) GAMMA* (RA 12 h 13 m   δ −17° 16′)   m 2.59 at the northwest corner, the brightest star of tiny Corvus, is a blue-white giant 450 light years distant and 1,200 *L* ☉. *ALGORAB (δ) DELTA* (RA 12 h 27 m   δ −16° 14′)   m 2.95, marking the northeast corner, is a wide double (separation 900 AU), magnitudes 3.1 and 8.4, 125 light years distant and easy for small telescopes. Although often listed as "yellow and lilac," the components' spectral types, B9 and dK2, indicate white and orange, the difference probably arising from chromatic aberration in the objective of the telescope.

NGC 4038 (RA 11 h 59 m   δ −18° 35′), popularly known as the Ring-tail Galaxy, is classed as a "peculiar" spiral. Lying about 3.7 degrees west-southwest from γ, it resembles a loop with a tapering tail (fig. 72). Long-exposure photographs reveal two elliptical masses joined together and two faint curved streamers arcing out from the point of their fusion. Astronomers believe that this strange image may represent two galaxies in collision, or perhaps one galaxy dividing in two. The streamers may be caused by mutual tidal action. *Peculiar* galaxies, those in which either intergalactic collisions or internal explosions have changed their shapes, constitute only 2 percent of all known galaxies. When such collisions do occur, they are surprisingly quiet, for the stars themselves do not collide. Stars are separated from one another by such immense distances that even when two galaxies interpenetrate and the density of their stellar population is doubled, no stars touch; they remain many light-years apart! However, in such an event, the atoms of the galactic gas/dust clouds collide repeatedly, forming pockets of very dense material from which new stars are born. Thus, a celestial "catastrophe" becomes an act of creation!

The Ring-tail is nearly 90 million light years distant, its diameter about 100,000 light years and its luminosity about 20 billion *L* ☉. Its Doppler shift indicates that it is receding at 910 miles per second. Peculiar galaxy NGC 4027, probably an associated system, lies 0.7 degrees to the west-southwest, its single filament pointing to the Ring-tail.

LEO

The first of Hercules' labors was to slay the gigantic Nemean lion, an anatomic marvel whose skin could not be pierced by iron, stone, or bronze. Invincible Hercules stopped up one entrance to the lion's double-mouthed cave, then wrestled with the beast until he choked it to death. A fitting memorial to Hercules' many superhuman deeds, Leo (The Lion) was placed in the heavens by the gods, where it has been known for

72. NGC 4038–39, the *Ring-Tail Galaxy,* in Corvus, a "peculiar" spiral system formed by two interlocking structures. A source of radio noise, it may be a dividing galaxy or two galaxies in collision. Two long, faint arcs of material curving out from the system's fusion point are visible in this photograph, taken with the 48-inch Schmidt telescope. (*Hale Observatories*)

thousands of years in the West and Near East as a constellation of the Zodiac. The Babylonians and Egyptians first perceived its stars in the form of a lion, and the Egyptian king Necepsos taught that the Sun rose near Denebola (β Leonis) at the creation; from this idea, Leo became the emblem of fire and heat to many ancient peoples. Five thousand years ago, the Chaldeans noted that at the time of summer solstice, when the sun reaches its highest point overhead (i.e., maximum declination), and the days are therefore longest, it was crossing the stars of Leo; owing to precession, of course, this is no longer so. Astrologers called Leo the "House of the Sun," from its Roman name, *Domicilium Solis,* and ancient physicians believed that medicine turned to poison when the Sun was "in Leo," and even forbade bathing at this time. Pliny states that the Egyptians worshiped Leo because the Sun's entrance among its stars marked the flooding of the Nile.

The lion is one of the most archaic symbols known, and despite its later masculine identification with light, heat, and bravery in battle, it was probably associated with the earlier matriarchies, agricultural tribes to which the lion was an important animal of the hunt, even as the bull came to represent more advanced, patriarchal societies, which practiced the domestication of animals. The lion was often depicted by pagan peoples as accompanying their chief goddesses, for example the lions of Atargatis, goddess of the Nabataeans, who was also the goddess of agriculture and fertility of the soil. The chief god was often shown with a bull, symbol of masculine power and virility, for with the advent of animal husbandry and the raising of cattle, the male rose to a position of dominance. The symbolic clash between lion and bull was variously depicted; on Ninevite cylinders, for example, Leo and a bull are locked in deadly combat. In its conscious meaning, such a struggle signifies the victory of light over darkness; but in the collective unconscious, it is a basic expression of male-female conflict, rooted in the clash between archaic societal forms. To this, we add that the symbolism of dreams and the subconscious, of ancient mythology, and the pictorial forms of the constellations, all stem from a common source, our deepest human feelings and mankind's earliest experiences.

REGULUS (α) *Alpha Leonis*
(RA 10 h 6 m   δ +12° 13')   m 1.36

Marking the base of the "Sickle," a distinctive shape that also includes η, γ, ζ, μ, and ε, Leo's brightest star was known from earliest times as the "Guardian of Heaven," ruling stellar affairs and keeping celestial order. This reputation gave it the Babylonian name *Sharru,* or "King"; its Indian title *Maghā,* "Mighty"; and the Persian *Miyan,* "Center." Copernicus derived Regulus from the diminutive form of its Ptolemaic name, *Rex* (or perhaps the *Regia* of Pliny), but the Roman *Cor Leonis,* "Lion's Heart," was always popular. Like Sirius, in classical days Regulus was thought responsible for the summer's heat. Euphrateans called it *Gus-ba-ra,* the "Red Fire," but this concept stemmed from its supposed association with the weather and bore no relationship to its color; for with a color index of −0.11 and spectral classification B7-V, α is a normal blue-white main-sequence star. It is 26 parsecs, or about 84.76 light-years, distant and has a luminosity of $160 L \odot$ and a surface temperature of about 13,000 degrees K. Regulus is a "visual triple," the system including a small companion, magnitude 8, class dK1, 177 seconds from the primary. This dwarf is resolvable in very large telescopes as a binary; the third, very faint (m13) star is also a dwarf, only about 0.25 $L \odot$. A fourth star of even larger magnitude has been noted at 217 seconds from the primary but not gravitationally bound to it. Because of α's position, very close to the ecliptic, it is called a "lunar star," for the Moon often appears to pass "near" it, that is, they are in nearly the same line of sight; therefore it is useful to sailors as a navigational guide.

DENEBOLA (β) Beta Leonis
(RA 11 h 46.5 m δ +14° 51')   m 2.14

At the opposite (eastern) end of the constellation lies Denebola, derived from *Al Dhanab al Asad*, "The Lion's Tail." Ulug Beg called it *Al Sarfah*, "The Changer" (of the weather), for it was believed to turn away the heat on rising and the cold on setting. The state-conscious Chinese named twelve groups surrounding it after officers and nobles of the empire, and called β, along with four small nearby stars, the "Seat of the Five Emperors." Euphrateans called it *Lamash*, the "Colossus," and in India it was the "Star of the Goddess Bahu," the Creating Mother. Because it marks the tail, astrologers believed β brought misfortune and disgrace, as opposed to the favorable character of Regulus.

Denebola, about 43 light years distant, is a blue-white main-sequence star about 20 $L \odot$, and surrounded by several very faint distant optical companions; the brightest of these, sixth-magnitude β603, is a normal class-A dwarf discovered by S. W. Burnham in 1879.

ALGEIBA (γ) Gamma Leonis
(RA 10 h 17 m   δ +20° 06')   m 1.98

Its name was probably derived from the Arabic *Al Jabbah*, "The Forehead," but don't look for it there, because Algeiba, easternmost star of the Sickle, lies on the bend of Leo's neck. Bayer, Riccioli, and Flamsteed correctly called γ *Juba*, Latin for "mane." Lying about 90 light years distant from us, γ is an excellent example of a double, its duplicity discovered by Herschel in 1782; with a period of six or seven centuries, the class-K and -G yellow-giant components, magnitudes 2.14 and 3.39, and about 90 and 30 $L \odot$, respectively, are separated by approximately 125 astronomical units. This distance is gradually increasing; thus our descendants at the beginning of the twenty-second century will be able to view the pair in small telescopes. Optically close (22') to its south lies fifth-magnitude *#40* Leonis, a class-F subgiant; and about 2 degrees northwest of γ is the radiant point of the Leonids, a spectacular meteor shower, the debris of the Tempel-Tuttle comet, whose orbit nearly intersects that of the Earth. The shower occurs annually around November 17, but it produces a particularly brilliant display every thirty-three years in the wake of the comet's passage.

Only 3 degrees 34 minutes north-northwest of γ is *ALDHAFERA (ζ) ZETA*, a white giant; 130 light years distant, it rests on the crest of the Lion's mane. With a small instrument, we may also view **#35,** an optical companion 5.5 minutes to ζ's north; almost on a line between γ and ζ, 18 minutes to ζ's south-southeast, lies **#39,** a close binary consisting of an orange and a red dwarf, magnitudes 6 and 11. To ζ's northwest, at the top of the Lion's head, lies *(μ) MU,* a reddish fourth-magnitude star, called "Rasalas" in some cataloges, from *Al Rās al Asad al Shamāliyy,* "The Lion's Head Toward the South."

To its southwest lies third-magnitude *(ε) EPSILON:* μ and ε were the Arabian *Al Ashfār,* "The Eyebrows," but since we know no lion with eyebrows, we think "face" would be more appropriate. The Chinese, however, called ε *Ta Tsze,* the "Crown Prince": not a bad guess, because this intrinsically brilliant star, about 340 light years distant from us, is a yellow subgiant, its luminosity 580 times the Sun's. Completing the Lion's head are *(κ) KAPPA* and *(λ) LAMBDA,* a magnitude-4.31 red star called Alterf, from *Al Tarf,* "The Extremity," signifying its position in the Lion's open mouth.

Traveling eastward from γ along the Lion's back, we find a third-magnitude class-A main-sequence star, **ZOSMA *(δ) DELTA*** (RA 11 h 11.5 m   δ +20° 48'), from the Greek for "Girdle," also called "Duhr," from *Al Thahr al Asad,* "The Lion's Back." An interesting historical curiosity is Flamsteed's accidental sighting of the planet Uranus in its field on December 13, 1690, which he recorded without realizing that he had discovered our then unknown seventh planet. Directly south of δ by 5 degrees 6 minutes lies *(θ)* **THETA,** sometimes called "Chort," from the title *Al H·arātān,* "The Two Little Ribs," referring to δ and θ together. About 90 light years distant from us, θ is a blue-white main-sequence star, similar in type to δ. Iota (Σ1536), to its south-southeast, 78 light years distant, is a yellowish binary of magnitudes 4 and 6.5, its period about 190 years, the primary a class-F2 subgiant.

The long-period variable ***R Leonis*** (RA 9 h 44 m   δ+11° 40'), 5 degrees west of Regulus, is the fourth of its type, discovered in 1782. Already known were the red giants Mira, R Hydrae, and Chi Cygni. R Leonis ranges between magnitude 5 at maximum and magnitude 10 or fainter at minimum, during an average period of 312 days. Suitable for smaller telescopes, it is a late-class-M giant of a particularly deep and intense red hue.

WOLF 359 (RA 10 h 54 m   δ +07° 19')   m 13.7, an extremely faint red dwarf, is only 7.75 light years distant, the third-closest star known. It lies somewhat west-southwest of the binary *(χ) Chi* and about 1.4 degrees northwest of the magnitude-5 white star Fl. #59. Its luminosity is only about 1/63,000 $L \odot$, and its absolute magnitude is +16.8. About the size of Jupiter, its proper motion is quite large, and it is of especial interest as a flare star. Compare Wolf 359 to Ross 614 B Monocerotis (p. 125), of similar radius and luminosity.

Leo's Many Spirals

South of λ, below the Lion's jaw, lies **NGC 2903 (56')** (RA 9 h 29 m   δ +21° 44'), a bright, many-armed, and elongated spiral galaxy (fig. 73) that resembles a crab more closely than does Cancer! The nucleus consists of eight intensely bright globules or knots that are probably HII regions.

In the Lion's head, north of the bright star γ, lies a small group of galaxies (fig. 74), **NGC 3190** (RA 10 h 15 m   δ+22° 08'), a spiral galaxy seen edge on, and its neighbors

73. *NGC 2903, a spiral galaxy in Leo.* A young (type-Sc) galaxy. Note the faint outer arms extending beyond the frame of this photograph, taken with the 200-inch telescope. (*Hale Observatories*)

74. *Group of four galaxies in Leo,* including NGC 3190 (upper center), a spiral galaxy seen edge on, showing a dark obscuration lane of dust and matter; to its right is L-shaped NGC 3187, a barred spiral; at the lower right is NGC 3185, also a barred spiral; and the elliptical NGC 3193 is at the upper left of this photograph, taken with the 200-inch telescope. (*Hale Observatories*)

NGC 3185 and *3187,* both barred spirals, and *NGC 3193,* a perfectly round "elliptical" galaxy.

The star fields east of Regulus are rich in spiral galaxies; notable are *M95 (NGC 3351)* and *M96 (NGC 3368),* discovered in 1781. M95 (RA 10 h 41 m δ +11° 58') is a bright barred spiral which, like others of its type, resembles the Greek letter θ in photographs; M96, to its east, has a large, bright nucleus and many faint spiral arms. This pair is receding from us at about 420 miles per second, at an approximate distance of 29 million light years.

M65 (NGC 3623) and *M66 (NGC 3627),* discovered in 1780, lying about 2½ degrees south-southeast of θ (RA 11 h 17 m δ +13° 20'), are accessible to small instruments. M66 is a bright, rather irregularly shaped spiral, its outer region streaked with long, dark channels of nonreflecting dusty material, and its arms display many clumps of stars. M65, more symmetrically shaped, appears oval or elongated because of its tilted position relative to our line of sight. The redshift, or rate of recession, for M65 and M66 is about 380 miles per second, and the approximate distance is 29 million light years. North of M66 is *NGC 3628,* an edge-on spiral.

The Classification of Galaxies

In 1926, Edwin P. Hubble (1889–1953), noted for his investigations of the redshifts of galaxies, introduced a structural scheme for their classification. He divided the galaxies into three main types: elliptical, spiral (normal and barred), and irregular structures, classed according to their shapes, and he represented them on his famous forked diagram (fig. 75). Although enlarged and revised, Hubble's system is still used. The elliptical galaxies range from the perfectly round E0 type to the lenticular E7. Next is type SO, which combines characteristics of elliptical and spiral systems, resembling highly flattened nuclei of spirals that have lost their arms. The spiral galaxies are subdivided into two broad classifications: normal S types (such as our Galaxy and the nebula in Andromeda), and barred spirals, classed SB. The normal spirals are further subdivided into three subclasses, a, b, and c, characterized by the degree to which the spiral arms are wound. The width of an arm, from one-tenth to one-fifteenth of the total galactic diameter, is related, with certain limitations, to this diameter; and the lengths of the arms also vary from short structures to those which wind completely around. The subclasses of barred spirals are as follows: SBa, which resembles θ (Theta); SBb, wherein the nucleus and the bar running through it are clearly defined and the spiral arms, which extend completely around, originate at the ends of the bar, almost at right angles to it; and the S-shaped SBc structure, which consists of a very bright nucleus with a bar running through it, from the ends of which two spiral arms extend about a quarter of the way around. Finally, the irregular galaxies, type I, show no definite spiral structure and no well-defined nucleus, as best exemplified by the Magellanic Clouds (pp. 389, 392).

Hubble's scheme suggests an evolutionary relationship among the various types of galaxies, and in fact he did believe that the ellipticals are the youngest, ultimately developing into spiral forms. According to his theory, the original round structures rotated very slowly, but this rotation speeded up because of gravitational contraction, and the systems became highly flattened, while, at the same time, exploding popula-

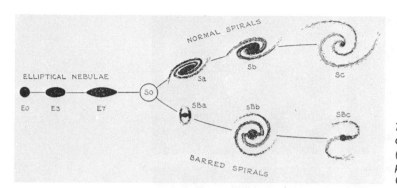

75. Hubble's classification of galaxies. (Yerkes Observatory photograph, University of Chicago)

tion-II stars (not known as such to him at the time) ejected their contents, rich in heavy elements, into space. From this material dust was formed and swirled by the rapid rotation into the outer regions, forming the spiral arm where the young population-I stars were born. Although Hubble's scheme is logical, observation seems to have refuted it; for, according to Hubble, ellipticals should be the youngest galaxies and we should expect to find far more ellipticals than spirals in the more distant parts of the universe; but instead we find the same number of very distant ellipticals as spirals (those in younger parts of the universe, from which light began its journey long ago). Another objection to Hubble's theory is that the amount of angular momentum in spherical and elliptical galaxies as we observe them now is insufficient to cause the amount of flattening required to account for the spiral structures.

Because of the weaknesses in Hubble's theory, other astronomers have suggested an opposing evolutionary scheme in which irregular galaxies and spirals end up as elliptical systems. According to this theory, galaxies begin as irregularly shaped distributions of gas and dust containing a fairly dense nucleus of population-II stars. As rotation speed increases, the outer regions develop into spiral arms, while the nucleus becomes spherical, rotating like a solid structure. After the gaseous material in the arms is collected into stars so that the spiral arms gradually disappear, a smooth elliptical structure is formed. Unfortunately for this theory, the foregoing process does not really lead to a smooth globular form from spiral structures. Because of severe difficulties presented by both theories, the Dutch astronomer J. H. Oort suggested that the various types of galaxies develop independently of each other, their forms determined by the original speed of rotation (or lack of it) in the protogalaxies, and the presence or absence of dust. In this scheme, the spherical galaxies had very little rotation to begin with, whereas spiral systems, which are highly flattened, began with a good deal of angular momentum. The important question of galactic evolution, however, is not yet fully answered.

LEO MINOR

The Greater Bear hath his "cub"; so also the "Greater Lion"! North of Leo but beneath Ursa Major's paws lies Leo Minor, the Lesser Lion, a modern constellation formed by

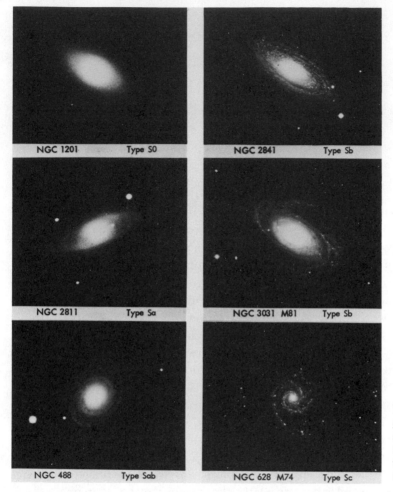

NGC 1201　　　Type S0

NGC 2841　　　Type Sb

NGC 2811　　　Type Sa

NGC 3031 M81　　Type Sb

NGC 488　　　Type Sab

NGC 628 M74　　Type Sc

76. *Typical examples of galaxies, elliptical and spiral, classified by Hubble. Compare this with the scheme shown in Figure 75.*

Hevelius. Lacking a mythological story to advise how he arrived at this rather ignominious location, we invite the reader to create one! The Borgian globe, however, shows the faint stars of Leo Minor as the Arabian *Al Ṭhibā' wa-Aulāduhā,* "The Gazelle with Her Young."

Fl. #46, of magnitude 4, is the principal star, single but with three distant optical companions. To its northwest lies β, a binary, the primary a late-solar-type (class-G) deep yellow star. About 1.33 degrees southwest of #46 lie *NGC 3395* and *NGC 3396* (RA 10 h 47 m　δ 33° 15–16'), an interacting pair of bright small galaxies, classed as a spiral (Sc) and an irregular "peculiar" barred spiral, of particular interest.

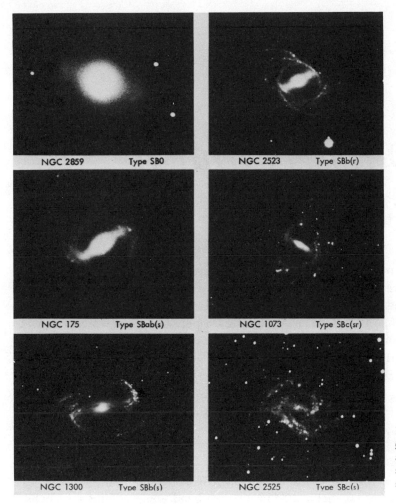

NGC 2859   Type SB0

NGC 2523   Type SBb(r)

NGC 175   Type SBab(s)

NGC 1073   Type SBc(sr)

NGC 1300   Type SBb(s)

NGC 2525   Type SBc(s)

77. Examples of barred galaxies, showing additional subclassfications of spirals.

SEVEN

Two Damsels and a Herdsman

As the nights grow warmer, we visit two great constellations, Virgo and Boötes, and the neighboring Coma Berenices and Canes Venatici. Because the circumpolar groups help us locate stars at various latitudes, we first undertake an orientation trip as we depart from Megrez (δ) and Phecda (γ), of the Big Dipper, and zoom straight down to Regulus, in Leo, passing the stars of Coma Berenices on the way. We zip along the ecliptic to Spica, the bright star in Virgo, and then we swing upward in a sweeping arc to brilliant Arcturus, in Boötes, returning to the Dipper via its handle.

COMA BERENICES

In modern charts, Denebola marks the Lion's tail. For many centuries, however, Leo had a much longer, fluffier tail, delineated by a rough quadrangle of faint stars behind his body, which was called *Al Halbah,* "The Tuft," by Arabians. As the tail follows the Lion, we now follow up our exploration of Leo (and his Cub) with a brief visit to the minor group first charted separately in modern times by Tycho Brahe as *Coma Berenices,* or "Berenice's Hair," after an ancient historical event to which it is traditionally related. This story concerns the long, golden hair of Berenice, wife of the Egyptian king Ptolemy III, who waged war on the Assyrians to avenge his sister's death at their hands. In gratitude for her husband's safe and victorious return from battle, Berenice had her beautiful hair shorn as a sacrifice to the goddess Aphrodite in the temple at Zephyrion. During a night of festivities and rejoicing, Berenice's tresses mysteriously disappeared from the altar in the sealed sanctuary of the temple. The royal couple were greatly distressed, and threatened to execute those priests who guarded the temple, should the hair not be returned. However, the Greek astronomer Conon of Samos saved the day by declaring that Berenice's sacrifice had so pleased Aphrodite that she had transferred the missing tresses to Leo's starry realm, so that all humanity might admire them!

"Berenice" is derived from the Greek word for "victory-bearing"; some philologists say that our word "varnish" is derived from the amber color of the queen's hair. Bayer called the constellation "Tricas," originating in the Low Greek for "tresses." Coma was

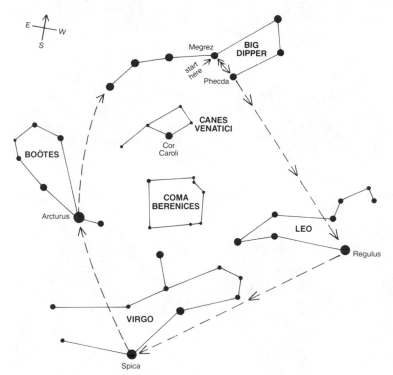

78. A round trip
(orientation chart for
spring).

also related to the constellation of Virgo, as a distaff held in the Virgin's hand.

Most of the stars of Coma Berenices are only of magnitudes 4 and 5. In the southeast corner of its irregular quadrangle lies *Alpha Comae,* a magnitude-4.23 class-F main-sequence double, whose fifth-magnitude components just miss forming an eclipsing binary by their nearly edge-on orbit. Interesting objects for binoculars or small telescopes include *#17,* a very wide double (m 5.4 and 6.7), white and bluish in color, lying south of Gamma Comae, in the northwest corner; and *#24,* a brilliant double of contrasting components, a magnitude-5.5 orange star and a magnitude-7.0 "emerald." This binary lies on the southern edge of the "quadrangle," several degrees to the west of Alpha.

Star Clusters

Six faint naked-eye stars, forming a semicircle or urn-shaped figure in the northwestern part of Coma, and some thirty fainter stars, all with the same proper motion, are members of the well-known *Coma Berenices Star Cluster* (often identified separately as the queen's hair). At only 250 light years, the Coma cluster is one of the nearest open clusters; its members include several spectroscopic binaries, the components ranging in type from blue-white to yellow main-sequence stars. The Coma Cluster's sparse population of about one star per 10 cubic parsecs gives it a small total mass; for this reason, astronomers suggest that it may be dispersing.

M53, NGC 5024 (RA 13 h 10 m   δ +18° 26'), 1 degree northeast from α and 65,000

79. *NGC 4314, peculiar spiral galaxy in Coma Berenices,* taken with two different filters. The inset (corner) shows details of its faint outer arms and apparent barred-spiral structure. Photographed with the 200-inch telescope. (*Hale Observatories*)

light years distant, is a compact globular star cluster which observers have always referred to as one of the most beautiful of its type.

The Virgo-Coma Galaxy Cluster

M85, NGC 4382 (RA 12 h 23 m δ +18° 28′), in the southern part of Coma, an "outlying" member of the Virgo galaxy cluster, was discovered by Mechain in 1781 as a "very faint nebula," which even our modern telescopes have been unable to resolve into stars! It is classed as an elliptical or an "SO" galaxy, i.e., a "hybrid" type combining the features of both elliptical and spiral galaxies. Receding at 450 miles per second, it is some 44 million light years distant, with its great mass equal to about 100 billion $\mathcal{M}\odot$.

M88, NGC 4501 (RA 12 h 29 m δ +14° 42′) is a member of the galaxy-rich Virgo-Coma galaxy cluster, one of its fine multiple-arm spirals, type Sb, suitable for the small telescope.

M98, NGC 4192 (RA 12 h 11 m   δ +15° 11′), one-half degree west of #6, is a nearly edge-on spiral of 130 billion solar masses, even more massive than the Black-Eye; and its absolute magnitude is about −21.

M99, NGC 4254 (RA 12 h 16 m   δ +14° 42′) is a type-Sc face-on spiral, nearly circular, its western arm reaching far out, and both arms marked by bright star clouds and nebulous regions; short segments radiating from a main arm on the northeast give it a many-armed appearance. The nucleus is small and starlike.

NGC 4314 (76¹) (RA 12 h 20 m   δ +30° 10′) is a "peculiar" barred spiral of type SB2, one of a group of galaxies in the northwestern part of Coma.

M100, NGC 4321 (RA 12 h 20 m   δ +16° 06′) is the largest spiral of the Virgo-Coma cluster, its coarse, prominent swirling arms reminiscent of a giant scarab or spider. Many secondary arms and segments radiate from its bright central nucleus. Light must travel for nearly three thousand years to traverse the thickness of a major arm; its diameter is about 110,000 light years, its total luminosity is nearly 20 billion times that of the Sun, and its mass is 160 billion solar masses. In December of 1979, a supernova erupted in this galaxy which has been extensively observed at visual, radio, ultraviolet, and X-ray wavelengths.

THE BLACK-EYE GALAXY, M64, NGC 4826 (RA 12 h 54 m   δ+21° 57′)   m 8, lies within the Coma "quadrangle," to the northeast of the Virgo-Coma cluster. It is distinguished for its striking central dust lane embedded in its inner spiral arms and giving it the unique appearance from which it derives its popular name.

NGC 4565 (RA 12 h 34 m   δ +26° 16′), the largest edgewise spiral, lies 1.7 degrees east of #17 and less than 3 degrees from the north galactic pole. One of the most impressive of its type, it displays a well-rounded nucleus and a prominent dark absorbing lane running "from tip to tip" (fig. 80). It is rather flattering to man's ego to reflect that Nature has tumbled the great galaxies at every possible angle to our line of sight, almost as though it intended to provide us with a complete view of each type, face-on to "profile," so that we may better understand their formation and structure! Even more overwhelming is the possibility that each of these vast stellar systems may house many reasoning, wondering, and awestruck observers like ourselves. Yet we bear in mind that there is but one universe, in which all systems are properly arranged, each in its place and all in balance.

The Coma Berenices Galaxy Cluster

An extremely remote and rich galaxy cluster lies in the northeast corner of Coma, about 2.3 degrees west of β. Until recent revisions, which indicate a far greater figure, its distance was estimated at nearly 400 million light years. It includes at least one thousand galaxies of every conceivable type; the brightest are *NGC 4889* and *NGC 4874*, the former a giant elliptical system and the latter of the hybrid "SO" type. The diameter of the total system is about 20 million light years, and the central core about 7 million light years. This great cluster is receding from us at the rate of 4,250 miles per second!

Using the latest electronic detectors and optical telescopes, scientists have reassessed the redshifts of this and of a neighboring cluster, indicating a distance somewhere between 5 and 9 billion light years. Distant galaxies like those in the Coma

80. NGC 4565, edgewise spiral galaxy in Coma Berenices; the largest of its type, photographed on an unfiltered red-sensitive plate with the 200-inch telescope. (*Hale Observatories*)

81. *Part of the Coma Berenices galaxy cluster,* showing NGC 4889, a bright elliptical galaxy, and NGC 4874, a hybrid "SO" type, in the same field but probably far "behind" it. The cluster is billions of light-years distant. Photographed with the 200-inch telescope. (*Hale Observatories*)

cluster are studied to test various theories on the evolution of galaxies, because they are so far away that we see them as they were when the universe was much younger. For this reason, they are also useful as tracers of the cosmological expansion.

VIRGO

Holding a palm branch in her right hand and an ear of wheat in her left, this universally recognized Maiden bewilders us with her many identities! In Egypt, Virgo appears on the zodiacs of Denderah and Thebes, and some writers associate her with Isis, the great Egyptian goddess, who, like Juno, was credited with forming the Milky Way—not because of a nursing problem, however, but with the kernels that fell from the magic ear of corn Isis was carrying as she fled from her evil brother Set, a fierce and warlike thunder god.

Virgo was the Chaldean Ishtar, "Queen of the Stars," called "Ashtoreth" in the First Book of Kings, the Sidonese goddess of one of Solomon's many foreign wives. In India she was *Kauni,* or "Maiden," mother of the great god Krishna; some of Virgo's principal stars seem to delineate the keel of a ship, which also gave her the title "Woman in a Ship" on the Sinhalese zodiac. The Turkomans called her *Dufhiya Pakhiza,* the "Pure Virgin"; a similar concept finds expression in an early Greek myth, in which she is Astraea, goddess of innocence and purity, the daughter of Themis, a Titaness representing justice and order. This was during the "Golden Age," when Greek gods walked among the people in a land of peace and plenty. To toughen men's spirits, Zeus brought a cold climate and great hardship to half of the world, an event that marked the coming of the so-called "Silver Age." Discontent and conflict soon arose among the populace, so that at the beginning of the "Iron Age," the gods abandoned human society to take up residence in heaven. When Astraea, the last celestial being to leave the Earth, saw that the people would not mend their evil ways, she rose slowly upward into the skies until she reached her present position along the ecliptic between Leo and Libra, where at night she still shows herself to mortal man as a constellation of the zodiac.

Virgo is also mythologically bound to the neighboring Boötes through its identity as the Athenian Icarius, whose daughter Erigone (Virgo) hanged herself in grief at his death and was transported with him to the skies.

In her most famous representation, Virgo is Persephone (Proserpina), the daughter of Zeus and Demeter (Ceres), goddess of the harvest. Kidnapped by Pluto, god of the underworld, Persephone was borne off in his chariot (the stars of nearby Libra!) to become the queen of Hades. Her grieving mother prevented all seed from sprouting, causing a never ending winter to encompass the Earth, until Zeus, alarmed at the loss of his agricultural tributes, ordered Pluto to return Persephone to Demeter; but the king of Hades claimed his marital rights on the basis of six pomegranate seeds that Persephone had swallowed while a captive in his realm. Zeus resolved this case by ordering that Persephone spend one third of each year in Hades as Pluto's wife, returning to the upper world for the other two thirds. On her daughter's release, Demeter joyously allowed the dormant plant life to awaken and cover the Earth with its verdancy. Since those events, the cycle is repeated each year as winter proclaims Persephone's banishment, and her return is celebrated in the coming of spring.

SPICA (α) *Alpha Virginis*
(RA 13 h 23 m   δ − 10° 54')   m 1.0

Its modern name derived from the Latin *Spicum,* this blue-white star was the Hebraic *Shibboleth,* the Syrian *Shebbelta, Chushe* of Persia, and the Turkish *Salkim,* all signifying Virgo's ear of corn or wheat. Among its numerous titles are also the desert name *Al Simāk al A'zal,* "The Defenseless" or "The Unarmed," and the Coptic *Khoritos,* or "Solitary," because it appears to the naked eye as though unattended by any companion.

The spectroscope, however, tells a different story about Spica, revealing its binary nature. The components, only 11 million miles apart, are a hot, massive (class-B) primary of 10.9 solar masses, and its somewhat cooler and less massive secondary (6.8 $\mathcal{M} \odot$), which partially eclipse one another during their very short (4-day) period of revolution. The luminous primary, also intrinsically variable, is not a young star, for it seems similar in type to both Beta Canis Majoris (Merak, of the Big Dipper), the classic example, and Beta Cephei, which are characterized by small magnitude variations occurring within a period of a few hours, caused by the pulsations typifying slowly expanding stars at the beginning of their departure from the main sequence.

Spica is associated with Hipparchus' discovery of the precession of the equinoxes, based upon his comparisons of its position in 150 B.C., and that of Regulus, with the recorded observations of the Alexandrian astronomer Timochares 150 years earlier. Scholars believe that a number of ancient Egyptian and Greek temples, erected at various intervals spanning a period of about thirty-five hundred years, were all oriented

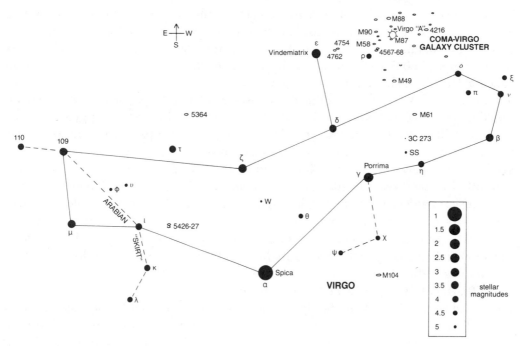

82. *The stars of Virgo.*

to Spica's setting, which, owing to precession, changed its position over the centuries. The adjustments that the builders of the later temples made to take account of this indicate that precession, as a matter of practical knowledge, may have been known to them long before Hipparchus.

Spica, Denebola, and Arcturus form a large, almost equilateral triangle; by adding the fainter Cor Caroli (of Canes Venatici), we have a prominent four-sided figure called the *DIAMOND OF VIRGO,* which helps to guide us as we explore this interesting and important region of the heavens.

In the "Doghouse"

Early Arabs saw the stars β, η, γ, δ, and ε as the outline of a kennel forming the eleventh lunar station *Al 'Awwā'*, appropriately "The Barker." Pale yellow β, magnitude 3.6, derives its intriguing name, "Zavijava," from *Al Zāwiah*, the corner, or angle, of the kennel (R. H. Allen thinks this inappropriate because γ marks this corner); but if ν, o, and π, the faint stars of Virgo's head, are included in the kennel's outline, we see instead a very nicely formed doghouse, and β marks the right corner of its *roof!* Incidentally, because of her faint head, the Romans knew Virgo as headless *Fortuna.*

Best-known of Virgo's major stars after Spica is *PORRIMA (γ) GAMMA* (RA 12 h 39 m δ −1° 11′) magnitude 2.7, equally distant from α and β. Porrima was named by the Romans after two ancient goddesses of prophecy; in Babylon it was the "Star of the Hero," and interestingly, it is associated with the first observation of the planet Saturn, recorded by the Babylonians on March 1, 228 B.C., with γ as the reference point.

Porrima, first observed as a double in the early eighteenth century, is an outstanding example of a visual binary whose components are nearly of the same spectral type (class F) and equal luminosity (7 $L \odot$); it is the first star for which an orbit was calculated (by Sir John Herschel). Its extremely eccentric system (0.88) results in an elliptical orbit which brings the components close together every 171 years, when they are only 279 million miles (3 AU) apart and difficult to observe separately. At their greatest separation they are 6,510 million miles (70 AU) apart. *Periastron*, their closest approach to one another, will next occur in the year 2007; the present apparent separation is 4.5 seconds of arc. One of the closer binaries, at a distance of 32 light years, Porrima is gradually approaching us at the tiny speed of 12 miles per second.

Another of the "kennel" stars is (δ), magnitude 3.4, lying about 7 degrees south-southwest of ε and marking the Virgin's right shoulder; the two were regarded as one of seven pairs of "twin stars" by the Euphrateans, while δ alone was named *Lu Lim*, "Gazelle." A red giant, it bears the unique distinction of a title inspired by modern astronomy: *Bellissima*, a reference to its beautifully banded spectrum, by the Italian astronomer Father Angelo Secchi (1818–78).

VINDEMIATRIX (ε) EPSILON, (RA 12 h 59.7 m   δ +11° 14′)   m 2.84, δ's Euphratean "twin," marking the tip of Virgo's outstretched right arm, was known as "Grape Gatherer" to the Romans, because it arose with the Sun shortly before the annual grape harvest. It was also called "Almuradin," possibly from the Arabic *Al Muridīn*, "Those Who Sent Forth." It may have earned the Euphratean title "Star Man of Fire" because of its warm color (actually related to its late-class-G spectrum). Although its spectral class is that of the Sun's, ε is a so-called "bright giant," with a luminosity of 50 Suns.

To the west of γ, fourth-magnitude *ZANIAH (η) ETA*, its name derived, like β's, from *Al Zāwiah*, "The Angle" (and, in its field, 6th-magnitude #13), helps form the side of the Arabian Kennel; but, to the Chinese, η was *Tso Chih Fa*, the "Left-Hand Maintainer of Law."

Stars of Virgo lying outside our doghouse, i.e., in the robes of the Maiden, include: *(θ) THETA*, magnitude 4.5, an attractive triple system northwest of α, the components a pale white class-A, or "Sirian," primary, a ninth-magnitude secondary at a separation of 7 seconds, and a magnitude-10 tertiary at 71 seconds; *(ζ) ZETA*, magnitude 3.4, another Sirian main-sequence star, about 11 degrees to the north of Spica and directly on the celestial equator, its luminosity 30 $L \odot$; and *(ι) IOTA, (κ) KAPPA* and *(φ) PHI*, on the train of Virgo's robe near the borders of Libra. According to Al Biruni, the three form the most fortunate Arabian lunar mansion, lying between an ancient lion and a scorpion (not our modern figures), but also marking the location of Mohammed's horoscope and coinciding with the birth date of Moses.

W VIRGINIS (RA 13 h 23 m   δ −3° 17′), lying about 3⅝ degrees southwest of ζ, is the prototype after which an important class of variable stars has been named. We told the story of δ Cephei and how the discovery of two populations of Cepheid variables made it possible to determine the distances of galaxies with great accuracy. Called Type II or "W Virginis" stars, these Cepheids lie at considerable distances from the plane of our Galaxy, within the halo of globular clusters surrounding the Galaxy as well as outside these clusters, and also toward the galactic center. Like the classical (population-I) Cepheids, they are very luminous, but with the following differences: their mean spectral classes lie between F2 and G2, while type-I Cepheids have a somewhat broader spectral range, extending into later subclasses. The periods of W Virginis stars range from two to seventy-five days, but most of them complete a light cycle in about seventeen days, whereas classical Cepheids have shorter periods. W Virginis stars' absolute magnitudes range from −0.5 to −2.5, while classical Cepheids reach absolute magnitudes of −4.4.

Interesting features of some type-II Cepheids are their bright hydrogen emission lines, which are most intense at maximum luminosity, and also the simultaneous appearance of two different absorption spectra at the moment when luminosity begins to increase. This phenomenon probably arises because the outer layers of these stars begin to contract while the material in their deeper layers is still expanding. The velocity curves of classical Cepheids are the mirror images (reverse) of their light curves. The velocity curves of the W Virginis stars differ from those of classical Cepheids in that the former do not duplicate their own light curves, although they follow them closely (see fig. 83).

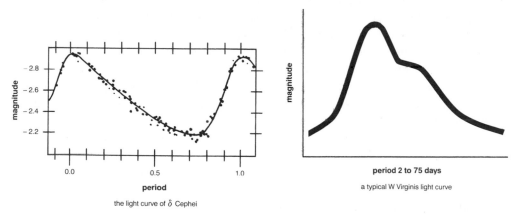

the light curve of δ Cephei

period 2 to 75 days

a typical W Virginis light curve

83. Comparison of the light curve of a classical Cepheid and that of a typical W Virginis star.

W Virginis itself lies 1,750 parsecs (about 5,707 LY) off the plane of the Milky Way, within the Galaxy, and not in a globular cluster; it is about 3,380 parsecs (11,000 LY) distant from the Solar System. With its variations in light, W alternately expands and contracts, its spectrum ranging from G0 Ib at minimum to about F0 Ib at maximum and its magnitude from 10.7 to 9.5. Its luminosity at maximum is approximately fifteen hundred times that of the Sun! The radial velocity of W Virginis varies from about 23 to 56 miles per second in approach; its proper motion is about 0.006 second per year.

All population-II Cepheids have high space velocities, as compared to classical Cepheids and population-I stars generally, but only *with regard to the Sun;* in their highly elongated orbits about the galactic center, W Virginis stars appear to be moving rapidly, because the Sun, moving in its almost circular orbit with higher orbital speed, either overtakes these distant clusters (in our line of sight) or leaves them lagging behind.

The Story of Quasars

Lying about 4.7 degrees northwest of γ, *3C 273 Virginis* (RA 12 h 26 m   δ +2°
19′)   m 12.8 is one of the earliest known examples of the most mysterious objects in
the universe. Called "quasars," "quasi-stellar radio sources," or "QSS's," they emit an
excess of ultraviolet and infrared radiation as compared to ordinary stars or galaxies,
and, when their radio emission is relatively weak but optical properties are similar to
those of the QSS's, they are known as QSO's or "quasi-stellar objects." Many aspects of
the true nature of these objects are still unexplained. Certain faint starlike images
visible on photographic plates for many years were mistaken for galactic stars until, in
1960, it was discovered that, unlike ordinary stars, they are sources of strong radio
waves. By 1963, the spectra had been obtained for two of these strange bodies: 3C 48,
in Triangulum, and 3C 273, in Virgo. These spectra baffled astronomers, because the
spectral lines (and those of similar "radio stars" subsequently discovered) do not
correspond to the patterns of any known atom, or even to one another!

In 1963, Cyril Hazard, in Australia, used an occultation of 3C 273 by the moon to
pinpoint the quasar's exact position and found it to be a double source of radio waves:
the bright nucleus, and the tip of a jet 160,000 light years long, streaking out from the
quasar's southwest (both visible in photographs). Surprisingly, much of the radio
emission has now been shown to originate in the jet itself, which is actually the trail or
track of an unimaginably violent explosion in the quasar's core, which erupted about 1
million years ago and is continuing to emit vast amounts of energy. Later in 1963,
Maarten Schmidt, of the Palomar Mountain Observatory, solved the problem of the
quasar's strange spectra when he noticed that the spacing of the lines of 3C 273
resembles that of the visible hydrogen spectrum (the Balmer lines); by assuming a very
large redshift for these lines, he accounted for their unusual wavelengths, which actually correspond to the hydrogen spectrum shifted toward the red or longer wavelengths by 15.8 percent. The spectral lines of doubly ionized oxygen and of neon and
magnesium were subsequently identified. Later the spectra of other quasars were also
matched up with the hydrogen lines by applying even larger redshifts. In accord with
the Hubble law of recession (p. 102), the redshift of 3C 273, about 16 percent, or
28,400 miles per second, places it at 3 billion light years from the Sun, and thus creates
another mystery: this distance and its apparent magnitude (13) indicate that 3C 273,
has a tremendous luminosity, nearly 30 trillion times the Sun's, or comparable to that
of three hundred giant galaxies. This is all the more amazing in the light of our
knowledge that the diameter of 3C 273 is only 7,500 light years. Other quasars with still
larger redshifts appear to be receding with still higher *relativistic* speeds, i.e., nearly that
of light, and they are therefore even more distant than 3C 273.

Most astronomers now believe that quasars are, indeed, very distant galaxies which
appear deceptively starlike in photographs; but some have suggested that they are
relatively nearby objects, and therefore not so luminous. In this case, an explanation
other than distance has to be found for their great redshifts; however, this has never
been done. In our account of the companion of Sirius, we explain the gravitational
energy loss (reddening) of a photon leaving the surface of a massive body of very high
density; a collapsing star approaching the black-hole stage can produce the observed

spectral redshift of a quasar. Yet this explanation for the redshift of quasars has been rejected by the majority of astronomers, for a number of reasons.

For one, if quasars were dense enough to produce strong gravitational redshifts, they could not produce their characteristic bright-line emission spectra. For another, quasars seem to share certain features with some types of galaxies: radio galaxies, for example, which display double sources of energy, giving them a "twin-lobed" appearance on radio contour maps, as do many quasars; and the "Seyfert" galaxies (discovered by Carl Seyfert in 1943), with their extremely large output of energy originating in small nuclei, inexplicable in terms of greater mass (for they contain fewer stars and have smaller diameters than normal galaxies).

Like quasars, the spectra of Seyfert galaxies also contain more emission lines than do ordinary galaxies. Thus there seems to be a sequence in galactic type, ranging from normal spirals, through Seyfert galaxies (and the still more luminous "nuclear," or N-type, galaxies with their very small, bright nuclei), to the quasars, which apparently exaggerate the unusual qualities of the Seyfert and N galaxies.

Another conjecture links quasars with galaxies as follows: within the small spatial area of a quasar, stars are packed very closely together, and collisions among individual stars occur frequently, thus contributing to the great luminosity of the quasar, by producing supernovae and, ultimately, energetic pulsars, which add somewhat to the luminosity. This theory, however, cannot account for the tremendous luminosity of such quasars as 3C 273 and has been discarded.

Furthermore, all quasars are variable, with relatively small periods, both in radio emission and in visible light; the luminosities of some quasars increase more than a magnitude in a few weeks. A galaxy-sized object, however, cannot possibly show such a rapid light-change, hence quasars must be much smaller than normal galaxies. On the other hand, if quasars are not galaxies, but much smaller, we face the question of how so much energy can be produced in such small bodies. Other theories suggest that quasars are contracting clouds of gas and dust; giant supercondensed pulsar-like objects, their own gravity the source of their energy; annihilating collisions between matter and antimatter; massive black holes in galactic nuclei, which are fueled by capturing the surrounding gas and stars; and "lagging cores" or material left over from the original expanding universe.

Quasars are apparently the most distant objects known, and we must remember that when we record many of them, we are looking billions of years into the past and viewing the universe in a much earlier state; such ideas as the "lagging core" theory are therefore particularly interesting. Recalling the theory of the original "Big Bang" in which the universe was produced by the explosion of a supercondensed and massive sphere, we wonder if quasars are tiny "universelets," concentrated globes of subatomic particles in their earliest explosive and expansive process. Indeed, was the universe really produced from one single condensed sphere? Might it not have originated in a *cluster* of tiny but unimaginably massive objects?

When the true nature of quasars is finally understood, however, it may be a rather sad day for astronomers, marking the end of a lot of speculative fun! Meanwhile, having added our own fanciful notions to the quasar-theory kitty, we return to sanity and prepare to visit the great cluster realm of conventional galaxies in Virgo.

A Treasure Trove of Galaxies

To the west of ε, centered upon the northwestern corner of Virgo and extending over the border of Coma Berenices, lies the vast **VIRGO GALAXY CLUSTER** (fig. 84), one of the largest aggregates of galaxies. At approximately 78 million light years, or 24 megaparsecs, this vast array of star systems is suitable for exploration with a smaller (at least 6-inch) telescope. Galaxy clusters were first detected by Sir William Herschel over a century ago, and since then they have been extensively photographed with large telescopes, and their details observed. F. Zwicky's thorough analysis of the compact Coma Berenices cluster revealed that in an area of 25 square degrees, cluster galaxies outnumber field galaxies by at least twenty-three to one, and this finding led astronomers to conclude that clustering among galaxies is the rule, rather than the exception.

Aggregates of hundreds to thousands of galaxies with a definite concentration toward one or more points, galaxy clusters may be dense and compact, often of a fairly well-defined spherical structure, or they may be less compact, i.e., open and irregular in structure. Dense clusters contain primarily elliptical or "SO" systems, with few spirals evident; open clusters, such as the Virgo cluster, contain all types of galaxies, ranging from the globular to the irregular. The Virgo cluster itself contains chiefly spiral galaxies, some elliptical galaxies and intermediate SO systems, plus a few irregulars. The dimensions of clusters depend on their compactness and on the total number of galaxies they contain; the Virgo cluster, one of the largest, with at least twenty-five

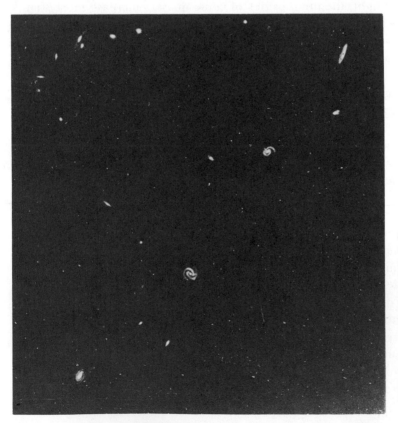

84. *The Virgo Galaxy Cluster,* a very large open cluster containing at least two thousand five hundred galaxies. This Harvard photograph shows several barred and normal spirals as well as some ellipticals. The speed of recession of the cluster is 1,150 kilometers per second. (*Yerkes Observatory photograph, University of Chicago*)

85. M87 (NGC 4486), a giant elliptical galaxy in Virgo, with its explosive jet clearly shown as it emerges from the bright nucleus and juts out to the northwest; this short exposure was taken with the Shane 120-inch reflector. (*Lick Observatory*)

hundred members, has an apparent diameter of 12 degrees. Even with its vast number of galaxies, the Virgo cluster cannot remain bound gravitationally without hidden mass; at least thirty times as much matter (the hidden mass) must be present to supply the gravitational glue to prevent the twenty-five hundred galaxies from dispersing.

We mention a few of its most prominent galaxies, representing various galactic types, as follows: lying near the Coma border, northwest of the optical double (ρ) Rho, is the giant elliptical *M87,* or *NGC 4486* (RA 12 h 28 m   δ +12° 40′),   ph m 9.7, classed "E0 peculiar," one of the most luminous galaxies known. Many globular clusters are scattered over its entire image. Intense radio waves were detected from M87; subsequently they were found to originate in a bright jet (now called "Virgo A") emerging from the northwest side of the galaxy (fig. 85). The jet's polarized blue light shows a continuous spectrum, synchrotron radiation resulting from intense magnetic fields caused by a gravitational collapse in the galaxy core, as in galaxy M82, of Ursa Major. Both M87 and its jet are strong sources of X-rays as well. The jet looks tiny on photographs, but remember, it takes light 4,100 years to traverse its length, and 400 years to cross its width; these figures indicate the vastness and tremendous force of the explosion that produced the jet.

Lying directly on the Virgo-Corvus border, 20 degrees south of the center of the Virgo cluster, and almost due south of γ (and χ), is an outstanding example of a bright spiral galaxy viewed edge on: *M104,* the *SOMBRERO GALAXY,* or *NGC 4594* (RA 12 h 37 m   δ −11° 21′), listed as type Sa or early Sb. Noteworthy features are a large nuclear bulge bisected by a heavy dust lane, in total effect resembling a round head wearing the Mexican hat of its title (fig. 86). The galaxy has an apparent diameter of 82,000 light years, its combined spectral type is G3, and its very great mass is 1.3 trillion $\mathcal{M} \odot$. The Sombrero's redshift, indicating a recession speed of nearly 700 miles per second, was discovered by V. M. Slipher in 1913, giving impetus to Hubble's research on the

86. M104 (NGC 4594), the Sombrero Galaxy, famous edge-on spiral, of historical importance in Hubble's investigations of the distances and nature of galaxies. A striking object because of its heavy and sharply defined dust lane and its large nuclear bulge; this photograph made with the 200-inch telescope. (*Hale Observatories*)

distances of nebulae which show that these objects are not local gas clouds, but extragalactic star systems, galaxies other than our own in an expanding universe.

Lying 8 degrees northwest of γ is the large face-on spiral *M61*, or *NGC 4303* (RA 12 h 19 m δ +4° 45') m 10, its diameter 60,000 light years and its mass 50 billion $\mathcal{M}\odot$. The arms are angular, so that in photographs it resembles a huge tarantula! A very luminous supernova was recorded in 1961 at this galaxy's eastern periphery.

Southeast of the main core of the Virgo cluster by about 2½ degrees is the compact barred spiral *M58, NGC 4579* (RA 12 h 35 m δ +12° 5'), of 160 billion solar masses.

North-northeast of M61, east of (o) Omicron, lies the bright elliptical *M49, NGC 4472* (RA 12 h 27 m δ +8° 16'), one of the largest and most massive known, its diameter 50,000 light years and its mass five times that of our Galaxy!

About 7 degrees west of ε and some 2 degrees northwest of fifth-magnitude ρ are the interesting spirals *NGC 4567* and *NGC 4568* (RA 12 h 34 m δ +11° 32'), called the "Siamese Twins," because their borders appear to touch, at least in our line of sight.

Another interesting galactic pair is *NGC 5426* and *NGC 5427*, lying on the 14th hour

87. *NGC 5426 and NGC 5427,* intertwined spirals in Virgo. Their connective bridge of interlacing spiral arms is clearly visible in this photograph taken with the 120-inch reflector. (*Lick Observatory*)

line at (δ +5° 49′ and 47′), due west of Iota, in the maiden's "skirt." A clearly visible bridge of matter connects these spirals.

More than 11 degrees to their north and slightly to the west (some 10° east of 5th-magnitude, reddish σ), is the galactic pair ***NGC 5363*** (RA 13 h 54 m   δ +5° 29′), an eleventh-magnitude elliptical with a small, bright nucleus, and ***NGC 5364*** (RA 13 h 54 m   δ +5° 15′)   m 11.5, a face-on spiral of delicate beauty.

88. *NGC 5364*, a spiral galaxy in Virgo. An Sb–c faint spiral galaxy of great beauty with rather loosely wound spiral arms that form a pattern of delicate perfection in this photograph taken with the 200-inch telescope. Note the thin, bright ring around the core, enclosing some spiral structure. NGC 5363, a bright elliptical galaxy 14½ minutes to its north is also shown. (*Hale Observatories*)

BOÖTES

First appearing in Homer's *Odyssey,* some twenty-eight hundred years ago, this constellation's ancient title was given various derivations by authorities; some said it came from the Greek words for "ox," and "to drive," thus the Wagoner or Driver of the Wain; but more recently, Boötes was associated with the "Hunting Dogs" Asterion and Chara (Canes Venatici) as they pursue the Great Bear around the north pole. He was *Arator,* the "Ploughman," a "Clamorous One" shouting to his oxen; or a Hunter urging on his hounds, as in early Rome, where he was *Venator Ursae,* the "Hunter of the Bear." In France, Boötes is *Bouvier,* or "Herdsman," watching over nearby circumpolar stars that constitute the herd (this a concept of desert dwellers); with the Arabians he was also *Al Ḥāmil Luzz,* "The Spear Bearer."

Mythological personalities were often subject to much confusion when assigned stellar identities; thus, according to Ovid, *Arcas,* the Little Bear, is more properly identified with Boötes, as "Arctophylax"—a name that shares its derivation from the

Greek *Arktikos* (bear) with Arcturus, the principal star. As mentioned, Boötes was also "Icarus" or "Icarius," the Athenian who taught the world how to make wine and whose daughter Erigone accompanied him to the heavens as Virgo.

Boötes is a large constellation, bridging an arc of 50° all the way from Virgo to the circumpolar zone. Poets made much of its sluggishness in setting, which is due to this great length and the closeness of its northern boundary to

> . . . that point where slowest are the stars,
> Even as a wheel the nearest to its axle.
> Dante's *Purgatorio*

ARCTURUS (α) *Alpha Boötis*
(RA 14 h 13 m   δ +19° 27')   m −0.06

An approximate transliteration of the Greek word for "bear guard," Arcturus' name was shared with Boötes in ancient days, but also with Ursa Major. Despite this confusion, long before our era the star was accorded great fame by poets, statesmen, farmers, seamen, and the general populace. Worshiped by early Egyptians, it was also the Chaldean *Papsukal*, or "Guardian Messenger," and the Arabian "Keeper of Heaven." Its risings and settings were closely observed, and engraved on ancient Greek calendars; and the Arcadians, in classical Greece, used its apparent motion relative to the Sun to regulate their annual festival. The physician Hippocrates believed that Arcturus, in connection with the weather, exerted various effects, both beneficial and harmful, upon the human body.

Often charted on the Herdsman's knee or marking the fringe of his tunic, Arcturus, a K2 giant, is easily identified by its golden color; Alcaid (the end star in the handle of the Big Dipper), some 30 degrees to the northwest, is helpful for orientation. The fourth-brightest star and one of the first visible in early twilight, Arcturus was observed with telescopes in the daytime in 1635 and again in 1669; glimpsed through the comet of 1618, it also shone brightly through the Donati comet, of 1858, the first comet ever photographed. Arcturus' large proper motion (2.29″ of arc annually toward its south-southwest) was observed by Halley in 1718, when he discovered stellar proper motions. The nineteenth-century astronomer and optician Bernhard Schmidt of the Hamburg Observatory said that Arcturus' color faded early in 1862 for a period of several years, an observation confirmed by Argelander and Kaiser of Leyden.

With a diameter of about 20 million miles, or twenty-five times that of the Sun, and with a luminosity of about 115 Suns but containing only some four solar masses, Arcturus has a very low density. About 37 light years distant and a member of the halo of population-II stars surrounding the galactic nucleus, it is presently moving through the galactic plane toward Virgo with a space velocity of 90 miles per second; its orbit, highly inclined to the plane of the Galaxy, takes Arcturus beyond the galactic plane, into the halo, and back again, so that some thousands of years hence it will be much farther from us and no longer visible. At present Arcturus is still approaching us, and its spectrum is therefore shifted toward the bluer wavelengths. However, Arcturus' Doppler effect also changes periodically as the direction of the Earth's motion changes,

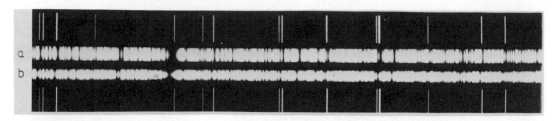

89. *Two spectra of Arcturus taken six months apart* to show orbital velocity of the Earth by means of the Doppler shift. (+ equals "in recession"; − equals "in approach"). *(Hale Observatories)*

for in its revolution about the Sun, the Earth is always approaching some stars and receding from others; we can therefore use Arcturus' Doppler shift (as well as that of the other stars) to measure the orbital velocity of the Earth (see fig. 89).

Two Chinese asterisms, "Officers," formed mostly by faint stars, flank either side of Arcturus, the "Emperor." The dignitary to the west, called *Yew She Ti,* includes *MUPHRID (η) ETA,* magnitude 2.69, lying about 4 degrees west of Arcturus; it was sometimes shown on the Herdsman's left leg and therefore called *Saak,* or "Shinbone." A type G subgiant, 32 LY distant, *almost* the standard distance of 10 parsecs (its apparent and absolute magnitudes therefore about the same), it shines with seven times the Sun's luminosity.

Tso She Ti, the eastern "Officer," includes *(ξ) XI,* magnitude 4.54, a yellow and red binary which lies about 8.5 degrees east of Arcturus and is only 22 light years distant from us; with a period of 150 years, its components are now near maximum separation, of 7.3 seconds of arc. About 10 degrees to the northeast of Arcturus, we find *IZAR (ε) EPSILON,* (RA 14 h 43 m   δ +27° 17′)   m 2.7, the Arabic "Girdle" or "The Belt of the Shouter"; its modern name of "Pulcherrima" was originated by F. G. W. Struve because of ε's beauty in larger telescopes as a deep golden and pale blue contrast double. The primary is a class-K giant, with about five hundred times the Sun's luminosity; 2.8 seconds distant from it lies its companion, a class-A main-sequence star with the luminosity of 45 Suns.

The Herdsman's brightest stars, α, β, γ, δ, and ε, form a simple kite, by which he is easily recognized even when viewing conditions are not optimum. The *BOÖTES TRAPEZOID* (which was called *Al Dhibah,* or "The Female Wolves," by early Arabs) marks the upper part of the Herdsman's body and is formed by β, γ, δ, and the much fainter μ. At the northernmost corner of the Trapezoid, marking Boötes' head or cheek, lies *NEKKAR (β) BETA* (RA 15 h 0 m   δ +40° 35′)   m 3.48. A golden giant, it is seventy times more luminous than the Sun and, at a distance of 140 light years, is slowly drawing closer to us, at 12 miles per second. *SEGINUS (γ) GAMMA* (RA 14 h 30 m   δ +38° 32′)   m 3.05, marking Boötes' left shoulder (southwest of β), is the closest Trapezoid star to Ursa Major, and is a white "Sirian" type, its luminosity seventy-five times the Sun's. *(δ) DELTA,* the right shoulder, the southeast corner of the Trapezoid (RA 15 h 14 m   δ +33° 30′)   m 3.47, is a late-class-G giant like β, and also 140 light years distant. Giving us a notion of how faint our own Sun would appear at that distance, δ's seventh-magnitude component, at the large separation of 195 seconds, is a type-G main-sequence star very similar to the Sun. The fourth member of the Trapezoid, northeast of δ, is *ALKALUROPS (μ) MU,* the Herdsman's "Club," or "Staff," and also called *Venabulum,* or "Hunting Spear"; μ is a yellow-white subgiant with a faint yellow companion at a separation of 108 seconds. This component is in itself a close binary consisting of two class-G stars similar to the Sun, magnitudes 7 and 7.6.

The Herdsman thrusts his upraised left hand far into the circumpolar zone, reaching a point a few degrees east of the Dipper's handle, near Alkaid. The stars of this hand are *(θ) THETA,* magnitude 4.1; *(ι) IOTA,* a triple of magnitudes 4.4, 4.5, and 8; and *(κ) KAPPA,* a blue-white visual and spectroscopic double, magnitude 4.6. With *(λ) LAMBDA* on the left arm, they were the "Whelps" of the trapezoidal wolves on early Arabic globes.

The Solar Triplets

Picture a system of three stars like our own Sun! This is *#44 (i Boötis),* about 7 degrees due north of β, consisting of a solar-type primary, magnitude 4.76, and a sixth-magnitude companion at 45 astronomical units; the latter is an elipsing binary formed by two dwarfs of class dG2, which, like the primary, are very similar to our Sun. With a revolutionary period of about 6½ hours, these dwarfs, separated from one another by a mere three-fourth of a million miles, are exchanging gaseous matter much as the tiny components of W Ursae Majoris are. Astronomers theorize that such systems are probably the forerunners of cataclysmic dwarf novae like SU Ursae Majoris and U Geminorum; therefore, thousands of years hence, this exchange of material will lead to periodic eruptions of the components' surface hydrogen.

CANES VENATICI

Boötes *(Venator)* holds his "Hunting Dogs" or "Hounds" *(Canes Venatici)* on a leash as they spring lightly after the Bear. *Asterion,* or "Starry," the northerly dog, is formed of many faint stars; *Chara,* or "Dear" (i.e., to his heart), the more southerly, contains α and β, the two brightest stars of this minor constellation. *COR CAROLI (α) ALPHA Canum Venaticorum, Fl. #12* (RA 12 h 53 m δ +38° 35′) m 2.9–2.95, "The Heart of Charles," was originally named after the executed Charles I of England, but it is popularly associated with King Charles II. An attractive double star 120 light years distant and accessible to the smaller telescope, its bluish and pale yellow components, magnitudes 2.89 and 5.6, are separated by 770 astronomical units. The primary, a metal-rich star classed "A0 peculiar," is the standard example of a *magnetic variable,* because in its spectrum certain unusual fluctuations occur, jointly owing to the star's pulsations and to the *Zeeman effect:* each spectral line of the atoms of a gas in magnetic fields splits up into a number of components. This effect is of great importance in astronomy, since it enables us to determine the strengths of magnetic fields in stellar atmospheres. The Zeeman effect changes periodically in stars of the Alpha Canum type, which have large magnetic fields, and these changes are in step with the stars' pulsations, because when such a star contracts, its magnetic field is compressed and becomes more intense; thus the Zeeman effect is also increased. The intensities of the spectral absorption lines also change regularly with periods very nearly equal to the period of rotation, which may be explained by assuming that the magnetic field is attached rigidly to the star and rotates with it. Other examples of spectral variables are Alioth, of the Dipper, and Iota Cassiopeiae.

Y CANUM ("LA SUPERBA") (RA 12 h 43 m δ +45° 43'), lying north-northeast of β, with its magnitude varying from 4.8 to 6.3, is a vivid red star named by Father Secchi, the nineteenth-century Italian astronomer. Characterized by the beauty of its multicolored rays, Y is a type-N carbon star, cool and gigantic, similar to R Leporis, "Hind's Crimson Star" (p. 124). With a difference of 9.5 magnitudes between the visual and ultraviolet wavelengths, Y's spectrum is very weak in the blue and ultraviolet because of strong molecular absorption by the triple atomic molecule C_3. The star is too distant for accurate determination of its parallax.

An Important Globular

At the southern border of Canes Venatici, between Cor Caroli and Arcturus but somewhat closer to Arcturus, lies *M3, NGC 5272* (RA 13 h 39 m δ +28° 38'), a well-known, beautiful globular cluster discovered by Messier in 1764. Its radiating streams of stars are concentrated into a very luminous nucleus, and these features are particularly impressive in larger telescopes (fig. 90). M3 is notable for its large percentage of population-II variable stars, called cluster-type variables or RR Lyrae stars, after the classic example of this type.

An important feature of a globular cluster is its luminosity function: the number of stars it contains in each absolute-magnitude range. The most complete data on the luminosity function were obtained for M3 by A. R. Sandage, who analyzed the evolution of stars in the open cluster M67 in Cancer (p. 161). He found that while most of the light from M3 is emitted by stars brighter than absolute magnitude +4, 90 percent of the cluster's mass is contained in fainter stars. This tells us that many superdense white dwarfs—Sandage estimated about fifty thousand—are present.

Globulars are not truly spherical, but ellipsoidal, with discernible flattening, which indicates that they are spinning. There is also a considerable amount of variation in the concentration of stars toward the cluster centers, which has led to a classification of globulars, from class I (the highest degree of concentration) to class XII, whose stars appear to be rather loosely distributed. M3 belongs to class VI, i.e., moderately concentrated.

We described the Sandage technique of superimposing the calculated evolutionary tracks of population-I stars on the observed C-M diagram of *open* clusters, as applied to M67; population-II stars may also be studied in this way by superimposing their evolutionary tracks on the C-M diagram of *globular* clusters, which Sandage did to analyze M3, in Canes Venatici.

The *WHIRLPOOL GALAXY, M51, NGC 5194* (RA 13 h 28 m δ +47° 27') m 8, southwest from Alkaid, at the end of the Dipper's handle, was the first spiral discovered; Lord Ross, at Birr Castle, first detected its spiral pattern in 1845; cosmologists believed it to be a new planetary system in formation, until Hubble and other astronomers demonstrated that such objects are a frequently occurring type of *galaxy* like our own. M51 is about 35 million light years distant, and its well-defined spiral pattern,

90. *M3 (NGC 5272)*, a globular star cluster in Canes Venatici, very luminous and notable for its large percentage of RR Lyrae stars (population-II variables); analyzed by Sandage on the C–M diagram. Photographed with the 200-inch telescope. (*Hale Observatories*)

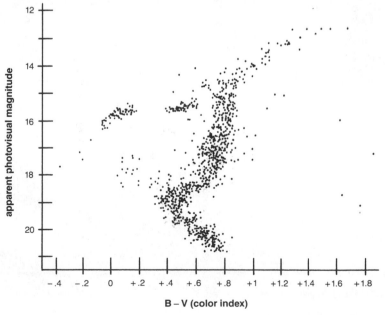

91. *Stars of globular cluster M3* plotted on the color-magnitude diagram.

92. *M51 (NGC 5194), the Whirlpool Galaxy in Canes Venatici,* the first spiral galaxy discovered, with its attached satellite system NGC 5195, looks like a vast, glowing figure "9" in this photograph taken with the 200-inch telescope. (*Hale Observatories*)

accentuated by dark, slender dust lanes, is visible only in larger telescopes (at least 12-inch) (fig. 92). It is as massive and luminous as the Andromeda galaxy, i.e., about 20 billion $L \odot$, with a total mass of about 400 billion $\mathcal{M}$. Attached to the northern end of one of its spiral arms is the smaller satellite galaxy **NGC 5195,** of undetermined type; its gravitational action may account for the visible distortion of one side of the parent galaxy.

A Galaxy That Is Consuming Itself

About 5 degrees to the west and slightly north of fifth-magnitude star #6 Canum (and about 12° WNW of Cor Caroli) we find one of the strangest objects ever observed in deep space, **NGC 4151, 165[1]** (RA 12 h 08 m   +δ 39° 41′), a Seyfert galaxy whose energy-producing mechanism scientists are only now beginning to understand. Classed either as a barred or a peculiar spiral, this somewhat elongated but almost face-on galaxy, one of the brightest Seyferts and also the most variable known, has been studied intensely since the International Ultraviolet Explorer satellite (IUE) was launched, in 1978.

The well-defined nucleus of NGC 4151 is much brighter than the galactic arms; to account for its luminosity, astronomers, using the Doppler shift, measured the orbital velocities of expanding carbon and magnesium clouds, and the distances of these clouds from the center of 4151 were also measured during an outburst of core variability in 1979. These measurements indicate that this extraordinarily luminous nucleus contains a highly contracted object with a mass equal to that of 100 million Suns, possibly an enormous black hole which is sucking in all neighboring stars, gas and dust. The turbulent, extremely hot accretion disk formed by such infalling material produces a tremendous output of radiant energy, enough to account for the quasar-like luminosity of NGC 4151. Since most astronomers now believe that quasars and Seyferts are closely related, an explanation of the common mechanism by which they radiate such extraordinary amounts of energy would add considerably to our knowledge of the origin and evolution of galaxies.

IV

Summer

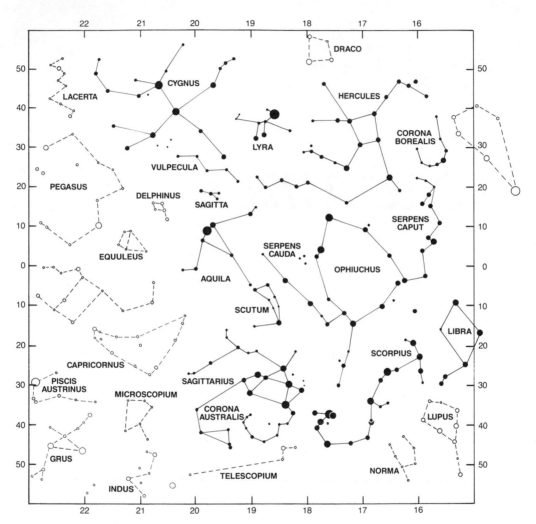

93. THE SUMMER SKIES

EIGHT

Three Tragic Myths

CORONA BOREALIS

. . . Lifting the folds, soft folds of her garments, baring her ankles,
Dashed into edges of upward waves that trembled before her;
Uttered, anguished then, one wail, her maddest and saddest,—
Catching with tear-wet lips poor sobs that shivering choked her:—
"Thus is it far from my home, O traitor, and far from its altars—
Thus on a desert strand,—dost leave me, treacherous Theseus? . . ."*

We told the story of Pasiphaë, queen of Crete, who became enamored of the bull
Taurus and bore him the monster Minotaur. Daedalus, the Athenian craftsman at King
Minos' court, builder of the artificial cow by which Pasiphaë seduced Taurus, was
probably the most remarkable—and useful—mythological character, fabled as the
inventor of the ax, awl, bevel, and saw, glue, the plumb line, and even a folding chair!
Following the dangerous Minotaur's birth, King Minos commissioned Daedalus to
build an underground maze at Cnossus for its restraint. Daedalus' labyrinth consisted
of numberless winding passages and turnings opening into one another without beginning or end, preventing any escape of the monster. Following a dispute with Aegeus,
king of Athens, concerning the death of Minos' son Androgeos, Minos exacted a
penalty from Athens consisting of seven Attic youths and maidens, sent every year to
Crete and fed to the Minotaur. Resolving to destroy this horrible beast, Aegeus' son
Theseus, famed for his heroic exploits, joined a group of victims en route to Crete.
When the handsome youth was exhibited together with his sacrificial party before King
Minos, beautious Princess Ariadne, the sheltered and innocent daughter of Minos and
Pasiphaë, saw Theseus and fell deeply in love with him. Ariadne and her sister Phaidra
secretly gave him a sword and a spool of silken thread, one end of which they fastened
to the entrance of the labyrinth. As he went to meet the Minotaur, Theseus played out
his spool of thread; quickly he slew the beast, then followed the slender filament back to

* *From* The Wedding of Peleus and Thetis—*Catullus, LXIV; translation by C. M. Gayley.*

the entrance of the maze. Ariadne released him and the intended victims from the labyrinth. That night, after boring holes in Minos' ship to prevent him from following, Ariadne and Theseus, who had promised marriage as a condition of her help, departed for Athens. On the high seas, Ariadne became ill (according to some versions, pregnant by Theseus), and he put in at the isle of Dia (later called Naxos), where she fell asleep on the white strand. When the unfortunate princess awoke, Theseus had set sail, abandoning her; in despair and grief she hanged herself. According to other versions kinder to Theseus, his ship was blown off course to Cyprus and, after the ailing princess was set ashore to rest, was driven out to sea again by waves so violent that he couldn't return for several months—when he sorrowfully learned that Ariadne had died in childbirth. Some also say that the goddess Minerva appeared to Theseus in a dream and commanded him to desert Ariadne (for those who accept the Freudian theory that dreams are wish fulfillments, this circumstance scarcely exonerates him).

In the most popular and satisfying account of Ariadne's ultimate fate, Bacchus (Dionysus), the god of wine, accompanied by a retinue of celebrants, arrived at the shores of Naxos, his favorite island, where he passionately wooed and won the sorrowing Ariadne. For a marriage present he gave her a golden crown encircled with gems; she lived very happily with her divine rescuer but, only a mortal, she eventually died, whereupon Bacchus hurled her crown of jewels upward into the sky. As it mounted, the gems grew brighter and were turned into seven stars, which since ancient times have been called *Corona Borealis,* the "Northern Crown," one of the most perfectly formed and easily recognizable constellations. When some fainter stars are included, Corona also looks like a complete hairpiece, and it was indeed called the "Coiled Hair of Ariadne" as a companion to "Berenice's Hair" *(Coma Berenices).* According to some authors, including the Roman poet Ovid, Ariadne herself became the constellation; this version of the myth inspired Hugo von Hofmannsthal's intensely poetic text for Richard Strauss's opera *Ariadne auf Naxos,* wherein Bacchus, welcomed by a delirious Ariadne as the supposed "god of death," promises her eternal life instead, as he transports her to the heavens.

Ariadne's tiara had numerous titles, among them the Roman *Corona Ariadnae* and *Diadema Coeli;* the Arabic *Al Fakkah,* or "The Dish," and Bayer's *Parma,* or "Shield," also of Arabian origin; the *Kāsah Shekesteh,* or "Broken Platter," as the Persians saw this semicircle of stars; and the Australians' *Woomera,* or "Boomerang."

GEMMA (α) ALPHA Coronae (RA 15 h 32 m δ +26° 53′) m 2.23, also called **ALPHECCA,** after *Al Nā'ir al Fakkah,* or "The Bright One of the Dish," is the major star of the Crown, about 75 light years distant and a main-sequence star like our Sun, but much hotter and forty-five times as luminous. A spectroscopic binary, its close eclipsing components at a separation of 17 million miles, produce small changes in magnitude as they eclipse each other; the secondary companion is a small golden star of class dG6. Gemma may be part of the Ursa Major moving cluster.

NUSAKAN (β) BETA, magnitude 3.66, northwest of α and 100 light years distant, is a spectroscopic binary; the primary is a magnetic variable similar to Cor Caroli. About 2.5 degrees to its north-northeast lies *(θ) THETA,* magnitude 4.2. Excluding θ, the mnemonic "B-A-G-D-E-I," which we used to memorize the principal stars of Casseopeia, also indicates the order of the jewels in Ariadne's Crown. Proceeding clockwise around the starry semicircle, we find *(γ) GAMMA,* a fourth-magnitude close double,

94. *Ariadne Abandoned by Theseus.* From a painting by F. Ehrmann. (*The Bettmann Archive*)

lying to the east of α and 140 light years distant, its interesting primary a pulsating short-period variable of the Delta Scuti type (p. 264). Fainter *(δ) DELTA Coronae,* magnitude 4.7, is a single object; fourth-magnitude *(ε) EPSILON* (northeast of δ) is a class-K spectroscopic binary; and north-northeast of it lies fifth-magnitude *(ι) IOTA,* completing the semicircle.

A Disintegrating Tiara

According to myth, the Diadem retains its form, enshrined forever in the heavens; this notion is a poetic illusion, alas, when viewed over thousands of years, for Ariadne's stellar jewels are pursuing individual paths, each moving at its own speed in a different direction from its neighbors. Thus the accidental optical placement of seven stars has created a fragile, impermanent wonder that we, only in our present age, are privileged to enjoy. Corona is not a true star cluster, therefore, because its members do not have a common proper motion; α and β have in fact already exchanged places in the past fifty thousand years. One hundred thousand years hence, the proper motions of all seven stars will have scattered them in various directions, dissipating the perfection of Ariadne's Crown and giving it the appearance of a crude dipper or, perhaps, the "sail" of Corvus; no longer a suitable ornament for her tresses!

Other Jewels

About 1.3 degrees northwest of β lies the well-known binary *(η) ETA* Coronae, 50 light years distant and observable with a telescope of at least 6 inches; the period of revolution is 41½ years, and the rather close separation of its solar-type components, of magnitudes 5.5 and 6, is increasing and will be widest in 1993.

Directly to the north of Iota lies *(σ) SIGMA,* magnitude 5.26, about 70 light years distant, an attractive and interesting binary system whose widely separated yellow components take more than a thousand years to orbit their common center of mass. The primary is also dual in nature, a very close spectroscopic binary with a period of about eight days. Twelve minutes to the southwest of σ lies a proper-motion companion, a very faint red dwarf designated LTT 14836. Thus the entire system actually consists of four objects. In addition, closer in to the main pair, there is a tenth-magnitude optical companion not sharing their proper motion.

Just over a degree east of σ are *(ν) NU 1* and *2,* constituting a fine naked-eye double whose deep yellow components are of fifth magnitude. Far to the northwest of Nu, and 5 degrees north-northeast of Theta, is the attractive optical double *(ζ) ZETA,* a fifth-magnitude blue-white star and a sixth-magnitude blue one.

A Topsy-Turvy Star

R CORONAE (15 h 46 m   δ +28° 19′), lying within the enclosure of the semicircle, about 2½ degrees to the northwest of ε, is the standard example of an irregular variable of an unusual and little understood type. In contrast to the behavior of U Geminorum stars, which are variables characterized by fairly large and rapid increases followed by slow declines, the R Coronae Borealis stars remain at maximum luminosity for very long periods of time compared to their periods of minimum luminosity. Their absolute magnitudes at maximum are always the same, but they vary considerably at minimum. R Coronae itself maintains an apparent magnitude of +6 for several years at a time and then suddenly undergoes a tremendous decrease in luminosity, bringing its magnitude to anywhere from 7 to 15 in a completely random fashion; most often, it reaches an apparent magnitude of +12.5. A very ancient (population-II) supergiant, about class F7, R is poor in hydrogen and maintains its stability while slowly changing its chemistry by transforming helium to carbon. As a member of the galactic halo, it lies far above the galactic plane, but its exact distance has not been determined. One theory suggests that R's luminosity changes arise from the emission of carbon clouds originating in its interior which temporarily surround the star's surface and block the light.

The Nova of 1866

T CORONAE, also known as the *BLAZE STAR* (15 h 57 m   δ +26° 4′), on the southern border of Corona and less than a degree south-southeast of ε, was the first nova examined by spectroscope. Discovered in May of 1866, it reached the second apparent magnitude, comparable to α. Within a week it was no longer visible to the naked eye and soon afterward reached its minimum luminosity; about a hundred days after maximum it brightened again to magnitude 8.5 and Sir William Huggins made his historic spectroscopic examination revealing bright hydrogen lines superimposed upon a solar-type spectrum. As predicted, T is a recurrent nova and again blazed forth, reaching the third apparent magnitude in February of 1946 and then rapidly declining, following the same pattern as it had eighty years previously.

Fast and Slow Novae

A so-called *fast nova* such as T Coronae completes a cycle of changes in luminosity, returning to its original prenova stage in a matter of months, or, at most, in a few years. The initial rise to maximum, a change of twelve to thirteen magnitudes, is extremely rapid, occurring within a few hours to a day or two. The nova then loses about three magnitudes rather smoothly in a few days, and continues to diminish in brightness quite slowly while undergoing a series of fluctuations, before the final smooth decline to initial luminosity. T Coronae, however, unlike other fast and recurrent novae, has a strong secondary maximum. *Slow novae,* on the other hand, which may take as long as a month to rise to maximum luminosity, show similar variations but over a number of years or even centuries.

T Coronae is also unusual because its outer layers are expanding at a rate of about 2,700 miles per second, the highest in our Galaxy except for Cassiopeia A and Tycho's Star (N 1572 Cassiopeiae). T's change in luminosity is vast, ranging from forty or fifty times that of the Sun at minimum to approximately 200,000 $L \odot$ at maximum. The bright hydrogen spectrum Huggins observed represents the expanding gases of the nova; at minimum, the spectrum is of type M. Astronomers now believe that its duality indicates the presence of two stars virtually in contact, a stable red giant and a compact blue dwarflike companion that periodically explodes, as material originating in the red giant falls upon its surface, although this theory has recently been challenged as an explanation for such recurrent flares (see U Geminorum, p. 151).

The Corona Borealis Galaxy Cluster

At the southwest corner of Corona, lying near the Corona/Boötes border (15 h 20 m δ +27° 50′), is an extremely distant group of more than four hundred galaxies, covering an area of the sky 1.5 degrees wide, and much too faint for viewing except with large telescopes (fig. 95). The amazing distance of this cluster is now estimated at between 1 and 1.3 billion light years, and according to its Doppler shift, its member galaxies are receding at about 30,000 miles per second.

95. *The Corona Borealis Galaxy Cluster,* photographed with the 200-inch telescope. (*Hale Observatories*)

HERCULES

Alcmene, the granddaughter of Perseus and Andromeda, was loved by Zeus (Jupiter). From this union between god and mortal was born Herakles (Hercules), the greatest hero of ancient Greece; in comparison with his exploits, those of our modern Superman pale into insignificance. As Queen of the Heavens and Patroness of Conjugal Law, Zeus's wife, Hera, was determined to destroy Hercules, and she dispatched two great serpents to his cradle—but the mighty infant strangled them barehanded. Amphytrion, king of Thebes, gave his stepson Hercules a broad education, but in an outburst of temper the youth killed his music teacher; to modify this wild behavior, the king sent Hercules to live among herdsmen in the mountains, where he developed into a skilled hunter. Maturing, he performed numerous deeds of valor, including the liberation of Thebes, for which he was rewarded by marriage to the Princess Megara; Juno, enraged at his good fortune, drove Hercules temporarily insane, causing him to kill his own children.

In expiation, he was placed under the command of his cousin Eurystheus, whom Juno favored; this service led to the famed twelve labors, two of which we described: Hercules' battle with the Nemean Lion (Leo) and his slaying of the Hydra.

Released from his servitude to Eurystheus, Hercules ended his marriage to Megara; not long after, he was again seized with a fit of madness and hurled his friend Iphitus to his death from the top of a city wall. Stricken soon afterward with a dreadful disease, Hercules could only atone for his crime by three years in bondage to Queen Omphale of Lydia (possibly the first feminist), who forced the belligerent hero to wear woman's dress and work at domestic duties, while she wore his lion's skin and wielded his club! This reversal of sexual roles did not prevent Hercules from accomplishing numberless valiant deeds, including his participation in the famed voyage of the Argonauts. When freed from his servitude to Queen Omphale, Hercules led a successful attack against Troy, and later conquered Elis. In the midst of these exploits, Hercules found time to celebrate the first Olympian games! Perhaps Hercules' greatest deed, however, was the freeing of Prometheus, bringer of enlightenment to mankind, chained to a rock and tortured by a vulture because he had displeased Jove.

Following expeditions against Pylos and Sparta, Hercules married Deianeira, daughter of Oeneus of Calydon. After causing the death of a youth related to the king, Hercules went into voluntary exile with Deianeira, arriving at the banks of the river Evenos; there, Nessos the Centaur, acting as ferryman, forgot his professional duties and attempted to kidnap Deianeira, but Hercules shot him with an arrow. Before dying, Nessos slyly gave Deianeira a mixture of his blood, promising it would revive her husband's love, should he stray.

In a campaign against Eurytos of Oichalia, Hercules slew the king and took his daughter Iole captive. Deianeira, smitten with jealousy, poured some of Nessos' supposed love-potion on a ceremonial robe that Hercules had ordered. When he donned the robe, fiery poison engulfed his limbs; in agony, he tried to remove the deadly garment, but his flesh was torn away. Hercules was transported home, where, in anguish, Deianeira hanged herself. Ascending Mount Oeta above Trachis, Hercules ordered the construction of a great funeral pyre. Renouncing his immortality so that his

96. Hercules and Omphale. Queen Omphale wearing his lion's skin, triumphs over a worshipful Hercules wearing feminine dress. Painted in 1861 by Gustave Boulanger. (*The Bettmann Archive*)

pain might end in death, he lay down upon the pile and ordered his friend Philoctetes to kindle the wood. The flames rose and consumed Hercules' body, but Jupiter transported his immortal soul to Olympus, where Juno, finally reconciled to Alcmene's son, gave him the hand of her daughter Hebe in marriage.

> But him, on whom, in the prime
> Of life, with vigor undimm'd, . . .
> Mournfully grating, the gates
> Of the city of death have forever closed—
> Him, I count him, well-starred.
> Fragment, *Dejaneira*

"Well-starred" he is indeed, a great constellation stretching upside down from Ophiuchus to Draco. Its origin, much older than the classical saga of Hercules, is shrouded in mystery; it has been known by the hero's name only in more recent times. In ancient Phoenicia it represented the sea god Melkarth; the Greeks called it *Engonasi* and the Romans, *Genuflexus,* "The Kneeler," also *Clavator,* or "Club Bearer" and *Saltator,* "The Leaper," adapted by the Arabians as *Al Rakis,* "The Dancer." After it was associated with Hercules, various figures in his life served as titles, e.g., Nessus (the

Centaur), *Aper* (the wild boar), Heros Tirynthius (where he was reared), and *Oetaeus,* after the mountain where his funeral pyre was erected.

With reference to the Apples of the Hesperides (Hercules' eleventh labor), the French and the Italians created the subconstellation *Ramo e Cerbero (ramo =* "branch"), or *Rameau et Cerbère,* combining the apple tree with its guardian dragon Cerberus as a symbol clutched in Hercules' left hand; but Cerberus was also regarded as the three-headed watchdog of Hades, which Hercules brought back in his twelfth labor.

RAS ALGETHI (α) *Alpha Herculis*

(RA 17 h 12 m   δ +14° 27')   m 3.1–3.9

Also called "Rasalgeti," its name derived from *Al Rās al Jāthiyy,* "The Kneeler's Head," this is the southernmost major star of the inverted figure of Hercules, only about 5 degrees to the west-northwest of α Ophiuchi. Together, the two stars were known as *Al Kalb al Rā'i,* or "The Shepherd's Dog," to the nomads of Arabia. The Chinese saw α Herculis as *Ti Tso,* the "Emperor's Seat"; since 1795, however, when Sir William Herschel discovered its variability, astronomers have noted it as one of the brightest irregular variables, a red giant that ranges from the third to nearly the fourth magnitude in an average period of about ninety days, with a diameter about four hundred times the sun's, or at least half the size of Betelgeuse. Alpha's distance, difficult to determine, may exceed four hundred light years; with an approximate luminosity of 830 Suns, and a mass only a few times the Sun's, this great star has a very low density; one of the coolest stars known, its temperature varies from 2,400 degrees K. to about 2,650. At an apparent separation of 4.6 seconds, a yellow companion (often described as blue-green) creates a beautiful contrast with the reddish primary; they are most comfortably viewed with a 100× telescope. The companion, dual in nature, is a very close (spectroscopic) binary. The primary of α Herculis is gradually losing material, which envelops the system in the form of an expanding gaseous shell, as in other binaries in which the primaries are red giants.

The Figure of the Kneeler

KORNEPHOROS (β) BETA Herculis, about 11.5 degrees northwest of α, marks his right shoulder; somewhat brighter than α, its apparent magnitude is 2.8, and the spectral type is G8 3. The nineteenth-century American astronomer Elijah Burritt called it *Kornephorus vel Rutilicus;* R. H. Allen thinks this title might be derived from *rutilus,* or golden red, which he believes inappropriate for β. Yet, with its late-G-type spectrum, "golden" or warm yellow seems an apt description. Ideler, however, believed that its rare name *Rutilico* was from the diminutive for *rutrum,* a sharp instrument that Hercules is shown carrying on Arabic globes. Beta, 105 light years distant, is about sixty-five times more luminous than the Sun.

Proceeding to the Kneeler's upraised right arm, southwest of β, we find fourth-magnitude *(γ) GAMMA,* and fifth-magnitude *MARFIK (κ) KAPPA,* also Mirfak and Marsic, from *Al Marfik,* "The Elbow," a yellow and orange optical double. The "Club," considered a separate constellation by Pliny, is marked by *CUJAM (ω) OMEGA,* perhaps helped along a little by nearby #29 and a few still fainter stars.

(δ) DELTA, his left shoulder (RA 17 h 13 m δ +24° 54′) m 3.2, north of α, is a white ("Sirian") star, and although single, has a faint optical companion, yellowish in hue and contrasting attractively with the primary. The two stars are separating.

A row of stars stretching to the northeast of δ delineates Hercules' outstretched left arm and the hand that grasps Cerberus the 3-headed dog. *LAMBDA (λ),* properly marking the upper arm or elbow, is mistakenly named Masym, from the Arabic *Mi'ṣam,* or "Wrist" (an oversight on the part of Bayer). In 1806, Sir William Herschel, investigating the Sun's motion among the stars, deduced that its apex† lay near λ; today, however, astronomers believe that our Sun is transporting us in the approximate direction of Vega, the bright star of Lyra.

Next along the left arm is *MU (μ),* marking the forearm, a triple star system of magnitude 3.42, of interest as a solar-type subgiant primary circled by a distant binary consisting of two faint red dwarfs. *XI (ξ),* at the end of the stellar row, together with nearby ν and o, form Hercules' hand.

The Keystone of Hercules

Four stars in the figure form this lopsided quadrilateral, a popular aid in identifying the constellation. At its southwest corner lies *(ζ) ZETA Herculis,* a visual binary consisting of a yellow subgiant and an orange (type-K) dwarf; their separation is 12 astronomical units, with an orbital period of 34.38 years. The class-G subgiant has already left the main sequence and is gradually becoming more luminous, thus serving as an evolutionary model for systems of this type. Only 30 light years distant, their separation will be largest in 1990.

At the northwest corner of the Keystone lies *(η) ETA,* magnitude 3.46, a class-G7 yellow subgiant, eleven times the Sun's luminosity. The northeast corner is occupied by *(π) PI,* magnitude 3.13, the type of cool, orange class-K star called a "bright giant," 750 times more luminous than the Sun. Faint *(ε) EPSILON,* magnitude 4, marks the southeast corner, completing the Keystone.

We may also utilize Eltanin, the red star in Draco's head, for orientation; about 5.5 degrees to its south-southwest lies *(ι) IOTA Herculis,* the left foot of the Kneeler; about 11 degrees to Iota's southwest is π of the Keystone. Look for *(ρ) RHO (#75),* the fourth-magnitude double immediately to the east and slightly to the north.

Two Interesting Binaries

Directly south of π lies the variable *#68* (RA 17 h 15 m δ +33° 09′), magnitudes 4.7 to 5.4, consisting of two hot blue-white giants separated by a mere 6 million miles, egg-shaped from their mutual tidal attraction. With an approximate period of two days, they partially eclipse each other with every revolution (see β Lyrae, p. 225). Almost directly on the 18th hour circle of right ascension and about 7.5 degrees south of ξ, we

† *The apparent point toward which it is moving.*

97. *Nova 1934 (DQ Herculis),* before and after its outburst. (*Yerkes Observatory photograph, University of Chicago*)

find *#95 (Σ 2264),* (δ +21° 36′) magnitude 4.42, a pair of white and yellow giants, spectral types A7 and G5, i.e., Sirian and solar, under observation since 1829, with great diversity of opinion, however, concerning their colors. The system is fairly distant at 400 light years, but the components, separated from each other by about 775 astronomical units, are easily viewed in the average 100× telescope.

A Christmas Nova

On the night of December 13, 1934, a third-magnitude star was observed to the east of ι and quite close to the Hercules-Lyra border, where previously it had existed as an obscure fourteenth-magnitude object. Within nine days, *NOVA 1934,* or *DQ HERCULIS,* reached a magnitude of 1.3, remaining at that level for a hundred days. It then dropped suddenly in luminosity and slowly brightened up again, remaining near maximum until the following May, when it very gradually faded to a postnova magnitude of +13.8 (fig. 97). This characteristic behavior made DQ recognizable as a *slow nova,* in contrast to T Coronae, the fast nova we have already discussed. As noted, slow novae like DQ (and the very similar nova T Aurigae of 1891) rise slowly to maximum, undergo fluctuations like those of fast novae, but occurring over much longer periods of time, and then decline in brightness much more gradually.

In 1954, DQ Herculis was found to be an eclipsing binary with an extremely short period: only 4 hours 39 minutes; the nova component, a bluish dwarf, is separated from its red-dwarf companion by only two hundred thousand miles, and each has only 0.25 solar mass. Their light curve is very similar to that of the dwarf eclipsing binary UX Ursae Majoris, but they differ from the UX system in their small masses. Although the components of UX are only 1.2 million miles apart, the separation of the low-mass dwarfs of nova DQ Herculis is more than six times smaller; this proximity, leading to an exchange of stellar material, probably causes the eruption of the blue component. A similar interaction may in time bring UX Ursae Majoris into the ranks of exploding stars.

A Spectacular Star Cluster

M13, NGC 6205 (RA 16 h 40 m   δ +36° 33′), about 2½° south of η, on a line between η and ζ, is well known as the great globular star cluster in Hercules, the finest of its kind in the northern hemisphere (fig. 99). Discovered by Halley in 1714, its erroneous early title was the "Halley nebula." With his vastly superior telescope, Herschel estimated a star count of fourteen thousand for M13, a result that astonished some observers of his time, who regarded it as a typographical error for four thousand!

98. *Nova Herculis* photographed in ultraviolet, green, and red light in 1951. Note the bright globules visible in red light (right). Taken with the 200-inch telescope using various filters and emulsions. (*Hale Observatories*)

99. M13 (NGC 6205), the Great Globular Star Cluster in Hercules, a very rich, compact cluster about 10 billion years old; more than 1 million stars lie in its core alone. Photographed with the 200-inch telescope. (*Hale Observatories*)

More than a century later, however, at the Mount Wilson survey, Harlow Shapley counted over thirty thousand stars brighter than the twenty-first apparent magnitude. Later estimates, still too small, placed this figure at one hundred thousand or more. The total number of stars in a globular cluster can be estimated only photographically, but a short exposure fails to record most of the faint stars, while a longer exposure results in a fused image of the densely concentrated stars in the center of the cluster. Although an exact calculation is not possible, we now know that more than one million stars probably lie in the central core alone of M13. More luminous than three hundred thousand Suns, the mass of the cluster may be 500,000 $\mathcal{M}\odot$. Its members, as with stars in all globulars, are exclusively population-II, including a number of luminous red giants, as well as faint stars that have evolved very slowly from the original main sequence.

Before a cluster can be analyzed in any detail its distance must be known. This can be obtained only indirectly by using some luminosity criterion for the individual stars in the cluster. The pioneer work in this field was done by Shapley, who obtained the distances of globular clusters by using both their RR-Lyrae-type variables (p. 227) and

100. A distant cluster of galaxies in Hercules, extremely rich in spirals, ellipticals, and barred galaxies. Note the "Siamese twin" spirals, upper center of photograph. Taken with the 200-Inch reflector. (*Hale Observatories*)

their brightest stars, since RR-Lyrae stars have about the same absolute magnitude, and the brightest stars are about 1.5 photographic magnitudes more luminous. Shapley also correlated the apparent diameters of globulars of known distance with their parallaxes to establish a distance scale based upon their apparent size. Since clusters differ in size, however, adjustments must be made to obtain an accurate result, and interstellar absorption must also be taken into account. With these corrections, we find that the great cluster in Hercules, with a linear diameter of 11 parsecs, or 35.86 light years, is 8,200 parsecs, or 26,732 light years, distant (current estimates vary), and it lies about 30,000 light years from the galactic center. The age of this very ancient globular is believed to be about 10 billion years.

M92, NGC 6341 (RA 17 h 16 m   δ +43° 12′), lying 9 degrees northeast of M13 and 6 degrees north of π, is also a splendid globular cluster, an easy object for the small telescope, at apparent magnitude 6.5.

LYRA

. . . "Ah, what!" she cried, "What madness hath undone
Me! and, ah, wretched! thee, my Orpheus, too!
For lo! the cruel Fates recall me now;
Chill slumbers press my swimming eyes . . . Farewell!"
                                        W. S. Landor

Whereas young Hercules was poor in music studies and hurled a lute at his teacher, Linus, with fatal force, Linus' half brother Orpheus, son of the sun god Apollo and the Muse Calliope, was the most unusual musician in the mythic world. His father, patron of music and poetry, taught him to play the lyre so beautifully that Orpheus' songs charmed the wild beasts and affected even the trees and rocks, winning the heart of the fair nymph Eurydice. Hymen, a divine personification of the marriage feast, attended the nuptials of Orpheus and Eurydice, but his torch smoked: a bad omen. Shortly after the marriage, Aristaeus, a love-smitten shepherd, surprised Eurydice in her garden; fleeing, she stepped on a snake and was fatally bitten. Eurydice was transported to the dark afterworld called Hades; in his grief, Orpheus sang his heart out to all who would listen, but none had the power to reverse Death's decree. The sorrowing bard then resolved to journey into the shades of Hades himself and win back his wife from its stern rulers. Presenting himself to Pluto and Proserpine, he sang with such compelling sweetness that even the ghosts of earthly wrongdoers paused in their torments to listen. Won over by his plea, the king and queen allowed Eurydice to return with Orpheus to the world of the living, on condition that he not look upon her until they had reached the upper air.

Orpheus agreed, and traced his way homeward from the Stygian realm, his wife following so silently that he began to doubt her presence. Almost past the borders of Hades, he violated his vow and turned around to look at her; she immediately vanished into thin air and was recalled to Hades. Unable to reach the underworld again, the broken-hearted poet remained seven months in a desert cave, pouring out his sorrow to the wild beasts and to the trees. His despairing strains attracted a group of Thracian maidens who were reveling in wine and dance; when he repelled their amorous advances, they stoned him to death, tearing his body limb from limb. Cast into the river Hebrus, Orpheus' lyre continued to play melancholy music as the waters flowed through its strings. Moved by these haunting melodies that even Death could not still, Jupiter placed the lyre between Hercules and Cygnus as a small but distinctive sky picture, highlighted by the brilliant star Vega, where

. . . still its force appears,
As then the rocks it now draws on the stars.
                                        Manilius

The Romans called it Lyra, but also *Apollinis* ("of Apollo," its first owner), and *Mercurii* ("of Mercury," its mythic inventor). It later acquired numerous musical titles,

101. *Orpheus and Eurydice.* The doomed couple flee the underworld as Cerberus (lower left) barks after them. From the painting by Ubaldo Gandolfi, Pinacoteca in Bologna. (*The Bettmann Archive*)

including "Cithara" and "Tympanum"; also *Fidicen* ("Lyrist") and *Canticum* ("Song").

The fourteenth-century Persian poet Hafiz named it the "Lyre of Zurah," and his countrymen called it *Sanj Rumi,* from which such European names as "Asange," "Mesanguo," and "Alsanja" evolved. The Bohemians called Lyra *Hauslicky na Nebi,* meaning the "Fiddle in the Sky"; the Teutons called it *Herapha,* and the Anglo-Saxons, *Hearpe* (harp), the same instrument imaged by early Britons in their *Talyn Arthur* ("Harp of [King] Arthur") and, in the 17th century, by Novidius of the "Biblical School," who called Lyra "King David's Harp." In ancient India, however, its stars were seen as an Eagle or Vulture, and three of them, Vega and its neighbors ε and ζ, as *Al Nasr al Wāki,* "The Swooping Eagle." Lyra then acquired other birdlike names, including *Aquila Marina,* or "Osprey," *Falco Sylvestris,* the "Wood Falcon," and, three centuries ago, its popular name *Vultur Cadens,* the "Swooping Vulture," pictured with the lyre in its beak!

The Alfonsine Tables (p. xx), however, show Galápago, the "Turtle," a reference to its origin as the little tortoise or shell from which Mercury first invented the Lyre:

". . . Why, here," cried he, "The thing of things
In shape, material and dimension!
Give it but strings and, lo! it sings—
A wonderful invention."
                                    James Russell Lowell

VEGA (α) *Alpha Lyrae*
(RA 18 h 35 m   δ +38° 44′)   m 0.04

Also called "Wega," derived from the Arabs' *Wāki* (see above), Vega, because of its brightness, often acquired the titles of its constellation, and was referred to as the "Harp Star." Because of the precession of the Earth's polar axis, Vega was the pole star some fourteen thousand years ago and therefore known in ancient Akkadia as *Tir-anna,* or "Life of Heaven"; the early Assyrians called it *Dayan same,* or "Judge of Heaven," for it then occupied the uppermost position (most northerly) of all stars! It still dominates our summer skies, appearing evenings near the zenith as a white star so dazzling that it is often called the "Arc Light" of the sky. Vega, the first star to be photographed (July 17, 1850, by the early daguerreotype process at the Harvard Observatory) is easily seen at various hours throughout the year.

About 27 light years distant, Vega is approaching us with a speed of about 8.5 miles per second, while our Sun is taking us on a space journey toward a point close to the Lyra-Hercules border at a velocity of 12 miles per second. However, Vega's approach will not endanger our remote descendents living thousands of years from now, for the great star and our own Sun will safely miss one another by many light years! Various estimates of the position of the *solar apex* (p. 217) differ slightly, but in any case we are not approaching Vega's exact position, and it is also very unlikely that our Sun will collide with some other star in Vega's vicinity, because Vega and the Sun lie within a sparsely populated spiral arm of the Galaxy; even within the "close quarters" of the galactic nucleus, such stellar collisions probably occur rarely, if at all, owing to the vastness of interstellar space.

Vega was the third star whose parallax was measured, this by the elder Struve (Wilhelm) in 1840 at the Russian National Observatory at Pulkova (61 Cygni by Bessel and Alpha Centauri by Henderson were the first two). Although inaccurate, these early measurements are of historic significance, because they were the first direct observations of the effect of the change in the Earth's position in space on the apparent position of a star, conclusive evidence that the Earth is a world in motion (see "Stellar Parallax," p. 13).

Vega's diameter is 2.8 million miles, or about 3.2 times that of the Sun (865,400 miles); it is an A-type star, and its temperature, 9,200 degrees K., is nearly twice the Sun's. However, while Vega is three times as massive as the Sun, its density is only two tenths that of the Sun.

Vega has two faint neighbors not bound to it gravitationally, a tenth-magnitude bluish star 1 minute of arc distant from it and a twelfth-magnitude star about 54 seconds distant.

Southeast of Vega, 1.7 degrees apart and lying on a line that forms the base of the Lyre, are the third-magnitude stars Sheliak and Sulafat, their names derived from the Arabic for "Tortoise," applied to the entire constellation. The westernmost member of this stellar pair is *SHELIAK (β) BETA,* (RA 18 h 48 m   δ +33° 18')   m 3.38 (variable), classified as "B8 peculiar." Its variability may be observed by comparison to Sulafat, of equal magnitude; every thirteen days, Sheliak drops to half its eastern neighbor's apparent brightness, with a secondary minimum some six days later. Beta Lyrae is the standard for an important type of eclipsing variable whose components are very close; from it astronomers have learned much about stellar evolution. John Goodricke, the discoverer of δ Cephei's variability, was also the first to observe the light variations of Sheliak, or β Lyrae, in 1784.

Close Eclipsing Binaries of the Beta Lyrae Type

Some very interesting phenomena occur in binary systems like that of β, the so-called "Lyrids," or "bright eclipsing variables," in which the components are quite close together. These effects, which are evident in certain unusual spectral features, cannot be accounted for in terms of the normal types of binary orbits. Velocity curves and the spectral lines that one observes from such binary systems can be explained by supposing that the two components are so close together that huge streams of matter pass between them, as Otto Struve pointed out after intense studies of β Lyrae.

β is losing material continuously because of gaseous currents between its two components, a class-B8 giant with a diameter about nineteen times that of the Sun with a luminosity of 3,000 $L \odot$, and a cooler, type late-A or early-F subgiant (not spectroscopically observed), its diameter about fifteen times the Sun's. These gaseous currents are indicated by emission lines of varying intensity in β's spectrum in addition to the normal spectra associated with its two stars. The primary contributes 90 percent of the system's light, but the F star is too faint to leave a deep impression on the spectrum. Only 22 million miles separates these stars from center to center, and as a result they have been flattened into egg shapes from their rapid rotation and mutual gravitational tidal action. They are so close together that a jet of hot material passes continuously from the B to the F star at 180 miles per second, whereas a cool stream of gas goes from the F to the B star. Owing to the eclipse, first one gas stream and then the other is cut off from sight, resulting in a variation in the bright-line emission spectrum.

As Struve observed, this system is losing mass continuously at the rate of 10^{22} grams per second, which causes the period of revolution (about thirteen days) to increase slowly. According to the mass-luminosity relation (p. 90) the more luminous primary should have the greater mass, but just the opposite is true, with the B star roughly of ten or eleven solar masses and the fainter, F star about nineteen or twenty times as massive as the Sun! This bizarre fact may be due to an "exchange of roles" in the evolutionary drama enacted by the components of β Lyrae, in which the massive F star has already passed through its giant stage and is chemically "older" than the primary. However, as it continues to acquire mass from the primary, the secondary B star may again become a giant, returning its excess mass to the primary. The primary will then become more massive and less luminous, as it contracts in size and undergoes the aging process already experienced by its partner in a reverse cycle of interactions.

Estimates of β's distance from us differ, because it is too remote for a determination of its trigonometric parallax; Sandage gives the distance as about 860 light years. β Lyrae also has two distant companions, a seventh-magnitude spectroscopic binary at 46.6 seconds of arc, or 12,000 astronomical units, and a ninth-magnitude star at 86 seconds. Two additional stars lie in Sheliak's field but are not gravitationally bound to the system.

Other Stars of the Lyre

Sheliak's eastern neighbor, which we mentioned, also meaning "Tortoise," is *SU-LAFAT (γ) GAMMA Lyrae* (RA 18 h 57 m   δ +32° 37') a third-magnitude type-B9 giant about 370 light years distant, its luminosity 525 L ⊙; there is a twelfth-magnitude optical companion at 13.8 seconds. The top of the Lyre's frame is occupied by a pair of much fainter stars; southernmost is the wide double *(δ) DELTA Lyrae 1* and *2*, magnitude 5.51 and 4.52 respectively, presenting a striking blue-red color contrast at a separation of 10½ minutes of arc, easily observed with binoculars. The components are surrounded by many fainter stars which probably constitute an open cluster, and provide an excellent field for the small telescope. Delta 2 is a luminous red giant; at 86 seconds of arc from it is an eleventh-magnitude optical companion, a close double with a 2.2 seconds separation.

About 1½ degrees to Delta's northwest, the upper corner of the frame is marked by *(ζ) ZETA,* also a double star, described as topaz green and worth observing with binoculars. To ζ's north, east of Vega, lies (ε); α, ζ, and ε form a triangle that was seen by Arabian desert dwellers as one of several *Athāfiyy,* or "Tripods," i.e., three stones on which the nomad placed his kettle! *(ε) EPSILON Lyrae* (1 and 2) is the famed "Double Double" (18 h 43 m   δ +39° 37')   m 5, first described by Sir William Herschel in 1779, a multiple system consisting of two very similar binaries separated from one another by perhaps two tenths of a light-year (nearly 13,000 AU), and in gradual revolution about their common center of mass. Because of this large separation, no relative change in their positions has been observed, but the period of revolution has been estimated at nearly 1 million years. The close components of Epsilon 1 (the northern binary) are separated by about 155 astronomical units, and those of Epsilon 2 (the southern), by 165 astronomical units. The brighter star of Epsilon 1 is in itself dual, a spectroscopic binary of uncertain period. The four chief components are of spectral class A, or "Sirian"; and the entire system, 180 light years distant, is approaching us at 17½ miles per second. Epsilon Lyrae may sometimes be seen as two separate stars with the naked eye, if the weather is very clear and one's eyesight exceptional; all four components are easily observable with binoculars.

About 4 degrees northeast of δ are the 4.5-magnitude stars *(η) ETA* and *(θ) THETA,* blue-white and amber, the latter a binary; in China, these stars bore the pleasant title *Lëen Taou,* ("Paths Within the Palace Grounds"). Fifth-magnitude (μ) MU, "The Swooping Eagle's Talons," *Al Aṭhfār,* lies about 2 degrees northwest of Vega. A fine star field surrounds it.

RR Lyrae and "Cluster Variables"

RR (RA 19 h 24 m δ +42° 41') m 9, lying far to the northeast of Vega, some 4 degrees southwest of Delta Cygni, is another important standard for a class of very old pulsating variables usually found in globular clusters such as M3 (p. 202), and M13 (p. 219). These so-called cluster variables lie in the galactic halo, their numbers increasing toward the center and away from the plane of the Galaxy. Showing the same peculiarities in spectral variation as type-II Cepheids, RR Lyrae stars have shorter periods, their light cycles ranging between 0.3 and 0.9 day. Balmer lines of hydrogen appear for a short time during the rapid increase in luminosity of an RR Lyrae star just before the break in its velocity curve. Two sets of absorption lines in its spectrum also occur for a short time while the brightness is increasing, one of these arising from surface material approaching us and the other from material that is rushing away, owing to the star's pulsations. The light curve of an RR Lyrae star shows that its luminosity varies by about one magnitude during any one cycle.

RR Lyrae stars constitute two groups: group I, those of periods under 0.45 day, which have smooth and symmetrical curves of small amplitude; and group II, those of longer periods (including RR itself), which exhibit sharp peaks following a sudden rise in luminosity and have greater amplitudes. RR's light period is 0.566837 day; its apparent magnitude varies from +7.1 to 8.0, and its spectral type ranges from A8 to F7. All RR Lyrae stars have average luminosities fifty to sixty-five times the Sun's, and their absolute magnitudes, according to recent data, range from about +0.3 to +0.6. Because the luminosities of these stars are similar, comparing their apparent and absolute magnitudes gives the distances of the clusters where they are found, as Shapley did with M13 and other globulars. RR Lyrae stars are less luminous than Cepheids, however, and therefore are too faint to be used for calculating the distances of external galaxies. With an absolute magnitude of +0.3, RR Lyrae itself, the brightest observed of this class, lies at a distance of about 900 light years.

The RR Lyrae stars in a globular cluster occupy the horizontal branch in the H-R diagram. They represent a rather short, unstable phase, about 80 million years, in the evolution of population-II stars, during which the periods of pulsation decrease from 1 to 0.3 day and luminosity increases somewhat. During this phase, in which energy is generated by the triple alpha process, they are in the core-helium-burning stage of their evolution, with the helium-burning core surrounded by a hydrogen-burning shell. (Refer to figure 91, the evolutionary tracks for M3).

THE RING NEBULA IN LYRA M57, NGC 6720
(RA 18 h 52 m δ +32° 58') m 9

Between β and γ, 60 seconds to the southeast of β, lies the most famous example of a widely observed, and until recently quite controversial, celestial object, the fascinating Ring nebula (fig. 103). The first to be discovered, this strange and beautiful nebular formation was found by the French astronomer Antoine Darquier, of Toulouse, in 1772. A few years later, Sir William Herschel made an intensive study of nebulae and

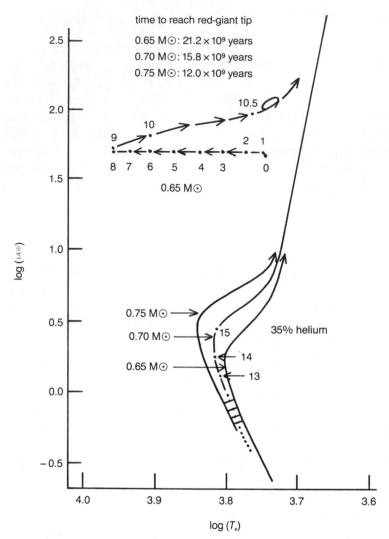

time to reach red-giant tip

0.65 M⊙: 21.2 × 10⁹ years
0.70 M⊙: 15.8 × 10⁹ years
0.75 M⊙: 12.0 × 10⁹ years

102. Evolutionary tracks of population-II stars, such as those found in globular clusters, plotted on an H-R diagram. The horizontal branch is shown by the dashed track, with figures from 1 to 8 representing time intervals of 10 million years each in a span of 80 million years. The red tip is at the points of the arrows, where a star becomes a red giant (according to Iben and Faulkner).

listed them in various categories according to his observations. He dubbed those with the appearance either of bright disks or smoke rings "planetary," because they reminded him of the disks or globes of planets; and their predominantly green color added to this illusion, because it was similar to the greenish tint of the planet Uranus (discovered by Herschel in 1781). He did not mean that this type of nebula is a planetary object; in fact, he imagined the characteristic "ring" to consist of many tiny stars. Subsequent investigation, however, revealed that it is actually gaseous in nature.

We mentioned the discredited element called "nebulium," which is really doubly ionized oxygen; it produces the spectral lines especially noticeable in planetary nebulae, giving them a characteristic green color. Huggins' discovery of this green spectral line revealed that planetary nebulae are gaseous and not composed of "a ring of stars," as Herschel speculated, or "minute stars, glittering like silver dust," as Father Secchi imagined; nevertheless, insofar as their origin and chemical composition are concerned, it is not incorrect to regard the so-called "planetaries" as very close "relatives" of certain types of stars—in fact, their offspring! According to recent theories, a young planetary nebula is the ejected product of an old star with a mass no greater than one and a half times the Sun's mass, in the first stages of its final evolution to a white dwarf. Material thrown off from its outer layers expands to form a gaseous "bubble" or shell, enclosing the doomed star like a luminous casket. Although the expanding stellar matter is dispersed throughout the nebula, in many cases it resembles a gigantic smoke ring or doughnut, with the hot white central star visible in the relatively dark area the ring encloses. This ringlike appearance is actually an optical illusion, however, owing to reflection, scattering, and ionization of its light at the outer boundaries of the nebula with relation to our line of sight.

The gaseous matter of a planetary is distributed in a thick shell transparent enough to allow the visible light from the central star to pass through it without much absorption, so that we can see the central star almost unobscured. The shell itself has a very complicated structure and is in a state of turbulence and internal motion resulting from the intense radiation of the central star. Velocity differences are also present within the

103. M57 (NGC 6270), the Ring Nebula in Lyra, famous planetary nebula, the first to be discovered. To view M57, it is advisable to use a telescope of at least 6 inches aperture; the larger it is, the more striking the results. Photographed with the 200-inch telescope on Palomar Mountain. (Hale Observatories)

nebula itself, ranging anywhere from 10 to 100 kilometers (6¼–62½ mi.) per second. This material may be expanding at speeds of about 20 kilometers (12½ mi.) to 30 kilometers (18¾ mi.) per second. Such a speed of expansion is produced by the great pressure of the ultraviolet radiation. Evidence also indicates that planetary nebulae are rotating, the rotation giving rise to additional internal motions.

Although the light emitted by the bright gaseous envelope of a planetary is dominated by the principal lines of the Balmer series of hydrogen, some forty other bright lines have been identified; infrared radiation is also present, probably emitted by heated dust particles within or near the expanding gaseous envelope. Thermal radio waves with a continuous spectrum have also been observed. This radiation probably arises from plasma vibrations in the expanding gaseous envelopes when these collide with interstellar gas clouds.

The ring of M57 is about twenty times as bright as its central star. Such brightness differences in some planetaries prompted Minkowski and Osterbrock in 1960 to suggest that the ring is actually the true shape in these cases: In fact, many differ greatly from M57 in appearance, resembling bright disks or irregularly shaped blobs. (Refer to the Owl Nebula, in Ursa Major, p. 54, and the Eskimo, in Gemini, p. 154).

The spectrum shows an expansion of about 12 miles per second for M57, indicating that the nebula is some fifteen thousand years old. Its distance is estimated between 1,800 and 2,000 light years. Assuming the smaller figure to be correct, we obtain a diameter of 0.2 parsec, or .47 light year (30,000 AU) and a total luminosity of 50 $L_\odot$ for the Ring, which contains less than one solar mass. A density of ten thousand ions per cubic centimeter has been given for it, which means that its material is extremely attenuated. This density may be determined by the following process: We note that the radiation in the continuous spectrum is produced by the recombination of free electrons and ions. The ultraviolet radiation from the central star first ionizes neutral atoms, thus ejecting electrons. These free electrons are then recaptured and emit continuous energy in the process; from the total continuous energy emitted in this way, we calculate the number of electrons per centimeter along the line of sight and this equals the number of ions.

The absolute magnitude of the hot central star of M57 is about +6, to be compared with the Sun's +4.71. The central stars of planetaries are extremely hot, their temperatures ranging from 30,000 degrees K. to about 400,000 degrees K., and some of them have very small diameters, which implies that they are approaching the white-dwarf stage. Planetaries fall near the blue end of the horizontal branch of the H-R diagram of a globular cluster (fig. 91, p. 203). This is confirmed by the single planetary nebula in globular cluster M15 (Pegasus), which appears to be one magnitude brighter than the RR Lyrae stars in that cluster. As mentioned, the RR Lyrae stage of a star is short; such periods of rapid change seem to be the rule for stars nearing the end of their lives. According to astronomer Donald Osterbrock (Lick Observatory, University of California), "The planetary nebula represents a relatively short-lived phase in the evolution of a star. In the final stages of the nebula, the nebular shell expands and merges with the interstellar gas, while the central star becomes a white dwarf."

This process apparently occurs only in the lives of single stars; the lone existence of a distended, irregular or long-period, red-giant variable with a mass less than 1.5 $\mathcal{M}_\odot$, burning helium in a shell around a carbon core, seems to be the prerequisite for the formation of a planetary nebula through an orderly ejection of the star's outer envelope. The combined action of intense radiation pressure and gas pressure resulting from a sudden increase in luminosity is the mechanism for this ejection of hydrogen and helium gases from the upper regions of the star's atmosphere. As more and more of the envelope escapes, the remaining hot, dense core reaches a stage at which it can remain in equilibrium as a white dwarf. Close binaries, on the other hand, discharge excess mass during their turbulent, "quarrelsome" later years by suffering more violent, nova-like explosions, and do not form planetary nebulae, although many do form temporary expanding gas shells.

The most interesting lesson to be learned from planetaries is the continuity of creation; the nebula returns to space the basic elements from which the dying star originally was formed: large quantities of hydrogen and helium mixed with carbon, which constitute the basic building blocks in the formation of future stars.

NINE

More Tales of Man and Beast

Journeying in a southerly direction, we visit our next group of constellations, most of them lying below the celestial equator and spanning an arc from about 14 hours 25 minutes to about 20 hours of right ascension; three of these figures—Libra, Scorpius, and Sagittarius—lie on the zodiac. Almost all are linked to stories that juxtapose traditional enemies—man, serpent, scorpion, and archer—in their heavenly abodes. Sagittarius, the Bow Stretcher, pursues the Scorpion, which, with extended claws (Libra), has already frightened its victim, Orion, into occupying the opposite side of the celestial sphere! Calm and unafraid, however, another great celestial giant dominates this scene, holding in his hands a long, writhing serpent, the symbol of immortality and of his profession of healing.

OPHIUCHUS (vel Serpentarius)

Mythology's great physician-surgeon was Aesculapius (here called *Ophiuchus),* son of Apollo and half brother to the poet-musicians Linus and Orpheus. Some say that Aesculapius' mother was Arsinoë, the daughter of King Leucippus of Messenia, whose other daughters, Hilaira and Phoebe, we remember, were carried off by Castor and Pollux. Many ancient authorities, however, claim that the Thessalian woman Coronis was the object of Apollo's affection. Coronis conceived Aesculapius, and, keeping her pregnancy a secret from her father, King Phlegyas, she exposed the newborn infant upon a mountain in Sclavonia, Peloponnesus; a she-goat and a dog suckled Aesculapius, and when their owner, a shepherd, found them with the child, he observed fiery rays about its head. Thereupon he spread the word about the countryside concerning the divine appearance of this abandoned child; hearing such miraculous news, a multitude of ill persons came to the child for relief, and many were cured.

According to a very widely supported second version of Aesculapius' early years, Coronis, unfaithful to Apollo, was slain either by his sister Diana or by Apollo himself, who was so angry with Corvus, the Raven who bore him the unwelcome tale of her deceit, that he turned the unfortunate bird from silver to black and sent it to Hades.

Mercury (according to the Greek writer Pausanias) or Apollo (according to the Greek poet Pindar) rescued the child Aesculapius from the ashes of Coronis' funeral pile and eventually gave him over to the Centaur Chiron, who taught his young ward the art of surgery. Aesculapius also became a skilled harpist, curing ills through music; skilled in divination and the preservation of his patients' future health, he studied the physical influences of celestial bodies on mankind, thus adding astrology to his activities as well. In a temple at Epidaurus, his image of gold and ivory is shown seated upon a throne, crowned with rays and holding in one hand a knotty stick that symbolizes the intricacies of medicine, while leaning with the other upon a serpent, the emblem of wisdom, healing, and regeneration. Plato, in fact, claims that Hippocrates composed his medical treatises from inscriptions in temples of Aesculapius.

The Aesculapius of mythology became a renowned physician, serving on the voyage of the Argonauts as ship's surgeon and even restoring to life several personages of ancient Greece, including Glaucus (Androgeus), the son of Minos of Crete; Tyndareus, father of Castor; and Hippolitus, the son of Theseus, whose severed limbs Aesculapius glued together after he was dragged to death by his own frightened horses. After this great healer tried to revive the dead Orion, Pluto complained to Zeus that Aesculapius was weakening the empire of Hades by reducing its population; whereupon Zeus, fearing that the doctor's art would become widespread among mortals, ended his medical career with a thunderbolt!

After his death, Aesculapius, pictured together with his serpents, assumed the status of a "potent" god because of his benefits to mankind, and was placed by Zeus in the heavens as a constellation, where, bounded by Hercules to the north, Aquila to the east, and Scorpius to the south, this great figure sprawls across the celestial equator. Known to the Romans under its modern name Ophiuchus, it was also called *Serpentarius* and *Serpentis Lator,* "The Serpent Holder." Among its numerous other mythic identities we mention only that of the Trojan seer and priest Laocoön, strangled along with his two sons by a serpent; this tragic struggle is immortalized in the lines of Vergil* and the famous Vatican statuary.

At about 5½ degrees to the east-southeast of Ras Algethi (α Herculis), we find the much brighter *RASALHAGUE (α) Ophiuchi* (RA 17 h 32.6 m δ +12° 36′) m 2.09, from the Arabic *Rās al Ḥawwāʻ,* the "Head of the Serpent Charmer." (In an early Arabic asterism, it was called *Al Rāʻi,* "The Shepherd," α Herculis was the "Shepherd's Dog," and a group of fainter stars scattered between α and ω Herculis constituted the "Flock of Sheep.") Alpha Ophiuchi was also the Chinese *How,* or "Duke." A white giant of the "Sirian" type, it is 18 parsecs, or nearly 59 light years distant, lying in the North Polar Spur, a prominent source of radio waves. A seventh-magnitude companion whose light is lost in that of the primary is indicated by periodic wavering in α's proper motion.

Rasalhague's unusual spectrum, with interstellar absorption lines, which are usually present only in more distant stars, was recently recorded by the International Ultraviolet Explorer (IUE) satellite. These interstellar lines are easily differentiated from the absorption lines of the star itself, because of the star's rapid rotation. Optical observations show only one neutral interstellar cloud in front of α; this cloud has the same velocity as the interstellar winds in various parts of our Galaxy. Such gaseous material

---

* *Publius Vergilius Maro (70-19* B.C.), *Latin poet.*

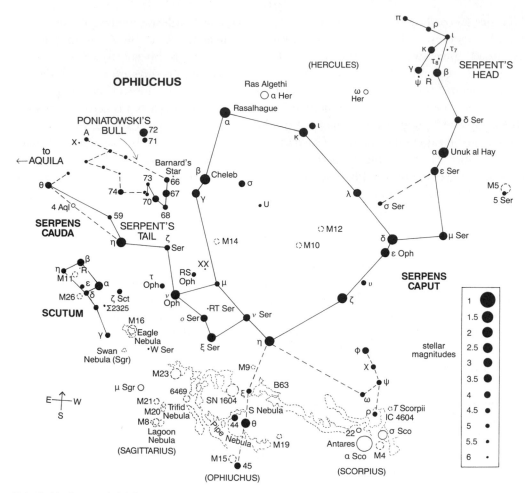

104. Ophiuchus and vicinity.

has been observed in front of the more distant Zeta and Chi Ophiuchi (in the southern part of the Giant's figure), as well as other stars of Ophiuchus and Scorpius; a comparison of the abundances of various elements, including iron and hydrogen in the clouds lying in front of the three stars, gives an estimate of the density of the gas in the nearby cloud (in front of α Ophiuchi), and the Doppler broadening in its spectral lines indicates that the gas is warm. The unusually high abundances of refractory metals in the cloud are believed to have been injected into the clouds by the high-velocity shock front of a supernova, possibly the supernova remnant of the North Polar Spur. It is believed that this remnant was extensive enough to have enveloped the Sun, which is immersed in it. Thus observations of α Ophiuchi have yielded data concerning the characteristics of interstellar material near the Sun. (See Hartmann's original studies of the K lines in Mintaka's spectrum, p. 100).

CHELEB (β) BETA Ophiuchi (RA 17 h 41 m   δ +04° 35′)   m 2.77, also "Cebalrai," is derived from *Al Kalb al Ra'i,* which means "The Shepherd's Dog" (the early Arabic title for α Herculis, above). Cheleb's distance is 125 light years and its luminosity is about

100 L ⊙. Roughly 10 degrees to the west-southwest of Rasalhague, forming a wide-based triangle with α and β, we find *(κ) KAPPA,* a magnitude-3.2 orange giant about 150 light years distant, and *(ι) IOTA,* only about 1¼ degrees to its northwest, which together formed the Chinese *Ho,* a dry measure. Ophiuchus' elbow is delineated by *MARFIK (λ) LAMBDA,* lying 10 degrees southwest of κ; a triple system, it consists of a binary of the white, "Sirian" type at 1 minute and a distant m-11 companion at about 2 minutes.

The northeastern corner of Ophiuchus' figure is marked by *(γ) GAMMA,* m 3.7, sometimes called "Muliphen," a single main-sequence object of the Sirian type. *DELTA* and *EPSILON,* at the western corner, form a contrasting naked-eye double (not gravitationally bound). Delta, a reddish giant of about 130 L ⊙, was known as "Yed Prior," from the Arabic *Yad,* meaning "hand"; and Epsilon, an orange giant of 35 L ⊙, was "Yed Posterior," or the "Hand Following After," as its more easterly position implies; with the Euphrateans, however, it was a "Man of Death"! These two stars mark Ophiuchus' left hand, which grasps the upper part of the Serpent.

About 8 degrees to the southeast of ε, near Ophiuchus' left knee, lies *(ζ) ZETA,* a magnitude-2.57 blue-white main-sequence star 520 light years distant, of about 4,000 L ⊙. About 10 degrees to ζ's southeast we find *SABIK (η) ETA,* the "Preceding One," a title it sometimes shared with ζ. A white magnitude-2.46 main-sequence star, η lies about 70 light years distant, its luminosity that of 40 Suns. Large telescopes reveal it as a close binary with a separation of about 0.2 second, or 20 astronomical units, the components completing a revolution every eighty-four years in a very eccentric orbit.

(ξ) XI, #40, magnitude 4.5, lying about 6 degrees south-southeast of η, is a close binary of spectral type F; together with η and 0, it forms the curved line of Ophiuchus' right leg, which was a "snake," however, to the Akkadians and the Copts (not to be confused with the modern Serpens). Interestingly, these stars were seen as a "Magician" by both the Sogdians and the Copts, and the Khorasmians called ξ and θ the "Serpent-bitten"; the long, winding curve of ε, ζ, η, and ξ was the Persian "Serpent Tamer." Perhaps all these ancient peoples of the East had a snake charmer in mind! Marking Ophiuchus' right foot, *(θ) THETA,* a variable star in the southern corner of the constellation, is of particular interest as a massive blue-white subgiant of the Beta Canis Majoris type (p. 134), showing tiny variations in magnitude as it begins to evolve away from the main sequence. Kepler's star, the supernova of 1604 (to be discussed), appeared immediately to θ's northeast.

It is easy to confuse the stars of Ophiuchus with those of Serpens Cauda, the modern Serpent's Tail, held in Ophiuchus' right hand, which is marked by the stars *(ν) NU* (RA 17 h 56 m δ −9° 46') m 3.34, and its fifth-magnitude neighbor *(τ) TAU,* about 2 degrees to the northeast. Nu Ophiuchi is a golden giant about 140 light years distant, surrounded by dark nebulae. (To the ancient Chinese, however, it was *She Low,* a "Market Tower"!) Tau is a close binary, both of its components F-type dwarfs.

The Other Taurus

Some 4 degrees to the east of γ lies the famous binary *#70 OPHIUCHI* (RA 18 h 2.9 m δ +02° 32') m 4.1, discovered in 1779 by Sir William Herschel; its orbital motion has been thoroughly studied because of its closeness, at only 16.5 light years. The components are class-K stars, one a main-sequence object and the other a dwarf of very low luminosity. Because of minute apparent deviations in their orbits, the presence of a third body in the system is suspected, possibly a planet about ten times Jupiter's mass, but the unseen planet's existence has not yet been confirmed.

The binary #70 Ophiuchi, however, was not always regarded as a star of the giant physician's realm! In former years it belonged, at least for a while, to *TAURUS PONIATOVII,* "Poniatowski's Bull," the French *Taureau Royal,* an asterism made up of stars lying between Ophiuchus' shoulder and Aquila. The little Polish Bull's brightest stars form the letter *V* and are reminiscent of the Hyades, in Taurus; its brief history began in 1777, when the Abbé Poczobut of Wilna, a Polish astronomer, asked the French Academy to honor King Stanislaus Poniatowski (the last Polish king) with a constellation. Although the new constellation became obsolete by the end of the nineteenth century, its stars being returned to Ophiuchus, Poniatowski's Bull is still found on some star charts. Its stars were also represented as a "triangular figure" on a thirteenth-century Arabian globe now resting in the Borgian Museum at Villetri. The Bull includes Flamsteed numbers 68 to 73 and some twenty to twenty-five other faint naked-eye stars; if added to these are fifth-magnitude A (HD #171802, SAO #123690), far to the northeast of #70, and variable *X OPHIUCHI* immediately to A's southeast (RA 18 h 36 m δ +08° 47′), this bull even has a tail! X is a long-period pulsating red giant of the Mira type, its magnitude varying from 6 to 9. In 1900, a close companion was detected at a separation of about 75 astronomical units, and this discovery led to the calculation of the mass of the red giant, apparently not much greater than that of the Sun.

Other Variables

U OPHIUCHI, in the heart of the giant's figure, southwest of β and γ (RA 17 h 14 m δ +01° 16′) m 5.70, is a short-period variable, a binary a thousand light years distant from us. Its components are two class-B main-sequence stars separated by about 5.5 million miles, their orbital period therefore only one and a half days. In U's field, less than a degree to the northeast, lies Z, a long-period reddish variable that never gets brighter than the seventh magnitude.

RS OPHIUCHI (RA 17 h 45 m δ −06° 42′), about 4° to the northeast of ν and very distant at 12,000 light years, is a widely observed recurrent nova, similar to T Coronae (p. 212), reaching naked-eye visibility at maximum luminosity, and ranging from the eleventh or twelfth magnitude to the fifth and even the fourth magnitude during its four recorded outbursts. The star changes color from white to red, shortly after maximum, owing to the dominant spectral emission lines of hydrogen. A rapidly expanding nova shell is indicated, and it was recently discovered that U is a binary: the primary, or erupting star, a hot subdwarf, and its companion, a large solar-type star, probably losing mass to the subdwarf.

Sometimes called the "Iron Star," *XX OPHIUCHI* (RA 17 h 41 m δ −06° 15′) is about 1.5 degrees west-northwest from RS and some 2½ degrees to the north-northeast of μ(magnitude 4.6). XX is an irregular variable that seems to follow no specific program, ranging from the ninth to the eleventh magnitudes at intervals of anywhere from ten to twenty years. Its changeable spectrum displays very strong absorption lines of ionized iron, as well as bright lines of hydrogen, but in 1925 it developed strong absorption lines of ionized titanium. The spectral class, normally B, changes to A at maximum luminosity. P. W. Merrill, who studied this unusual star, saw indications of an expanding gaseous shell in the spectrum; generally it behaves like R Coronae (see "A Topsy-Turvy Star," p. 212).

A Supernova Comes to the Aid of Science

In 1604, a "new" object blazed forth in the southern part of Ophiuchus, less than 2 degrees east-southeast of ξ (the middle star of the ancient Akkadian "Snake"). It soon reached an apparent magnitude of −2.25, exceeding at that time even Jupiter's brilliance, and it remained visible to the naked eye from October until March, but after that it could not be observed, because the telescope had not yet been invented. Kepler's student Brunowski notified him of the appearance of this supernova, the fourth and most recent to be observed in our Galaxy in the past thousand years. Now called **KEPLER'S STAR**, the **SUPERNOVA A.D. 1604** or V483 (RA 17 h 27.6 m δ −21° 26′) provided valuable ammunition for Galileo, who challenged Aristotle's axiom that the heavens are "incorruptible." Galileo demonstrated that the unknown star is more distant than the planets, thereby introducing the first correct concept of stellar distance, an accomplishment reminiscent of Tycho Brahe's determination that the supernova of 1572 is farther away than the moon! Like Tycho's Star (pp. 60–9), Kepler's Star (a population-II star) was a type-I supernova, which may attain a luminosity 100 million times that of the Sun, reaching an absolute photographic magnitude of about −16.

In 1941, using the 100-inch telescope at Mount Wilson Observatory and red sensitive plates because of obscuration clouds in that region of the heavens, astronomers discovered nebulosity near the position assigned to the star by Kepler. If a stellar core in the form of a neutron star has survived the vast explosion, such a remnant has not yet been positively identified.

Another Runaway

Slightly west-northwest of the magnitude-4.5 variable #66, which marks the tip of the Polish Bull's western horn, we find a faint red dwarf whose position will be out-of-date as quickly as we state it. This is **BARNARD'S STAR** (LFT 1385) (fig. 105), also "Barnard's Runaway Star" (RA 17 h 55.4 m δ +04° 24′) m 9.53, at 6 light years the second-nearest star, heading toward the north at the very large proper motion of 10.29 seconds per annum, the largest proper motion of any known star, indicating a transverse speed of about 60 miles per second relative to the Sun. Since the transverse velocity is one component of the pace velocity, the space velocity of Barnard's Runaway Star relative to the Sun is also very large: about 103 miles per second (see "Proper Motions of Stars," p. 40). A star's radial velocity is found by measuring the Doppler shift of its spectral lines; as it approaches us with a radial velocity of 87 miles per second, Barnard's Star is expected to reach a point less than 4 light years distant about eight thousand years hence, before its journey through the Galaxy takes it away from us again.

With a luminosity only 0.0004 that of the Sun, an estimated mass of 16 percent $\mathcal{M}\odot$

105. *LFT 1385, Barnard's Star* in Ophiuchus, with the greatest known proper motion, shown on August 24, 1894, and May 30, 1916. Compare its location (arrow) relative to closest bright star and to the little reversed-Hyades-like figure above it to note the positional change. North is toward the top. (*Yerkes Observatory photograph, University of Chicago*)

and a diameter of about 140,000 miles, the star probably has a density about forty times that of the Sun. An interesting feature of Barnard's Star is the astrometric variations in its motion, which indicates at least one unseen companion at a separation of 4 astronomical units and with only 0.0015 $\mathcal{M}\odot$, or about one and a half times Jupiter's mass, therefore believed to be a planet. Perhaps those stars with Jupiter-like companions may possess complete planetary systems, no member of which can ever be visible to us even in the largest Earth-based optical telescope. As it speeds toward us, Barnard's Star may be carrying with it a retinue of fascinating globes, as different from one another as are the moons of Jupiter and Saturn. Will our descendents, eight thousand years in the future, be able to travel to explore these planets? However, if remote-controlled vehicles equipped with the most advanced electronic optical devices were sent out far enough in advance to meet the Barnard system at its closest approach, a new and fabulous chapter might be written in the annals of space exploration!

Some Globular Clusters

For Charles Messier, the month of May 1764 was particularly productive in his discovery of globular clusters, although he did not understand their nature. The first of these was *M9 (NGC 6333),* lying in the curve of the Akkadian Snake, or, in more modern terms, on Ophiuchus' leg (RA 17 h 16 m   δ −18° 28'), about 3.5° southeast of η; in 1784, Sir William Herschel's great telescope revealed its starry contents. We now know it as one of the globulars surrounding the galactic core, lying only 7,500 light years from it, and 26,000 light years from our Sun, its luminosity about sixty thousand times the Sun's.

About 4 degrees west-southwest of θ, the star that marks Ophiuchus' right foot, and some 7 degrees east of α Scorpii (Antares), we find the globular *M19, NGC 6273* (RA 16 h 59.5 m   δ −26° 11'), a flattened, or "oblate," cluster, lying 20,000 light years distant from us but 13,000 light years from the center of our Galaxy.

Within the giant's torso, some 8½ degrees northeast of ζ, is *M10, NGC 6254* (RA 16 h 54.5 m   δ −04° 02')   m 7, a rich, bright cluster variously estimated as between 16,000 and 22,000 light years distant, its stars easily resolved with a moderate-sized telescope. About 3.4 degrees to its northwest lies *M12, NGC 6218,* somewhat dimmer and less condensed. As in the other cases, it was Sir William Herschel who first resolved these Messier "nebulae" into individual stars. Not quite 8 degrees south-southwest of β lies *M14, NGC 6402* (RA 17 h 35 m   δ   03° 13'), also a globular but somewhat elongated in shape, and more remote at 70,000 light years than M10 and M12; of note is Nova 1938, one of the only two known examples of novae in globulars.

Nebulae and Star Clouds

> . . . In Heaven's dark surface such this Circle lies
> And parts with various Light the Azure skies . . .
> Manilius

The eastern branch of the Milky Way is interrupted within the northeastern regions of Ophiuchus; here the Great Rift, an enormous dust lane, cuts through the bright star fields on its southerly course. To the south of the constellation, near the galactic center in our line of sight, lie dust-streaked spiral arms, concealing parts of the luminous core. We see brilliant star clouds crossed by an intricate network of twisting obscuration lanes, and, silhouetted against the star clouds, wonderful dark formations that stir the imagination. One of these, resembling a gentleman's pipe and visible to the naked eye, lies 2 degrees southeast of θ; a draft on this pipe, however, would be the longest in history, for the *PIPE NEBULA,* (fig. 106), bowl and stem included, spans several hundred light-years!

About 1½ degrees north-northeast of θ is another striking formation: the famed *S NEBULA, B72,* whose dark monogram shows up best in photographs. Moving about 10

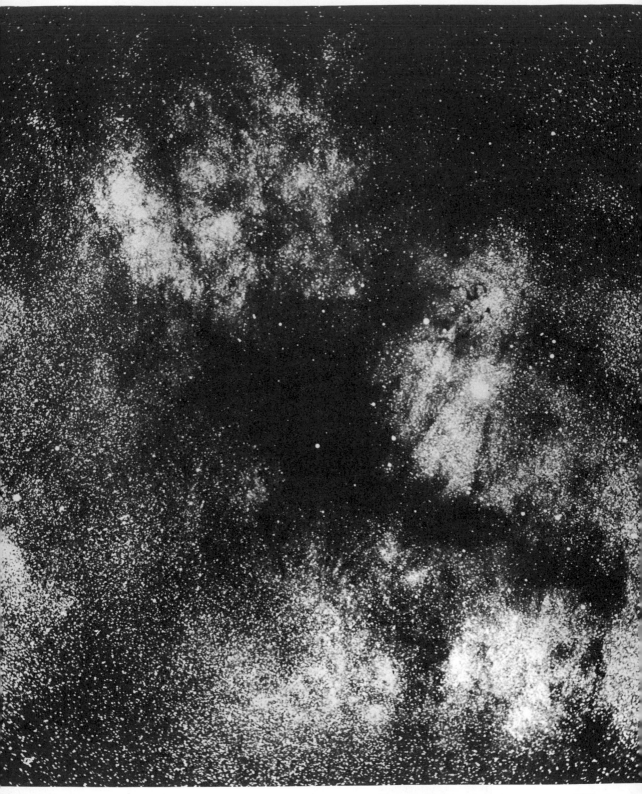

106. *The Pipe Nebula in Ophiuchus;* dark cloud Barnard 78 forms its bowl, while the stem includes B59 and B65–67. Surrounding it are bright star clouds of the Milky Way. The tiny S nebula is clearly delineated just to the right of the bowl's rim. (*Yerkes Observatory photograph, University of Chicago*)

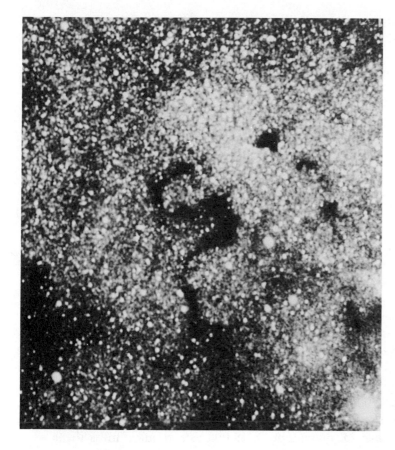

107. Enlargement of B72, the S-Nebula, a dark monogram lying in the bright star fields next to the Pipe Nebula. (*Yerkes Observatory photograph, University of Chicago*)

degrees westward and using the red supergiant Antares in Scorpius as a guide, we locate *(ρ) RHO* Ophiuchi, a magnitude 4.6 blue-white star lying about 3 degrees to Antares' north-northwest. Rho is embedded in the very faint diffuse nebula *IC 4604* and in neighboring clouds of nebulosity, from which numerous dust lanes point their fingers eastward toward ξ, bordering on the mysterious dark obscuration cloud *B63.*

SERPENS

. . . Heaven is beautified by His winds
And His Hand pierces the fleeing Snake . . .
Job, 26:13

The combined figures of Ophiuchus-Serpens deserve nomination as the most confusing in the heavens. Originally forming one constellation, their stars have been charted separately for many years, and the ancient Snake is now divided into two parts as well, *SERPENS CAPUT,* the Head of the Serpent, which lies to Ophiuchus' west, and *SERPENS CAUDA,* the Tail, to his east. Serpens crosses Ophiuchus as his left hand (δ and ε Ophiuchi) grasps its neck, and the Serpent Holder's right hand (ν and τ), holds its

twisting Tail. Universally known as a Serpent, this constellation's identification harks back to biblical times; the Hebrews, Hittites, and many other nations regarded it as a snake, which, together with Ophiuchus, formed the Euphratean *Nu-tsir-da,* the "Image of the Serpent." In ancient Arabia, however, several stars of Serpens, Ophiuchus, and Hercules were imaged as a pastoral scene. Serpens alone later became a snake *(Al Ḥayyah)* to the Arabians as well, after they adopted the Greek astronomy and its constellations.

Serpens Caput

Some 28 degrees of right ascension separate α Ophiuchi and *UNUK AL HAY (α) ALPHA Serpentis* (RA 15 h 41.8 m δ +06° 35'), lying to Rasalhague's west-southwest; also called "Unukalhai" or "Unuk," from *Unk al Ḥayyah,* the "Neck of the Snake," it was known as *Cor Serpentis,* the "Heart of the Serpent," to astrologers. A vibrant heart at magnitude 2.65, α is a gold-tinted giant of class K about 79 light years distant; with fourth-magnitude λ, 1° to its northeast, it was the Chinese *Shuh,* named after a feudal state. Other stars in α's field include a twelfth-magnitude optical companion lying almost directly to its north at a separation of about 58.2 seconds.

The back of the Serpent's uplifted head is marked by *(β) BETA* (which is almost due north of α) (RA 15 h 44 m δ +15° 35') m 3.67, and *(γ) GAMMA,* magnitude 3.9, some 2⅔ degrees to β's east. Known to the Chinese as *Chow,* a famous imperial dynasty, β is a type-A subgiant 95 light years distant, gravitationally bound to a yellow dwarf about 30.8 seconds (900 AU) to its west. There has been no noticeable change in their relative positions since the first measurement, by F. G. W. von Struve, in 1832. A second pair, the binary *Roe 75,* magnitude 9, 48,000 astronomical units to its west, forms a multiple system with β 1. Fourth-magnitude *(δ) DELTA,* at the bend of the Serpent's neck (RA 15 h 32 m δ +10° 42'), lying northwest of α and southwest of β, is a close binary consisting of two F-type dwarfs, fourth and fifth magnitudes; the primary is also a δ Scuti variable (p. 264).

About 1.2 degrees east-southeast of β lies the long-period variable *R SERPENTIS* (RA 15 h 48 m δ +15° 17') m 5.7–14, a distended red giant of the Mira type, about 600 light years distant and suitable for small telescopes.

Not quite 8 degrees southwest of α lies the spectacular globular cluster *M5, NGC 5904* (RA 15 h 16 m δ +02° 16', fig. 108), comparable with the great M13, in Hercules, and M3, in Canes Venatici. Discovered by Gottfried Kirch in 1702 and rediscovered by Messier in 1764, its great mass of stars was first resolved by Sir William Herschel in 1791. No more than 27,000 light years distant, but more than 10 billion years old, M5 was born when the Galaxy was young! Its roughly five hundred thousand stars, the brightest of which average magnitude 13.97, therefore belong to the old population-II generation. Among them are at least ninety-seven variables, mostly of the RR Lyrae type; the yellow binary in M5's field is #5 *(φ) PHI* Serpentis.

Serpens Cauda

The stars of Ophiuchus and the Serpent's Tail intermingle in the southwestern part of Cauda; ν, ξ, and o Serpentis (a class-A variable of the δ Scuti type) form a triangle,

108. M5 (NGC 5904)
globular cluster in
Serpens, more than 10
billion years old.
Thousands of stars are
crowded into the compact,
bright nucleus of this
spectacular cluster.
Photographed by K.
Cudworth with the 41-inch
reflector on June 19,
1977. (Yerkes
Observatory photograph,
University of Chicago)

with ν Serpentis at the long axis, lying directly between η and μ Ophiuchi. Some 5¹/₃ degrees to the north of ν Ophiuchi (p. 235)—do not confuse the latter with ν Serpentis —lies (ζ) *ZETA Serpentis,* magnitude 4.6. About 5¹/₃ degrees to ζ's east-northeast and some 9 degrees northeast of ν Ophiuchi, we find (η) *ETA Serpentis* (RA 18 h 19 m   δ +2° 55′)   m 3.25, near the northwest border of Scutum; η is a class-K giant or subgiant 60 light years distant, about 6 $L \odot$. The Chinese gave these two stars the appealing name of "The Heavenly Eastern Sea."

Lying within the western edge of the Milky Way, some 11 degrees to the northeast of η and marking the tip of the Serpent's Tail, is the fine wide double *ALYA (θ) THETA Serpentis* (RA 18 h 54 m   δ +4° 8′)   m 4.05, its name, like that of α, derived from *Al Ḥayyah,* the Arabic "Serpent." About 130 light years distant, its gravitationally bound components are both white, Sirian-type stars, and at a separation of 900 astronomical units, they form a binary that is exceptionally good for binoculars or a very small telescope.

Those with powerful instruments, however, will find it interesting to look for the erratic variable *W SERPENTIS* (RA 18 h 07 m   δ − 15° 34′) some 9 degrees east of ξ; W lies in the extreme southeastern region of Cauda, near its common border with Sagittarius. Catalogued as an eclipsing binary with a period of about fourteen days, its unusual and irregular light curve is accompanied by an intriguing change in its spectrum, in which the absorption lines are transformed into emission lines at the time of W's minimum luminosity. It is also an intrinsic variable with two different periods: a short period with a range of 0.2 magnitudes and a long period (about 270 days) with a range of about 0.6 magnitudes. The components are two F-type giants, six or eight times the Sun's diameter, separated by 14 million miles, with the primary star emitting a gaseous stream at high velocity to form an expanding shell, as it receives gaseous matter from the secondary, resulting in a rapid evolution of the system.

About 1 degree north-northwest from o (#56) Serpentis and some 5 degrees south-southeast of μ Ophiuchi lies the peculiar variable *RT SERPENTIS* (RA 17 h 37 m   δ −11° 55′)   m 16, discovered as a magnitude-13.9 star in 1909 and slowly brightening until it reached the ninth magnitude, in 1913, where it remained for over a decade. Its gradual decline took it to magnitude 14 in 1940 and ultimately back to magnitude 16 in 1963. RT's A-type spectrum showed the characteristic bright lines of a nova only late in

109. *M16 (NGC 6611),
open star cluster in
Serpens Cauda.*
Surrounding it is a vast,
luminous nebula crossed
by dark formations of dust,
creating a fantastic scene.
Photographed in red light
with the 200-inch Palomar
telescope. (*Hale
Observatories*)

its decline and no sign of the blue-shifted lines that indicate the high expansion velocities of most novae, possibly indicating a milder outburst than that of a true nova.

Some 3½ degrees northeast of W Serpentis, and about 6½ degrees southeast of ν Ophiuchi, near the border of Scutum, we find *M16, NGC 6611* (RA 18 h 16 m   δ −13° 48′), a large open cluster with a 25-minute field containing about sixty stars, many of them class O and B giants, about 8,000 light years distant. Surrounding the cluster is a vast, diffuse, and luminous nebula whose "faint" light even Messier, in 1764, observed. Modern photographs reveal a fantasy land of the heavens, both bizarre and wonderful (fig. 109). Dark dust-cloud formations are seen projected against the bright background cloud, whose tail- and wing-like structures give it the name Eagle Nebula; however, the center of interest of this marvelous scene is a great "throne-like" structure upon which stands the gowned figure of a young woman holding a sharply defined soft black boot! To the northeast side of the nebula, a gesturing figure clad in a short black ballet skirt and poised on the top of what appears to be a gnarled, twisted mushroom or even a palm tree, watches the "Princess." However, some may see instead a turbaned figure with immensely long legs, or even "Aladdin's genii rising from the magic lamp," in the words of Robert Burnham, Jr., who has named the entire cosmic spectacle the "Star-Queen Nebula."

Radio emission has been detected from M16, and a mass of 12,500 $\mathcal{M}_\odot$ has been estimated for it, corresponding to a diameter of 25 light years for the bright center and 70 light years for the nebula in its entirety. Studies show that all stars in the cluster hotter than or bluer than class B8 are pre main-sequence objects, that is, have not yet evolved to the main sequence. Dark globules believed to be newly forming protostars, and its very luminous blue giants, indicate that M16 is very young, the average age of its members about eight hundred thousand years.

LIBRA

> . . . I bear the Scales, when hang in equipoise
>   The night and day.
>
>   Longfellow

Libra, the "Balance," or "Scales," a faint trapezoidal figure lying on the zodiac between Virgo and Scorpius, is now visible in the summer evening skies, owing to precession, but it marked the autumnal equinox (the beginning of autumn) some three thousand years ago, when the Sun "rose in Libra." Thus the equal length of day and night at the beginning of fall was associated with the heavenly Scales by ancient peoples, and it symbolized the balance of justice in human affairs as well as the balanced arrangement of Nature's laws. Libra's uncertain origins are traced by some authorities to Egypt, where it appears on the planisphere in the temple at Denderah; others identify it with various sky pictures of the Chaldeans, including the archaic "Altar" or "Solar Lamp" carved on Euphratean gems and tablets; α and β Librae also suggested a "Chariot Yoke," or *Sugi,* to the Chaldeans, who believed that these stars influenced their crops.

Libra is associated with the Hebraic *Moznayim,* or "Scales," the "Balance" in which

Belshazzar, Nebuchadnezzar's son, was "weighed" and "found wanting" (Daniel, 5:27); it was also the Persian *Terāzū,* or "Scales." Mythology depicts Libra as "Pluto's Chariot," in which Persephone (Virgo) was transported to Hades, but Libra's stars were included by the early Greeks with those of Scorpius as the latter's "Claws," or *Chelae.* At the time of Julius Caesar, however, the "Claws" again became a separate constellation, its modern name, *Libra,* Latin for "weight" or "measure," first appearing in the Julian calendar, while Roman poets also called it *Jugum,* from the Greek *Zugon,* "Yoke" or "Beam" (of a scale).

This small sky picture has been subdivided into Northern and Southern Scales, but the main figure is a simple quadrilateral formed by α, ι, γ, and β, extended southward to σ on some charts. At the western corner of the "tetragon," Libra's brightest star formerly marked the Scorpion's "Southern Claw," in Arabic *Al Zubān al Janūbiyyah,* from which is derived its marvelous modern name, ***ZUBENELGENUBI (α) ALPHA*** Librae (Fl. # 8 and 9) (RA 14 h 48 m   δ − 15° 50′)   m 2.8. Another name is *Al Kiffah al Janūbiyyah,* "The Southern Tray of the Scale," which became *Kiffa Australis* a century ago. Alpha is a wide double consisting of a class-A3 "hydrogen" star whose spectrum also shows the presence of metallic elements, and a fifth-magnitude class-F white subgiant gravitationally bound at a separation of 3 minutes 51 seconds and suitable for the small telescope. Number 9, the primary, is also an extremely close spectroscopic binary, the components only .01 second apart, as astronomers verified photoelectrically using the 36-inch reflector at Kitt Peak in 1966 during an occultation of α by the Moon. If you wish a fair test for the small (2½-in.) telescope in resolving the components of a close double, you may direct it toward *(μ) MU* Librae, magnitude 5.5, some 2 degrees to the north-northwest of α.

The northern corner of the quadrilateral is marked by ***ZUBENESCHAMALI (β) BETA*** (RA 15 h 14 m   δ −09° 12′)   m 2.61, from *Al Zubān al Shamāliyyah,* "The Northern Claw," and also called *Kiffa Borealis,* the "Northern Scale Tray." A late-type-B main-sequence star, β is nearly 145 times as luminous as the Sun; there is a mystery, however, concerning its brightness in former times, for although its present magnitude is larger than that of α, Erastosthenes, in the third century B.C., called β the brightest of *all* stars in Scorpius and Libra, which implies that it exceeded even Antares (α Scorpii), while Ptolemy, in the 2nd century A.D., said that β Librae was as bright as Antares.

Above the scale beam, forming the top of the old Euphratean "Altar," is a slightly arced row of faint stars; to their north-northeast, about 3 degrees west-northwest of β, lies the interesting variable *(δ) DELTA* (RA 14 h 58 m   δ −08° 19′), some 200 light years distant, an eclipsing binary of the Algol type. Its components, a bright class-A star and a fainter, unseen G-type component, are separated by only some 4½ million miles; they complete one revolution about their common center of mass in a mere 2⅓ days. During an eclipse of the bright component, δ drops in apparent brightness from about the fifth to the sixth magnitudes in six hours; following a secondary eclipse, in which the primary occults the fainter star, almost twice as much additional light is reflected from the uncovered surface of the fainter component after the primary's passage.

At the southern corner of the quadrilateral lies the quadruple system *(ι) IOTA,* Fl. #24 and 25 (RA 15 h 09 m   δ − 19° 36′)   m 4.66; #24, sometimes described as "pale yellow," is a class-A star designated "peculiar" because of unusually strong spectral lines of silicon. A very close double, its components lie a mere 0.1 second apart, so that smaller instruments reveal but a single object; however, at 58.6 seconds (4,600 AU) from the primary pair, we find faint (m9.7), reddish Fl. #25, which is also a close double, its components only about 150 astronomical units, or 1.9 seconds, apart. Fl. # 24 and 25, the widely separated major pairs, present no problem for the moderate-sized telescope.

About 6 degrees to the southwest of ι, and a magnitude brighter, is the star *(σ)* *SIGMA* Librae (RA 15 h 01 m   δ −25° 05′) m 3.31, which perhaps more suitably defines the quadrangle's southernmost point but thereby creates a less symmetrical geometric figure. On older charts it was often designated as γ Scorpii, marking the creature's extended southern claw; but it was already known as Fl. #20 Librae. Ending a three hundred-year-old dispute, Benjamin Gould, in the nineteenth century, assigned it to the realm of the Scale with its present Greek letter. The Chinese, however, knew σ as *Chin Chay,* the "Camp Carriage!" A red giant fifteen times the Sun's luminosity and 60 light years distant, this stellar carriage is approaching us at the leisurely speed of 2.5 miles per second.

Less than 5 degrees east of γ Librae (the star that marks the eastern corner of the quadrilateral), and over 3 degrees to the southwest of ξ Scorpii, lies a classic shell star, *#48 LIBRAE* (RA 15 h 55.4 m   δ −14° 08′) m 4.85, about 640 light years distant and 450 times as luminous as the Sun. A type of irregular variable known as a "Pleione" star (see p. 146), #48 is classed as B or A "peculiar" owing to the unusual appearance of its emission spectrum, which includes sharply defined bright hydrogen lines along with very wide and ill-defined helium lines. Like the other stars of its type, #48 Librae is apparently stable for long periods of time, often ten years or more, but then its spectrum reflects periods of strong activity, during which the lines undergo drastic changes in position, intensity, and shape. Its large rotational equatorial velocity (about 240 mi./sec.) has generated a thick expanding atmosphere, or shell, of turbulent gases about the star (see ζ Tauri, p. 140). During #48 Librae's most recent period of major activity, in 1967–68, the spectrum also indicated that it is surrounded by many such gaseous rings. The puzzling dynamic processes that have produced them invite further scientific observation of this mysterious star.

SCORPIUS (SCORPIO)

As winter approached and the days shortened, the ancients (some three thousand years ago) noticed an ominous sky figure, a great curved scorpion formed of many brilliant stars, which appeared at dawn in autumn and forecast darkness and winter cold. Even the mighty Orion feared this invincible creature, his undoing, in fact, on Earth, for, each morning, they saw him flee before its celestial rising! Regarded as causing plagues, storms, and war, it became the astrological "birthplace of Mars," Manilius' *Martis Sidus* ("House of Mars"). Known to us as Scorpius, it is among the oldest delineated constellations and may have originated pictorially with Euphratean astronomy about 5,000 B.C., as one of six original zodiacal signs. Its great age as a recognized constellation is indicated by the prebiblical Epic of Gilgamesh, which tells of deadly monsters, half scorpion and half man. Scorpius was the Akkadian *Girtab,* the "Seizer" or "Stinger," and its dreaded stellar realm was called the "Place Where One Bows Down." Known to the Israelites as *'Aḳrabh,* the "Scorpion," its stars, some say, became the serpentine emblem of the tribe of Dan. Shown on the Denderah zodiac of Egypt, the heavenly Scorpion was also recognized in Persia, Turkey, India, and possibly by the American Mayan culture. Greek mythology identifies Scorpius as Orion's slayer, as we have told, and also as the monster that frightened Apollo's horses during Phaëthon's disastrous attempt to drive the Sun chariot across the sky. According to Maori tradition, however, Scorpius' tail is the magic fishhook the folk hero Maui used in capturing a fish so great that it raised the island of New Zealand up from the ocean!

Lying between Libra, to the west, which formerly possessed its claws, and Sagittarius the Archer, to the east, who threatens it with drawn bow, the Scorpion projects only its head and claws above the ecliptic, immersing the rest of its body safely in the deep waters of the Milky Way! The Sun, on its apparent path across the sky, passes through this constellation's northern portion in but nine days, narrowly qualifying Scorpius for membership in the Zodiac Club.

ANTARES (α) *Alpha Scorpii*
(RA 16 h 26.4 m　δ −26° 19′)　m 1

The glowing red heart of the Scorpion rivals in color the planet Mars, and thus this great red star was the Greek *Anti-Ares,* i.e., "similar to Mars," which was contracted to its universal name, "Antares." Early Euphratean inscriptions also associate Antares with Mars, giving it such titles as *Dar Lugal,* the "King" (after the god of lightning) and *Kakkab* Bir, the "Vermilion Star." Some say, however, that the name "Antares" originated with Antar, a warrior-hero of Arabian poetry, and the orientalist Beigel equates it with the Arabic word *antar,* meaning "shone." Antares was also *Al Ḳalb,* "The Heart," an Arabian lunar station; *Jayestha,* or "oldest," and *Rohini,* "Ruddy," a Hindu lunar station; and the Sogdian *Maghan Sadwis,* which means "The Great Saffron-Colored One." This star represented the Egyptian goddess Selkit and was seen at sunrise through the portals of her temples at the autumnal equinox about 3700–3500 B.C.; whereas several early Greek temples are said to have been oriented toward its rising or setting at the vernal equinox.

Red Antares is appropriately compared to the great Betelgeuse, its counterpart in Orion. With a diameter of some 600 million miles (according to recent estimates), Antares is nearly as large as Betelgeuse, and the "Scorpion's Heart," with a color index of $+1.83$, is only slightly less red than the "Armpit of the Central One" $(+1.85)$. This redness, as we have explained, indicates the advanced evolutionary stage of these two supergiants; thus the peoples of India may have chosen a more appropriate title than they realized when they designated the lunar station marked by Antares as "Oldest"! Chronologically, however, Antares, like Betelgeuse, cannot count itself more than 100 million years old, a short life span for a typical star, the result of consuming its fuel too rapidly because of its great mass, about 10 or 15 $\mathcal{M} \odot$. At a currently estimated distance of about 520 light years, and with an absolute magnitude of -5.1, Antares has a slightly variable luminosity of about 9,000 $L \odot$. Like Betelgeuse, it pulsates irregu-

larly, indicated by slight periodic changes in its radial velocity. Antares has also been identified as a weak radio source.

A faint (m-6.5) bluish companion (which looks green because of color contrast with the primary) circles Antares at a separation of 3.0 seconds. This dwarflike star was discovered in 1819 during an occultation of Antares by the Moon; it may be seen with a moderate-sized instrument under favorable conditions, and its emergence at the dark limb of the Moon during an occulation is of especial interest. Apparently a shell star, as indicated by metallic emission lines in its spectrum, its luminosity of 50 $L \odot$ seems inadequate for its B4 spectral class.

GRAFFIAS (β) *Beta*
(RA 16 h 2.5 m δ −19° 40′)   m 2.55

At the base of the Scorpion's left claw, over 8 degrees northwest of Antares, lies β, its name, Graffias, apparently derived from the Greek for "crab," because of an early folk belief that scorpions are generated from crabs! The Arabians included it in their lunar mansion, *Iklīl al jabhah,* the "Crown of the Forehead," the derivation of β's occasional name "Iclil." With Antares, it was the Chinese *Ta Who,* the "Emperor," announcing his laws to a gathering of stellar sons and courtiers, an appropriate concept when we consider β's luminosity, of 2,700 Suns, along with the great luminosity of Antares! A spectral-class B V (main-sequence) star, β is also a wide double; the fifth-magnitude components, at a separation of 13.7 seconds, are suitable for small telescopes. A faint second companion lies about 0.5 second from the primary, visible only in great telescopes; to complicate matters further, the primary is also a spectroscopic binary, its components, almost in contact, requiring about a week to complete one revolution about their common center of mass. The fifth-magnitude secondary member of the wide double is also suspected of being an extremely close double; thus the entire system of Graffias, which is at least quadruple, may even consist of five component stars.

Other Major Stars

In the northwestern portion of Scorpius lie two important systems: the first, *JABBAH (ν) NU,* about 2 degrees east-northeast of β and often compared to the "double-double" ε Lyrae, consists of two widely separated binaries in an almost common orbit, a fifth-magnitude type-B2 subgiant and a sixth-magnitude type-A subgiant containing unusually strong spectral lines of silicon for its class. Each of these stars actually consists of a close pair; the sixth-magnitude binary, discovered by S. W. Burnham in 1894, is extremely difficult to resolve. Nu lies within the faint diffuse nebula IC 4592.

The second of these multiple systems, situated near the far northern border of Scorpius, about 8½ degrees north of β, is (ξ) *XI* (RA 16 h 01.6 m   δ −11° 14′)   m 4.17, apparently called "Graffias" in earlier times, until the American astronomer Burritt assigned this title to β Scorpii. The duplicity of ξ was discovered in 1782 by Sir William Herschel, who resolved its two golden supergiants, separated by about 18 astronomical units and completing a revolution about their common center of mass every 46.69 years. A distant orange dwarflike companion circles them in retrograde motion less than once every thousand years. At 4 minutes 43 seconds, somewhat closer than the dwarf, the binary Σ *1999,* itself consisting of two type-K dwarfs, is also a physical part of the Xi system, giving it a total of five gravitationally bound stars.

About 3 degrees south-southwest of β, defining the Scorpion's forehead, is **DSCHUBBA *(δ) DELTA,*** magnitude 2.5, also forming the Hindu "Row" or "Ridge" along with β and π. This same row of stars (δ with β to its north-northeast and π to its south) was seen by the Euphrateans as *Gis-gan-gu-sur,* the "Tree of the Garden of Light," perhaps an archaic antecedent to the "Tree of Life" in the Garden of Eden; it was also the Persian *Nūr,* "Bright," and the Coptic *Stephani,* or "Crown," as well as the Arabic *Iklil al Aḳrab,* the "Scorpion's Crown." Intrinsically a bright member of this "crown," δ Scorpii, at 590 light years, has the luminosity of 3,300 Suns; its spectral type is the same as β's. Between δ and α, only about 2 degrees from the latter, lies the third-magnitude white and blue double *(σ) SIGMA,* called *Al Niyat,* "The Outworks of the Heart," as though protecting Antares. Also a spectroscopic binary, σ is a Beta Canis Majoris variable (p. 134). About 1 degree to its south-southeast, look for the globular cluster M4, the nearest to the Solar System.

Continuing to the southeast on our journey along the Scorpion, we pass third-magnitude *(τ) TAU,* like β and δ a class B0 V star, 750 light years distant, its luminosity 3,300 $L \odot$. The next bright star on this route, lying about 9 degrees south-southeast of Antares, is *(ε) EPSILON,* a golden subgiant of the second magnitude, which marks the beginning of Scorpius' Tail; the exact spectral type is K2 III or IV, and it lies 65 light years distant with a luminosity of about 45 $L \odot$. For those as far north as the 50th parallel of latitude, ε rises only about 6 degrees above the southern horizon. RV Scorpii, a Cepheid and visual double, lies some one and a half degrees to the east-northeast.

Proceeding southward farther along the Tail, some 3½ degrees south and slightly to the east of ε, we find the wide naked-eye double *(μ) MU* Scorpii Nos. 1 and 2 (RA 16 h 48 m δ −37° 58') m 3.12 and 3.56, lying on an east-west line. According to a Polynesian legend, these stars, called *Piri-ere-ua,* the "Inseparable," were a little girl and her smaller brother, fleeing ill treatment by their parents, the brighter stars λ and υ, to the east. Appropriately, μ 1 and μ 2 have the same proper motion in their headlong flight, 0.03 second per annum, as they approach us at about 15 miles per second. Although indeed "inseparable" to the naked eye, these type-B bright giants, lying 520 light years distant, are actually separated from one another by 5 minutes 46 seconds, or .88 light year! Mu 1 is also an eclipsing binary of the Beta Lyrae type, with a small variability (.03 magnitude); the components, class-B giants of fourteen and nine solar masses, respectively, are separated by a scant 6 million miles and therefore take only about a day and a half to complete their period of revolution. For observers at 50 degrees latitude north, μ Scorpii barely rises a degree or two above the southern horizon.

Some 4½ degrees still farther along the Tail, lying to the south-southeast on a line with ε and μ, is (ζ) *ZETA 1 and 2* (RA 16 h 51 m δ −42° 17') m 3.7, a naked-eye orange and blue contrast pair, optically aligned to the east-west, as are the components of μ; the Zetas, however, are not gravitationally bound. In fact, they are separated from one another by a vast distance, with ζ 2, the eastern member, at the nominal distance of 155 light years from our Sun, and ζ 1, a bright class-B supergiant with emission features, probably some 5,700 light years distant. Thus our Sun and ζ 2 are much more "neighborly" than are the Zetas to each other! A probable member of bright cluster NGC 6231 (1/2° to its north), ζ1, apparently losing mass and very turbulent, may have a luminosity higher than 100,000 L ☉, with an absolute magnitude of −8, and its spectrum suggests that it radiates even more energy per second than the great Rigel! These interesting stars of ζ Scorpii, however, are not visible as far north as 50 degrees latitude.

As we journey around Scorpius' curved tail, we pass *(η) ETA,* a third-magnitude white supergiant, about 3 degrees east-southeast of ζ and only 50 light years distant from the Sun. Another 4 degrees farther along to the east, we find *SARGAS (θ) THETA,* magnitude 1.87, marking the base of Scorpius' upturned stinger; with other stars at the end of the Tail, it formed the Khorasmian asterism *Khachman,* the "Curved." The bright stars from ε to υ were also known as a "Tail," or *Wei,* in Chinese astronomy, but this figure was probably a part of the "Azure Dragon," one of four divisions of the Chinese zodiac. Theta is an F supergiant, about 650 light years distant, with a luminosity of some 5,800 L ☉. Above 45 degrees north latitude, one cannot view it.

Some 3½ degrees to the northeast of θ lies the optical double *(ι) IOTA 1 and 2.* Third-magnitude ι1 is an F bright supergiant, its luminosity about 60,000 L ☉, and quite distant at 3,400 light years. Iota 2, magnitude 4.8, a spectral class-A star, lies to its east; both Iotas have companions at separations of several thousand astronomical units, probably not gravitationally bound. For observers at 50 degrees north latitude, ι nestles tantalizingly just below the southern horizon; however, at favorable times, observers at 40 degrees north latitude will be able to view the entire constellation.

The luminous second-magnitude subgiant *(κ) KAPPA,* about 1.4 degrees northwest of ι, is of interest as a variable of the Beta Canis Majoris type. Another 2½ degrees to its northwest our journey is completed at the tip of the Scorpion's stinger, marked by two stars: *SHAULA (λ) LAMBDA* (RA 17 h 30 m   δ −37° 04′)   m 1.62, from the Arabic *Al Shaulah,* "The Sting," or from *Mushālah,* meaning "raised," a type-B subgiant classed as a Beta Canis Majoris star because of minute short-period variations, and *LESATH (υ) UPSILON,* magnitude 2.71, lying immediately to Lambda's west, its name derived from *Al Las'ah,* also "The Sting"; υ is a type-B3 Ib so-called "faint" supergiant, its luminosity, however, some 1,900 L ☉. Because Scorpius' stinger is indeed "raised," these stars are visible to observers as far north as 50 degrees north latitude.

Most of the stars of Scorpius, except β and ξ, lie within bright portions of the Milky Way, and a number of them, including Antares, β, δ, $μ^1$ and $μ^2$, π and σ, belong to the *Scorpio-Centaurus association* (or *cluster*), a great aggregation of mostly young B-type stars about 500 light years distant, covering some 90 degrees of the southern sky and having a common motion through space (averaging 15 mi./sec. away from the Sun); its counterparts in the north are the Ursa Major cluster and the Perseus cluster. The convergent point of the Scorpio-Centaurus association is near the star β Columbae. Shell stars and spectral variables are included in the group, which is about 20 million years old; Antares, although it has evolved very rapidly, fits into this age category.

The Dutch astronomer Jacobus C. Kapteyn (1851–1922) announced the existence of this great stellar association in 1914; ten years earlier, Kapteyn first discovered a dynamic pattern in the motions of the stars of our Galaxy called the "preferential motion of the stars," and he then introduced his "two star-stream" hypothesis (later discarded) that two streams of stars in the Galaxy, intermingling with each other, move in two distinct directions as given by two convergent points. His calculations of the proper motions of class-B southern stars, however, ultimately revealed the velocity relationship of the members of the Scorpio-Centaurus group.

A RARE NOVA

Near Scorpius' northeastern border, about 1.3 degrees northwest of χ (Chi) Ophiuchi, on May 20, 1863, Pogson recorded the appearance of a ninth-magnitude nova that faded to the twelfth magnitude within a week. Since then, *NOVA U SCORPII* has flared up three more times, the last of its sudden increases in luminosity occurring in 1979. Reaching the ninth magnitude at its brightest, U takes only a few weeks to fade to the fifteenth magnitude and fainter; between such outbursts it normally remains at the eighteenth magnitude, making visual observation difficult. Astronomers suspect that most novae are recurrent, but examples of novae with short periods, like U Scorpii, are rare. Two other members of this unusual group are novae RS Ophiuchi, T Pyxidis, T Coronae, and WZ Sagittae.

A Mira-Type Variable

To the north-northeast of ε and about 6½° southeast of Antares lies the long-period variable **RR SCORPII** (RA 16 h 53 m δ −30° 30'), ranging from the twelfth to the fifth magnitudes. Its late, type-M spectrum gives it a deep red color, rich in bright hydrogen lines. Its luminosity is about 250 $L \odot$, and at a distance of about 600 light years, R is approaching at about 28 miles per second.

Discovery of the First Galactic X-Ray Source

In the northern corridor of Scorpius, not far from its border with Ophiuchus and about 5½ degrees northeast of β Scorpii, lies **SCORPIUS X-1, or V818** (RA 16 h 17 m δ −15° 31') m 11–14, a powerful source of X rays, radio waves, and optical radiation. Its discovery, in 1962, made during the routine flight of an Aerobee rocket, was of great consequence to the new science of X-ray astronomy, because X-1 was the first galactic X-ray source to be identified with an optically visible stellar object. Observations made during the flight of a NASA rocket in 1966 pinpointed the exact position of the source, where a hot blue star resembling a nova remnant was soon discovered. Spectra obtained by Sandage for this star, a short-period irregular variable, show strong emission lines of hydrogen and helium, and highly excited nitrogen and carbon; dark interstellar lines of ionized calcium place the object at about 1,600 light years distant, with a visual luminosity of about one Sun.

Most of X-1's radiation lies in the X-ray end of the spectrum, which is a thousand times more luminous than the visible part. Astronomers have tried to explain this great disparity by proposing that Scorpius X-1 is a binary; Shklovsky, in 1967, suggested that the X-ray source is a neutron star embedded in a hot shell, and the companion a cool dwarf. Studies by Gottlieb, Wright, and Liller in 1975 revealed a small periodic change in magnitude; Cowley and Crampton, at Kitt Peak, found a matching variation in the radial velocity, suggesting two components, each less than 2 $\mathcal{M}\odot$, a normal star and an extremely dense one, perhaps a rapidly rotating neutron star that is acquiring material from its close companion in the form of an ejected gas stream superheated by its acceleration in the powerful gravitational field of the neutron star. The semimajor axis of their orbit (half of their maximum separation) is less than four hundred thousand miles. Current studies of the radio structure of Scorpius X-1, which mimics an extragalactic source, indicate that the dynamics of objects like X-1 are very similar to those of the peculiar X-ray binary SS433 (p. 272) and the processes are similar to those in active galactic nuclei.

Star Clusters

Only 1.3 degrees west of Antares is the bright globular cluster *M4, NGC 6121* (RA 16 h 20.6 m δ −26° 24′)   m 6.38, one of the nearest globulars to our Solar System; dark nebulae lying nearby make it difficult to estimate its distance; figures range anywhere from 1.75 to 3 kiloparsecs (5,705–9,780 LY). A small instrument shows only a granular disk, but if the telescope's objective is 4 inches or larger, it reveals a loosely structured, very large cluster, with curved rows of stars streaming from its center; forty of its members are RR Lyrae variables. Observers with very large telescopes may note *NGC 6144*, only one-half degree northwest of Antares, a very rich but more distant globular of the tenth magnitude.

Again proceeding from Antares, this time about 4 degrees to its northwest, we find the globular *M80, NGC 6093* (RA 16 h 14 m δ −22° 52′)   m 8, an extremely rich and condensed cluster about 36,000 light years distant and best viewed with very large telescopes. M80 contained *T SCORPII (NOVA 1860)*, one of the few novae ever sighted within a globular cluster; T flared up close to the center of the cluster, reaching the seventh apparent magnitude, which gave it an absolute magnitude of −8.5, or 200,000 $L\odot$ at the time of its discovery, on May 21, 1860, by A. Auwers at Berlin. A week later, Pogson sighted it independently and observed its rapid fading, which occurred within only eleven days. T Scorpii has never reappeared.

On the southern border of Ophiuchus, directly upon the Scorpio-Ophiuchus border, about 7 degrees southeast of Antares and slightly over 1 degree to the northeast of RR Scorpii, we find *M62, NGC 6266* (RA 16 h 58 m δ −30° 03′), which Shapley called "the most irregular globular cluster." About 26,000 light years distant and lying near but above the galactic hub, it was discovered by Messier in 1771; if viewed with small instruments, it even resembles one of his comets!

About one-half degree north of ζ, at the bend of Scorpius' tail and suitable for southern observers only, is *NGC 6231* (RA 16 h 50.7 m δ −41° 43′), a galactic cluster so bright that it is visible to the naked eye under clear atmospheric conditions; it provides an exceptional sight for the small telescope. Most of its members are O and B supergiants, including two intensely luminous stars, one of them comparable to Rigel; it also contains two Wolf-Rayet stars and several P Cygni variable shell stars, as well as ζ

l itself. About 5,700 light years distant, NGC 6231 is apparently the nucleus of a much larger group of O and B stars, called the *I SCORPII ASSOCIATION,* which defines part of a distant spiral arm lying toward the galactic center; the richest portion of this association, just north of NGC 6231 and often shown on charts as "H 12," is connected to the cluster by a visible chain of stars. One degree to the north of H 12 is the small, compact cluster NGC 6242; between it and H 12, immediately northwest of the latter, is faint nebula IC 4628. The entire I Scorpii association is surrounded by a ringlike formation, some 300 light years in diameter, understood to be a great mass of ionized hydrogen like the Orion nebula. Such giant H-II zones (shells of ionized hydrogen around stars, as previously defined), the spawning grounds of new stars, are also found in external spiral and irregular galaxies.

A line drawn through the stars λ and υ, at the stinger's tip, points to the galactic cluster M7, some 5 degrees to the northeast of λ, with M6 lying 3½ degrees to the northwest of M7. These clusters apparently were the Arabian *Tāli' al Shaulah,* or "That Which Follows the Sting," in ancient times. *M6, NGC 6405* (RA 17 h 36.8 m δ −32° 11'), a large irregular open cluster that was described by the French astronomer Flammarion as ". . . three starry avenues leading to a large square," counts most of its members as class-B main-sequence stars, except for BM Scorpii, a K-type yellow giant of the sixth magnitude. Suitable for the 6-inch telescope, this cluster is computed to be only 100 million years old; estimates of its distance vary from about 1,300 to 2,000 light years.

M7, NGC 6475 (RA 17 h 50.7 m δ −34° 48') is an exceptionally large and bright naked-eye group of moderately concentrated stars, only 800 light years distant and spectacular in binoculars but lying a few degrees from the southern horizon for those of us at 50 degrees latitude north. H-R diagrams demonstrate that the bluer stars of various open clusters are somewhat off the main sequence (as opposed to yellow and red stars, which are still on it); this tendency may be observed in all those members of M7, brighter than apparent magnitude 7.5, most of them B-type stars that have begun to leave the main sequence. Based upon the evolution of its stars as shown on color-magnitude diagrams, M7 is about 260 million years old, more than twice the age of M6.

CORONA AUSTRALIS

To the immediate east of Scorpius' tail and tucked in between the southern portions of Sagittarius is *CORONA AUSTRALIS,* the Southern Crown, a faint replica of its famed counterpart in the north. In early days it was sometimes associated with Sagittarius as *Corona Sagittarii* and as the "Bunch of Arrows" held in the Archer's hand. It was the fifth-century *Parvum Coelum,* meaning "Little Sky" or "Canopy," and Lalande's *Sertum Australe,* or "Southern Garland;" in early Arabia it was *Al Ḳubbah,* or "The Tortoise," but also a "Tent" and an "Ostrich's Nest" for birds now in neighboring constellations. Its biblical associations are Caesius' "Crown of Eternal Life" and Julius Schiller's "Diadem of Solomon."

The brightest of its stars, *(α) ALPHA* Coronae Australis, the Latin *Alfecca Meridiana,* second from the northern end of the row, and *(β) BETA,* to its south, are only of magnitude 4.1. Gamma, to Alpha's northwest, is a binary only 39.1 light years distant; its fifth-magnitude components are 2.7 seconds apart, suitable for instruments of moderate size. (For observers 50° north, α and γ just peep above the southern horizon.) (ε) *EPSILON*, some 2½ degrees to Gamma's west, is a class-F dwarf eclipsing binary (magnitude range 4.7–4.9) of the W Ursae Majoris type.

Southern viewers may look for the optical double, η¹ and η², at the southern end of the Crown. To their northeast lies (ζ) *ZETA*, m 4.8, a type A main sequence star about 221.7 LY distant, whose hydrogen spectrum shows broad blurred lines, indicating that ζ is in rapid rotation.

Diffuse Nebulae

The double nebula *NGC 6726 and 6727* (RA 18 h 58 m   δ −36° 57′), lying in the region of γ, contains *TY Coronae Australis*, an erratic variable ranging from magnitude 8.8 to 12.5, embedded in the northeast portion of the nebula. A little to the southeast is *NGC 6729*, resembling a tiny comet and containing *R Coronae Australis*, an irregular class-F5 variable with an erratic pattern of light changes (magnitude range 9.7 to 12) which affect the appearance of the nebula (see Hubble's Variable Nebula, p. 128). In the same field lies *S*, a class-G dwarf with an emission-line spectrum similar to T Tauri's.

SAGITTARIUS

In the Archer's figure, man and beast are combined. Half horse and half man, he belongs to the mythological race of Centaurs, descendants of the miscreant Ixion, who dared to covet Hera, the Queen of Heaven, and was seduced by a phantom cloud disguised as Hera, which Zeus sent to entrap him. The offspring born of this deception was Kentauros, outlawed by god and man, who bred with the mares of Thessaly, thus producing the Centaurs. Most of these half-human creatures became the friends and advisers of mankind, but there were exceptions; one of these, fierce-looking Sagittarius, was transferred to the heavens some 15 to 30 degrees south of the celestial equator, where he perpetually aims his bow and arrow westward at Scorpius. Sagittarius was often confused by Greek and Roman authors with the constellation Centaurus, lying farther to the southwest, which represents the Centaur Chiron, Aesculapius' teacher. Mythology tells us that the mild-mannered Chiron, beloved by Diana and Apollo, invented the constellations and placed Sagittarius in the sky to guide the Argonauts in their expedition to Colchis. Historically, however, the present Sagittarius as a constellation apparently originated with the Euphrateans; on cuneiform inscriptions, this stellar figure personifies the Archer Nergal, a god of war; because of its zodiacal position, it was also the "Smiting Sun Face" and "Dayspring." It appeared on the zodiac of India three thousand years ago as a "Horse's Head" or "Horseman," whereas the human part of the Archer became a fan of lions' tails belonging to the wife of an Indian ruler!

Earlier Greek titles were translated by the Romans into the present name, but the form "Sagittary" was widely used. Other titles were *Semivir,* the "Half Man," and

Cornipedes, or "Horn-Footed," but numerous ancient peoples saw this constellation simply as a "Bow." In fact, its various parts have been pictured in differing ways: the central figure is our modern Teapot, more recognizable, we believe, than the ancient Centaur. In ancient Arabia, however, the stars constituting the Teapot's spout were a group of "Going Ostriches," and those of our handle, the "Returning Ostriches," all of them visiting the Milky Way; their Keeper was λ (at the top of the lid)! The handle, however, is also our inverted "Milk Dipper," which was the nineteenth Chinese lunar station called *Tew,* a "Ladle" or "Measure." Our Teapot's spout, their eighteenth station, was seen as a "Sieve." The Chinese complained:

> In the south is the Sieve
> Idly showing its mouth . . .
> But it is of no use to sift;
>
> In the north is the Ladle
> Raising its handle to the west . . .
> But it lades out no liquor!

Major Stars

RUKBAT (α) ALPHA (RA 19 h 24 m δ −40° 37′), from *Rukbat al Rāmī,* the "Archer's Knee," only of fourth magnitude, is much fainter than the stars of the Teapot, lying far southeast of the Teapot in a triangle of stars just east of Corona Australis. Nearby, to α's south, is *ARKAB (β) BETA 1* and *2* (RA 19 h 22 m δ −44° 27′), the Arabic "Tendon" (tying calf to heel), also called the two *Surad* (desert birds), a very wide naked-eye double; β 1, the northern companion, is also a magnitude-4.3 binary, its components a common-proper-motion pair of classes B and A, their separation 28.4 seconds of arc. Iota, to the east, completes Sagittarius' southern triangle, or foot, all of it hidden below the horizon for those of us at 50 degrees north latitude.

AL NASL (γ) GAMMA (RA 18 h 2.6 m δ −30° 26′) m 2.97, "The Point," marks the head of the Arrow, and also the tip of the Teapot's spout. On the Borgian globe it was *Al Wazl,* "The Junction," where arrow, bow, and the Archer's hand meet. A class-K giant, γ lies about 125 light years distant, with 85 *L* ☉; its somewhat variable radial velocity (13 mi./sec. in recession) hints at the presence of a very close invisible companion. Seen against the background of the Milky Way, γ lies just south of a bright region indicating the direction of the galactic core.

Where the spout touches the lid, we find *KAUS MERIDIONALIS (δ) DELTA,* magnitude 2.71, Arabic and Latin for "Middle (of the) Bow," also called *MEDIA.* With γ and ε, its neighbors of the spout, it was the Akkadian *Sin-nun-tu,* a "Swallow." Media is an orange giant, 85 light years distant and with a luminosity of 60 *L* ☉; a fourteenth-magnitude bluish companion at 25.8 seconds is probably not gravitationally bound. Some 4½ degrees south of δ is *KAUS AUSTRALIS (ε) EPSILON* (RA 18 h 20.9 m δ −34° 25′) m 1.81, the "Southern (part of the) Bow" and the brightest star in Sagittarius; like those of Draco, the Greek letters for the stars of the Archer bear little relationship to their magnitude order! A late-B-type subgiant, ε lies 125 light years distant, with

250 L ⊙; at 3.31 minutes to the north-northwest, a companion star of the same spectral type is an easy object for binoculars.

About 2.6 degrees south-southwest of ε is *(η) ETA,* a third-magnitude red giant some 90 light years distant and forty times the Sun's luminosity; a ninth-magnitude white companion has the same proper motion at 3.6 seconds of arc, a separation corresponding to about 100 astronomical units. Moving to the eastern corner of the Teapot, at its lower juncture with the handle, we find *ASCELLA (ζ) ZETA* (RA 18 h 59 m δ −29° 57′) m 2.61, from *Axillis,* the "Armpit of the Archer." This is also the southernmost star of the "Milk Dipper." Lying about 140 light years distant from us, ζ is a binary consisting of two class-A2 components, one a giant of MKK classification III and the other apparently a main-sequence star; their mean separation, only 0.4 second of arc, or about 23 astronomical units, is presently decreasing as they revolve around their common center of mass every twenty-one years. About 2½ degrees to the north-northeast lies *(τ) TAU,* a golden giant, easternmost of the "Milk Dipper" group.

Some 2.6 degrees northwest of τ and marking the northern corner of the inverted "Dipper" is *NUNKI (σ) SIGMA* (RA 18 h 52 m δ −26° 22′) m 2.12, at the juncture where Sagittarius' right hand grasps the end of the arrow and the bowstring. Nunki first appeared in the Euphratean *Tablet of the Thirty Stars,* a Babylonian astronomical work, as "The Star of the Proclamation of the Sea," which refers to a nearby domain of the heavens traditionally regarded as "watery" (see Chapter Eleven, p. 293). R. H. Allen cites the reference as another indication of the Euphratean origin of Greek astronomy. A blue-white main-sequence star 300 light years distant from our Sun, Nunki shines with eleven hundred times the Sun's luminosity!

Where the upper part of the handle joins the Teapot, just 2 degrees west-southwest from Nunki, lies *(φ) PHI,* a class-B8 giant 590 light-years distant and as luminous as some sixteen hundred Suns. Our next star, *KAUS BOREALIS (λ) LAMBDA* (RA 18 h 24.9 m δ −25° 27′) m 2.80, meaning the "Northern Bow," is auspiciously placed at the top of the lid, where it also marks the Milk Dipper's handle. A yellow-orange giant about 70 light years distant from us, λ lies on the eastern branch of the Milky Way, so that numerous images of stars in the galactic core fill its background. About 5 degrees to Lambda's northwest is the naked-eye double *(μ) MU 1 and MU 2* (RA 18 h, 12 m δ 21° 03′) m 4 and 6, marking the bow's upper extension. Lambda and Mu may have been associated with the Akkadian goddess Ishtar. MU 1, the primary, is an eclipsing binary and a luminous supergiant, class B8p, with variations in magnitude of a tenth of a degree. Several faint optical companions lie in its field.

Moving north-eastward from the Teapot, some 5 degrees north and slightly to the east of Nunki, we find an interesting naked-eye double, *(ξ) XI 1 and XI 2,* magnitudes 3.5 and 6, its white and orange components not gravitationally bound. To their south-southwest lies fifth-magnitude *(ν) NU 1 and NU 2* (RA 18 h 10.8 m δ −21° 05′), the Arabic *'Ain al Rāmī,* or "Archer's Eye," also a naked-eye double. More than sixteen hundred years before Sir William Herschel coined the word "binary," Ptolemy, in his *Syntaxis,* designated ν's reddish components as a "nebulous double star," an early use of this term. Although ν1 and 2 are not physically bound, ν 1 is, in fact, a close binary.

About 2½ degrees to the east of ξ is third-magnitude *(π) PI* (RA 19 h 10 m δ −21° 01′), sometimes called *Al Nā' ir,* "The Bright One," marking the back of the Archer's head; the viewer needs access to an observatory to "split" this extremely close triple 250 light years distant.

A Lyrid Variable

In the far-northeast region of Sagittarius, about 6 degrees north-northeast of Pi (and directly north of the optical double Rho), lies *(v) UPSILON, #46 Sagittarii* (RA 19 h 18.9 m δ −16° 03′) magnitude 4.3–4.4, an eclipsing binary of the Beta Lyrae type, which displays the two characteristic unequal minima but has an unusually long period for such a star: about 138 days. The components are a normal type-B or early-A star and a very turbulent class-F shell star, possibly a giant similar to P Cygni (p. 281), whose emission lines are lost through broadening. The spectrum of the visible star shows unusually strong metallic lines for a B-type star, and its temperature is computed at about 13,000 degrees K. At about 8,000 light years, the combined luminosity of Upsilon may exceed 80,000 $L \odot$! If so, it is one of the most luminous binaries in the Galaxy.

The exchange of mass between the two components of a binary system is an important phenomenon, and, as we have seen, has a very significant bearing on their evolution. The peculiar characteristics of Upsilon Sagittarii may derive from a close exchange of material between its components, altering their pattern of evolution. Such close binaries are also of great interest because of the many X-ray sources associated with them; the exchange of mass in these systems probably plays an important role in the generation of the X-rays they emit (see Scorpius X-1, p. 252).

About 4½ degrees southeast of Zeta lies the irregular variable *RY SAGITTARII* (RA 19 h 13.3 m δ −33° 37′), its magnitude ranging from 7.2 to 14. RY is unusual because it remains at peak brightness for a year or two and then fades very suddenly to minimum luminosity, seemingly the reverse pattern of a normal eruptive star, which flares up suddenly and fades slowly. Its peculiar spectrum indicates a deficiency of hydrogen and an overabundance of carbon, implying a late stage in its evolution, in which it may already have lost its hydrogen-rich envelope and is generating its energy through the C-N cycle and the triple-alpha reaction. The star's erratic fading is believed to be caused by the presence of an expanding opaque shell, much as in the case of R Coronae Borealis, the "topsy-turvy" star, which RY Sagittarii seems to resemble.

Famous Nebulae

About 4.7 degrees west and slightly north of Lambda is the *LAGOON NEBULA, M8, NGC 6523* (RA 18 h 1.6 m δ −24° 20′) m 5, lying near the galactic equator† and therefore in a field rich with stars of the Milky Way. With dimensions of eighty by forty minutes (1.3 × 0.7°), M8 is visible to the naked eye; moderate magnification reveals an irregular, cloudlike form, divided by the dark obscuration channel that gives it its name (fig. 110). Two luminous stars occupying the western half of the nebula contribute strongly to its illumination, and in the eastern half lies the open star cluster NGC 6530, which probably was formed only a few million years ago; its members, like those of the

---

† *(Not to be confused with the celestial equator!)*

Trapezium, in Orion, are young, class-B subgiants and pre-main-sequence T-Tauri stars. Scattered across the glowing clouds of M8 are numerous tiny dark globules similar to those of the Rosette Nebula, in Monoceros, nebular clouds 7,000 to 10,000 astronomical units in diameter, believed to be very young protostars that have not yet contracted to normal stellar size or begun to radiate light.

Only 1^1/2 degrees north-northwest of the Lagoon Nebula is another celestial wonder, the small bright **TRIFID NEBULA, M20, NGC 6514**, an irregularly shaped glowing structure of ionized gas; its southern portion, trisected by three dark obscuration rifts or channels of nonilluminated gas and dust, looks like a "pansy face" in long time-exposure photographs (fig. 111). With dimensions of about twenty by fifteen minutes, or about one-fifth those of the Orion nebula, the Trifid lies some 5,216 light-years distant. Centered within the "trifid" design is the triple star **HN 40,** a flashing blue giant, which discloses six components to very large telescopes; a small instrument, however, is advantageous for viewing the entire nebula, an impressive sight. Open cluster **M21, NGC 6531,** of magnitude 7, lying only 0.7 degree northeast of the Trifid nebula, contains about fifty stars in a compact group covering a twelve-minute field, the brightest of them young, hot class-B giants.

An Outstanding Globular of the Southern Skies

Roughly 8 degrees to the east of the Lagoon nebula is the great globular cluster **M22, NGC 6656** (RA 18 h 33 m   δ −23° 58′)   m 6, probably the third-largest globular known, with an apparent diameter of about seventeen minutes, or half that of the Moon. It may contain five hundred thousand stars, among them thirty-two variables, including many RR Lyrae stars. Its distance, often given as 3.1 kiloparsecs, or 10,106 light-years, is currently estimated as some 9,600 light years, with interstellar absorption by gas and dust taken into account.

110. M8 (NGC 6523), the Lagoon Nebula in Sagittarius, visible to the naked eye, a very large, bright nebula crossed by a dark obscuration channel. In the eastern half (left) may be seen NGC 6530, a very young open star cluster. Photographed in red light with the 200-inch Palomar telescope.

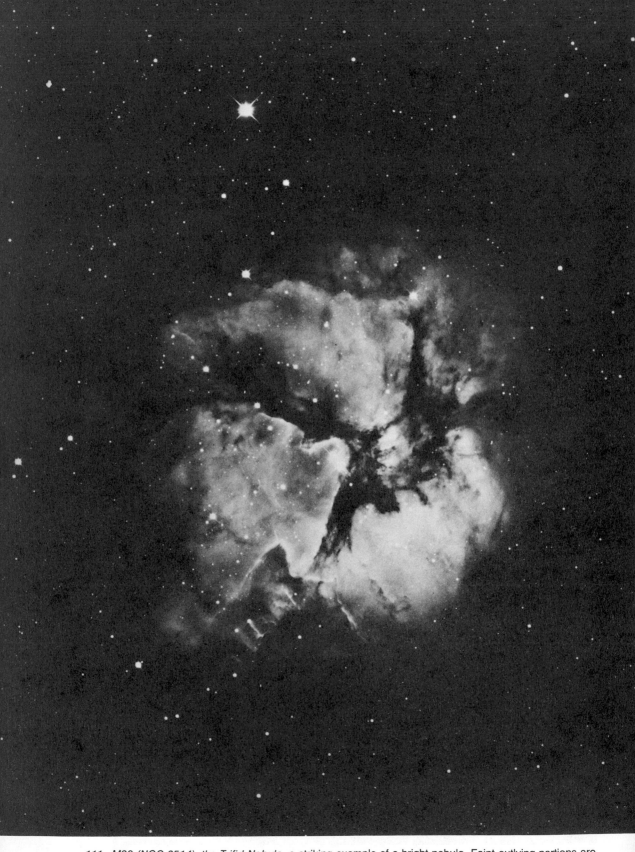

111. *M20 (NGC 6514), the Trifid Nebula,* a striking example of a bright nebula. Faint outlying portions are visible which greatly increase its apparent dimensions. The distinctive dark gaseous channels of the nebula are accentuated in this dramatic photograph, taken with the Shane 120-inch reflector. (*Lick Observatory*)

Moving our instruments back toward the west, some 4.7° northwest of the double star Mu and about 4 degrees north-northwest of the Trifid, we find the open cluster *M23, NGC 6494* (RA 17 h 54 m   δ − 19° 01′)   m 7, which contains about one hundred stars in a thirty-minute field, suitable for the small telescope.

As we follow the galactic equator, which cuts across the eighteenth circle of right ascension in a line moving toward the northeast, we encounter a multitude of deep-sky objects. Very important among these is *M24*, the **SMALL SAGITTARIUS STAR CLOUD**, with open cluster *NGC 6603* (RA 18 h 15.5 m   δ − 18° 27′), about 6 degrees northeast of the Trifid nebula and 3 degrees north-northeast of Mu; millions of Milky Way stars form the brilliant cloud, marked toward its northern borders by two dark nebulae, like a pair of black eyes (fig. 112). The tightly compressed small cluster, four minutes in diameter, 16,000 light years distant, and containing about fifty stars, lies within the northern portion of the cloud. About 3½ degrees east-southeast of M24 is the open cluster *M25, IC 4725*, containing fifty loosely scattered stars of magnitudes 6 to 10; the most luminous are of spectral class B4, but two solar-type members are also present, as well as the classical population-I Cepheid variable *U SAGITTARII*, which ranges from the sixth to the seventh magnitudes in about a week, changing from type F5 to G1 as it fades to minimum luminosity. It also has a faint companion at 1 minute 6.5 seconds, itself a close double. All classical Cepheids play an important role in the analysis of the Milky Way; most of them, however, lie at great distances in the plane of the Galaxy, out near the spiral arms, and are often found in globular clusters. U Sagittarii's presence in an open cluster is therefore unusual, an indication of the cluster's great age; at the relatively small distance of 2,000 light years, U is a useful instrument for accurately determining the Cepheid period-luminosity relation.

Near the northern border, about 2 degrees north-northeast of the small star cloud, and 2½ degrees south of spectacular nebula M16, in Serpens, is the prominent bright nebula *M17*, the **OMEGA**, or **HORSESHOE, NEBULA**, in Sagittarius, also called the **SWAN NEBULA, NGC 6618** (RA 18 h 18 m   δ − 16° 12′), some 5,700 light years distant. Messier's description, "A train of light without stars, 5′ or 6′ in extent, in the shape of a spindle . . ." is apt for the possessor of a small instrument. In large telescopes, its glowing form fills an area about twenty-six by twenty minutes, and the faint outer portions extend much farther, almost rivaling the Great Nebula, of Orion. Observers differ as to the "Swan's" appearance, some seeing a figure 2 or a check mark, whereas Sir William Herschel was reminded of the Greek letter (Ω) Omega (not unlike a horseshoe), originating its popular name; to the present authors, the "neck" seems rather short for a swan, and long-exposure photographs suggest a cocker spaniel, its forelegs tucked against its chest as it swims through the dark, starry waters (fig. 113). A degree south of the Swan (or Spaniel!) is *M18, NGC 6613*, a coarse open cluster of twelve stars, with an apparent diameter of seven feet.

Far out to the northeast lies *NGC 6822, BARNARD'S GALAXY* (RA 19 h 42 m   δ − 14° 53′), at 1.7 million light years one of the closest dwarf galaxies of the local group and suitable for medium-sized telescopes; in its irregular form lie hundreds of thousands of stars, among them luminous blue giants and Cepheids, the total stellar luminosity approximately 50 million $L \odot$. Barnard's Galaxy is of historical importance as one of three galaxies in which Hubble discovered Cepheid variables, thus establishing the stellar content of such so-called "nebulae."

112. *M24,* the small Sagittarius star cloud, with open cluster NGC 6603. The cloud contains millions of Milky Way stars. To its southeast, open star cluster M25 is seen glowing against a vast dark wedge-shaped region composed of underilluminated gas and dust. Note the dark nebulae in the northern part of the star cloud. Photographed with the 20-inch Astrographic telescope. (*Lick Observatory*)

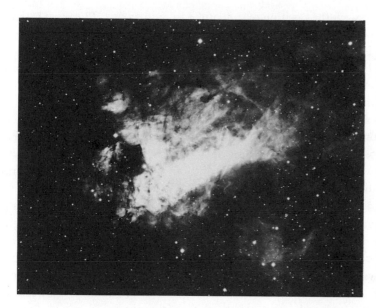

113. *M17 (NGC 6618),*
the Omega Nebula in
Sagittarius (Swan nebula),
very large and bright;
composed of thick,
gaseous clouds that
reflect the light of nearby
stars. In this print, south is
to the top; a dark
obscuration cloud
immediately to the west
(left) of our "spaniel"
creates the illusion of
"muzzle" and "neck"
(inverted, the "doggy"
effect is similar).
Photographed in red light
with the 200-inch
telescope. (*Hale*
Observatories)

Returning to the central figure, the "Teapot," we mention globular cluster **M69,** **NGC 6637,** about 2 degrees north-northeast of ε and 36,000 light years distant from us, notable because it is one of the most metal-rich clusters known; however, it can be effectively viewed only in large telescopes.

The Heart of the Milky Way

Like a burst of steam from our fanciful Teapot, just north of γ (the star at the tip of the spout), glows the **GREAT SAGITTARIUS STAR CLOUD.** A spectacular object of the Milky Way, it consists of many millions of stars lying near the central portion, or hub, of the Galaxy, some 30,000 light years from our Sun. Star clouds extend almost unbrokenly from Cassiopeia to the southern part of Cygnus, where dust clouds of the Great Rift begin to divide the Milky Way into its eastern and western branches; the eastern branch continues on through Aquila to Sagittarius, where the star clouds are densest. The Great Rift, comparable to the obscuration channels or equatorial bands of edge-on spiral galaxies, cuts across these luminous stellar masses a few degrees west of dark nebula **BARNARD 86** (RA 18 h 0 m   δ −27° 50′), situated within the great Sagittarius star cloud. Far behind the obscuring interstellar gas and dust of the Great Rift lies the unseen nucleus, or galactic core; astronomers, by studying the infrared light, radio waves, and X rays it emits, have pinpointed its location (RA 17 h 42 m   δ −28° 59′), about 4 degrees west-northwest of γ and 1.2 degrees south-southwest of the yellow fourth-magnitude Cepheid X Sagittarii.

Karl Jansky, in 1932, first detected extraterrestrial microwave radiation (later found to originate in the Milky Way), and his work was continued by radio astronomer Grote Reber, who discovered more intense radio signals coming from three specific regions, including that of Sagittarius. In 1951, infrared radiation was detected that was later revealed as coming from an elongated area lying along the galactic equator and the

Great Rift and coinciding with the position of the source in Sagittarius. The most intense region of this radiation was named **SAGITTARIUS A** and is now regarded as identical with the galactic core. It emits radio waves originating in interstellar (HI) regions of neutral hydrogen (p. 113); this emission is called "21-centimeter radiation," because the hydrogen electrons in such regions emit photons with wavelengths of 21 centimeters as they jump down to the lower energy level, the ground state (owing to a change in direction of their spin axes caused by atomic collisions). From the intensity of this 21-centimeter line, astronomers have determined the temperature and density of the HI region in the galactic center; and by analyzing it at various galactic longitudes they have obtained important information about the rotation of the Galaxy and its spiral structure, and the distribution of hydrogen in the arms.

Sagittarius A is also a strong source of synchroton radiation (p. 141), as well as a source of highly energetic gamma rays and of thermal radiation (caused when high-speed electrons are accelerated as they pass near protons). Encompassing an area no more than 50 light years in diameter with a density of some 1,700 $\mathcal{M}\odot$ per cubic parsec, it is crammed with stars, each separated by only a few hundred astronomical units from its neighbors. In addition to the enormous amount of radiation emitted by these closely packed stars, the core has two rotating arms of gas at opposite sides of the center, moving away from it at differing speeds, and an inner molecular ring with a mass of about 100,000 $\mathcal{M}\odot$, surrounding the central disk at a distance of about 1,000 light years and expanding at 100 kilometers per second. The two gaseous streams and the inner ring indicate that an explosion may have occurred some 10 million years ago in the galactic core. Such occurrences are common in Seyfert galaxies and in much more eruptive star complexes like M82 (p. 50); increased knowledge concerning the nucleus of our own "peaceful" Galaxy may suggest that it has a much closer bond with those violent star systems than hitherto imagined!

SCUTUM

North of Sagittarius, just east of Serpens Cauda, and immediately to the southwest of Aquila lies the small constellation Scutum (*Scutum Sobiescianum*, or "Sobieski's Shield"), a modern figure formed by Hevelius in 1690, out of seven fourth-magnitude stars. Scutum represents the coat of arms of heroic John (Jan) Sobieski III of Poland, who distinguished himself in resisting invaders of his own land as well as in stopping the Turkish advance on Vienna in 1683. Known by the Chinese as *Tien Pien*, the "Heavenly Casque," Scutum also resembles a dipper.

The Shield (or dipper) has a northeast-southwest orientation. In the southwestern corner of the Shield (or dipper), we find *(δ) DELTA Scuti* (RA 18 h 39 m   δ −09° 06′)   m 4.74, the standard for certain short-period pulsating variables with exceedingly small luminosity ranges; in contrast to Cepheids, their maximum luminosities do not coincide with their greatest velocities of expansion, but instead follow no definite rule. Delta Scuti stars are population-I giants of class A or F, and their spectral types remain constant throughout their light cycles, which range from about an hour to four hours; δ Scuti itself has the longest period, 4.65 hours.

114. Star cloud in the region of Sagittarius, the richest part of the Milky Way, containing millions of stars lying near the hub of our galaxy. Scattered across the cloud are several dark nebulae; strands of interstellar gas and dust cover the bright regions seen in the lower half of this photograph taken in red light with the 48-inch Schmidt telescope. (Hale Observatories)

If you have a moderate-sized telescope, look for the galactic cluster **M26, NGC 6694,** about 0.8 degree east-southeast of δ. It has a rather distinctive appearance, because of its resemblance to a tiny jeweled horseshoe (not to be confused with the more famous Horseshoe nebula, to its south, in Sagittarius).

In 1795, the British observer E. Pigott (who first noticed the variations in R Coronae) discovered an unusual star in the northern part of the Shield, 1°south of β Scuti: the semiregular variable **R SCUTI** (RA 18 h 44.8 m   δ −05° 46′), which soon became a favorite of star watchers who enjoy the unexpected. Its magnitude ranging from about 4.7 to 7.8, with no set pattern except for a pronounced fading every fourth or fifth minimum, R confounded these earlier observers with its erratic behavior. In the present century, after photography revolutionized the study of variable stars, R's light changes were recorded by major observatories; it is now classified as the standard for certain variables that are among the largest and most massive known, a subclass of the RV Tauri stars that lack the superimposed long light cycles of the RV Tauri stars.

A supergiant with a luminosity about eight thousand times the Sun's, R's approximate cycle is 100–146 days, with a spectral range from class G5 at maximum to class K

115. M11 (NGC 6705), an open star cluster in Scutum, lying in a Milky Way star cloud; photographed by Ritchey with the 40-inch refractor. (*Yerkes Observatory photograph, University of Chicago*)

or even M at minimum. The explanation of its complicated light pattern is that the rate of expansion and contraction differs for each gaseous layer within this great star. Estimated at anywhere from 2,500 to 3,000 light years distant, R Scuti is departing from us about 26½ miles per second.

About 1 degree to the southeast of R Scuti lies the rich galactic cluster **M11, NGC 6705** (RA 18 h 48 m　δ −06° 20′), which Sir John Herschel called a "glorious object" and the English astronomer W. H. Smyth likened to a "flight of wild ducks." Fan-shaped, and glittering with the light of at least four hundred Suns as well as numerous unseen or peripheral members, M11 appears only as a misty patch in binoculars, but even a small telescope reveals many star points; the greater the magnification, however, the more exciting the results (fig. 115). Middle-aged, M11 was formed about 500 million years ago; some 5,500 light years distant, it has a mass of about 2,900 $\mathcal{M}\odot$, and its luminosity is about 10,000 $L\odot$. The diameter has been calculated at about 15 light years, and with approximately eighty-three stars per cubic parsec, its core is so dense that individual stars may lie no more than a light-year apart.

TEN

Reserved for Great Birds

The Eagle and the Swan were sacred birds of such antiquity that it is impossible to describe their origins or to define their mythological identities with exactness: along with Lyra, they may represent the three Stymphalian birds, man-eating creatures with beaks, wings, and claws of iron, that Hercules slew in his sixth labor, at Lake Stymphalus, in the Arcadian woods. The nineteenth-century British ornithologist D. W. Thomson suggested that Aquila, rising in the east immediately after Cygnus but setting before it, symbolized the Eagle attacking the Swan but defeated by it; if so, this Greek fable stems from an astronomical fact: the Swan's more northerly location (near the circumpolar zone), causes its stars to remain above the horizon longer than the more southerly stars of Aquila the Eagle, for the northern latitude.

AQUILA (with Antinoüs)

Our saga of Hercules' adventures left off with his marriage in heaven to Hebe, Jupiter and Juno's daughter, but the story has a sequel. Hebe became the goddess of youth, and her duty was to appear before the gods as cup bearer; for this ceremonial she wore a flowered garland. After her marriage to Hercules (or, according to another version, after she fell while serving at a feast, causing embarrassment to Jupiter), Hebe gave up this office. Seeking her successor, the gods noticed Ganymede, a young Trojan prince of exceptional virtue and physical beauty; while among his hunting companions on Mount Ida, he was seized by Jupiter's faithful minister the eagle (or by the god himself, as some tell, disguised as an eagle) and borne aloft to the heavens; there Ganymede replaced Hebe in attendance upon Jupiter, pouring red nectar from a golden bowl. Jupiter declared the eagle to be prince of birds and carrier of his thunder; known by the Romans, however, as *Ganymedes Raptrix* and *Servans Antinoüm,* on star charts it is the constellation Aquila, flying eastward toward Delphinus and crossing the Milky Way. Although no longer officially recognized, the asterism Antinoüs, which represents Ganymede, is correctly depicted just below the Eagle, being carried unharmed in its talons.

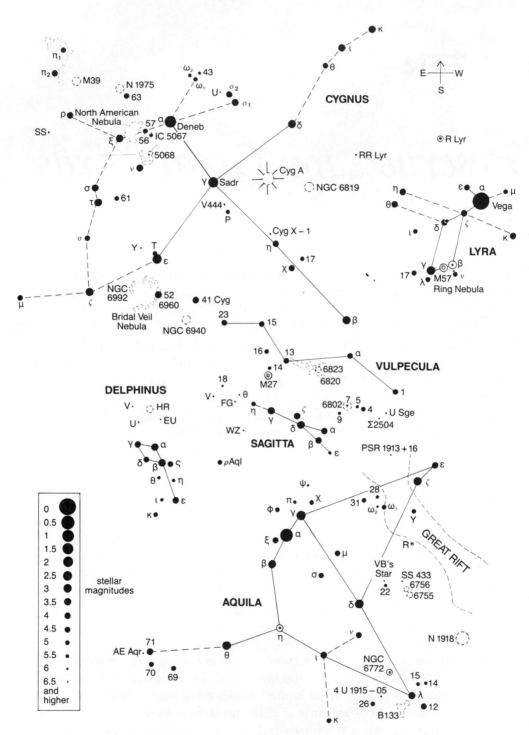

116. *The Eagle and the Swan*, with vicinity.

As early as 1200 B.C., an eagle, supposedly Aquila, was carved on a Euphratean stone of the heavens. Named *Dios Ornos,* or "Bird of Zeus," by the Greeks, and, similarly, *Jovis Ales* by Roman poets, Aquila was also known as *Jovis Nutrix,* the "Nurse of Jove"; but when it fought with Jupiter against the Titans, the versatile bird bore his armor as *Jovis Armiger.* The Arabians, too, pictured an eagle: *Al Oḳāb,* "The Black Eagle"; native Australians, however, knew it as *Totyarguil,* a legendary figure killed while bathing; and the Chinese pictured Aquila as Confucius' "Draught Oxen"!

Three Footsteps

"Altair walks hand in hand with two bright companions through the Milky Way," wrote Martha Martin in her charming book *The Friendly Stars.* Called the "Family of Aquila," the three stars Altair, Alshain, and Tarazed lie on a 5-degree arc; so like an eagle's outstretched wings, it was the *Vultur Volans,* or "Flying Vulture," of the Greeks and Romans, the Turks' *Taushaugjil,* or "Hunting Eagle," and the "Flying Grype" in Old England. Other names for this asterism is the *Çravana,* or "Ear," and *Açvattha,* the "Sacred Fig Tree," of India, and also the Three Footsteps with which Vishnu strode through the heavens, symbolized by a trident. Aquila is generally a summer constellation for evening viewers.

ALTAIR (α) ALPHA Aquilae (RA 19 h 48 m   δ +8° 44′)   m 0.8, its name derived from *Al Naṣr al Ṭāïr,* or "The Flying Eagle," marks the Eagle's neck or the junction of the upper wing with its body. Twelfth-brightest in the heavens, Altair was the standard first-magnitude star in Pogson's photometric scale (p. 15). In past times, it was used to determine lunar distances at sea, and it was a fundamental reference star for Flamsteed in his solar observations. Sixteen light years distant, Altair is a class-A main-sequence star, only one and a half times the Sun's diameter, with a luminosity of nine Suns. It rotates unusually rapidly, about 160 miles per second at the equator, the period of rotation only 6½ hours, the star's poles flattened. With brilliant Vega, in Lyra, and the relatively fainter Deneb, in Cygnus, Altair forms the well-known Summer Triangle, a landmark of the heavens.

ALSHAIN (β) BETA, southernmost of the three stars constituting the Family of Aquila, lies about 2½ degrees to the southeast of Altair. Forty light-years distant, β is a fourth-magnitude pale orange star, its name derived from the Persian title *Shahin tara zed,* the "Star-Striking Vulture." A late-solar-type subgiant, its luminosity four times the Sun's, it has a faint red-dwarf companion some 160 astronomical units distant from the primary but only 1/300 the Sun's luminosity and therefore not very suitable for the small telescope.

TARAZED (γ) GAMMA, the northernmost "Footstep," is brighter than β! A luminous orange giant, its name also derived from the Persian *Shahin tara zed,* γ lies about 340 light years distant from us, optically near dark regions of the Milky Way, the obscuring dust clouds B142 and B143, with their distinctive westward-pointing prongs.

The Scale Beam

A second group of three stars, δ, η, and θ, only of the third to fourth magnitudes, are spaced about 7 degrees apart, forming a line across Antinoüs' outstretched arms, that was known as *Al Mīzān,* or "The Scale Beam," in early Arabia. Of these stars, the supergiant *(η) ETA Aquilae* (RA 19 h 49.9 m   δ +00° 52′) is notable as a bright Cepheid variable of magnitudes 3.7 to 4.5. About 1,300 light years distant, it varies from class F6 to G4, and its light changes are easily observed with the naked eye under good viewing conditions; β, less than 6 degrees to its north, may be used for comparison.

A Cool Star

Together with ε, *(ζ) ZETA,* a close binary, marks the Eagle's tail in the far northwest corner of Aquila; 5½ degrees to its south lies the long-period variable *R AQUILAE,* a pulsating red giant similar to Mira, in Cetus. One of the coolest stars known, R's temperature ranges from 2,350 degrees K. at maximum to 1,890 degrees K. at minimum; the period of its light cycle has decreased from 350 to 300 days in the past eighty years, also an unusual feature.

Zeta and R lie on the opposite banks of a long, dark celestial "river": the **GREAT RIFT,** a nebulous obscuration channel or band that visibly divides the eastern and western branches of the Milky Way and is most apparent in the northwestern regions of Aquila, where it crosses the Eagle's tail.

Vain Birds Within the Eagle

Marking Antinoüs' left foot is *(λ) LAMBDA* (RA 19 h 03.6 m   δ −04° 58′)   m 3.44, a rapidly rotating blue-white main-sequence star near the Aquila-Scutum border; two degrees to its southeast lies the dark nebula *B133,* a dense cloud of gas and dust that totally blots out rich fields of background stars.

Lambda and *(κ) KAPPA* (which is the southernmost Greek-lettered star in Aquila) were imaged by the Arabians as *Al Thalīmain,* "The Two Ostriches"! With ι, δ, and η, Kappa also forms a well-defined banner called *Yew Ke,* or the "Right Flag," by the Chinese.

NOVA AQUILAE 1918 (V603)

On the night of June 8, 1918, one of Nature's most spectacular events rewarded diligent sky watchers. Near the Eagle's western borders, deep in the Milky Way (RA 18 h 46 m   δ +00° 31′), a nova appeared, so luminous that it far outshone Altair, and it became the brightest star in the northern heavens, next to Sirius. Formerly of the eleventh magnitude, *NOVA AQUILAE 1918* (see fig. 117), a "fast" nova, rose to an apparent magnitude of −1.4 by June 9, shining with the luminosity of 440,000 Suns! Unlike Nova Herculis, it did not remain long at maximum, but slowly faded to below naked-eye visibility by the following spring. Seven years later it had faded to the eleventh magnitude, with oscillations in the light curve during its decline. Today, Nova Aquilae is a dense, bluish type-O star; now known to be a close binary, its period is only 3 hours 20 minutes.

During maximum luminosity, spectroscopic analysis indicated that many gas shells

117. *Nova Aquilae 1918,* shown before and after its outburst. (*Yerkes Observatory photograph, University of Chicago*)

were being ejected from the star at high speeds; of great interest was the nova's expanding nebulous shell, seen as a minute disk by means of long-exposure photographs a few months after maximum. A year and a half after maximum, the nebulosity surrounding Nova Aquilae 1918 was comparable in size to the disk of the planet Neptune. Year by year this expanding shell increased in angular diameter at a substantially uniform rate—about 2 seconds per annum—until 1941, when it was too faint for further observation. The symmetry of the shell suggested that the initial disturbance or explosion was so deep-seated that it affected all parts of the surface layers with more or less equal intensity. Before it disappeared, the appearance of the shell invited comparison with planetary nebulae (pp. 227–31), whose relatively small numbers and great distances suggest a possible relation with novae.

Novae have also appeared within the borders of Aquila in the years 1899, 1905, 1925, 1927, twice in 1936, and in 1943 and 1945; none of these became more than faintly visible to the naked eye. It is estimated that nova-like outbursts may occur thirty or forty times a year in our Galaxy.

Exotic Binaries

Important discoveries in astronomy often seem to involve the duplicity—or multiplicity—of stellar objects, a fact we have frequently encountered in our exploration of the heavens. A single object proves to be double or even quadruple, and an erupting star reveals the presence of an invisible companion as the cause of its frenetic activity. This pattern of duality continues with the latest revelations from space: two challenging and mysterious stellar items within the realm of the Eagle. The first of these, lying near the Aquila-Sagitta border about 3½ degrees northeast of ζ, is *PSR 1913+16,* the first pulsar known to be gravitationally bound into a binary system; modern identification techniques also give its position. Pulsars are the rapidly rotating and highly condensed cores of exploded stars such as the central star of the Crab Nebula. Astronomers have studied binary pulsar PSR 1913+16 for several years using observational data consisting of some fifteen hundred measurements of pulse-arrival times obtained at the Arecibo Observatory, in Puerto Rico. These studies were undertaken to determine if the binary pulsar emits gravitational waves in agreement with the principles of general relativity. If the binary does radiate gravitational waves, it is losing energy and, hence, its orbit is decaying, that is, the two members of the binary are approaching each other and the period of the motion is decreasing. Early observations showed that the pulsar moves in a highly eccentric, elliptical orbit about another compact object of comparable mass, which, as deduced from the gravitational redshift and the advance of the periastron, is about 1.4 solar masses. According to general relativity, gravitational radiation causes the pulsar and its companion slowly to spiral closer together, at .00757 second per century, a value that conforms to the observed rate.

"Thus, Einstein's sixty-year-old prediction of the existence of gravitational radiation seems to have been fulfilled, and the dynamic aspect of gravitation has at last been experimentally verified," states Professor J. H. Taylor, of the Royal Astronomical Society.

SS433

The second of these modern discoveries in Aquila, and the most enigmatic, is the eccentric radio star known as **STEPHENSON-SANDULEAK 433**, or **SS433**, originally discovered twenty years ago by Bruce Stephenson and Nicholas Sanduleak, of Case Western Reserve University. In the fourth catalog of Earth's orbiting satellite *Uhuru*, this unusual emission-line star is numbered *4U 1908+05*—coordinates that place it a little over 4 degrees to the northwest of δ, about 1½ degrees to the west-northwest of Fl. #22, and also just northeast of the two galactic clusters NGC 6755 and NGC 6756. Centered in a cloud thought to be the debris of an old supernova remnant at the site of an intense patch of radio emission, SS433 emits an exceptional pattern of radiation in the visual, radio, and X-ray wavelengths. In X-ray emission, it tells the now familiar story of a binary in which the components are rapidly exchanging mass, but its visible spectrum includes bright emission lines that change wavelengths from night to night.

This unusual feature of SS433 aroused the curiosity of the scientific world after astronomer Bruce Margon, of Washington University, described the unusual lines, which were under observation by three independent groups of researchers. Their conclusions were that the spectral lines are emitted by neutral hydrogen but are enormously shifted in wavelength because the hydrogen is traveling at speeds over one-quarter the speed of light. They observed that each line, owing to the Doppler effect, has both a bluer component and a redder one, which implies that the star includes both approaching and receding gas. From this fact, Andy Fabian and Martin Rees, of Cambridge University, and, independently, Mordehai Milgrom, of the Weizmann Institute, in Israel, deduced that this gas is linked in the form of two jet streams whizzing in opposite directions from their stellar source at the relativistic speed of 80,000 kilometers per second. From spectroscopic observations, Fabian and Rees also concluded that the modulation of the Doppler shifts is cyclic, with a period of about 164 days. On his own (without this information), Milgrom postulated a rotating beam axis, inclined to our line of sight. Precession was considered the cause of this rotation, implying that SS433 has a close, gravitationally bound companion. Margon and G. O. Abell, utilizing extensive observations by University of California astronomers, elaborated Milgrom's suggestion, and out of these data evolved the so-called "kinematic" model of SS433.

Observations of the more normal emission lines of the star reveal that they have very slight Doppler shifts, indicating a source that approaches and then recedes at speeds up to 75 kilometers per second in a 13-day cycle; i.e., a binary whose components revolve once in 13 days about their common center of mass. With this discovery, astronomers tried to account for the enormous energy SS433 radiates—about 1 million times that radiated per unit time by the Sun—in ejecting the approaching and receding gas. In the model first advocated by Fabian and Rees, which was accepted by many astronomers, one of the binary's components loses a large amount of material to an extremely dense companion (a degenerate star), which flings the excess matter out into space in the form of two high-speed jets. According to C. Sarazin, of the University of Virginia, about a thousand times more mass is transferred than is normal for X-ray binaries. These gas jets also wobble, perhaps because of tidal effects within the binary, or, as suggested by other theorists, relativistic effects that warp space near the dense component. If the latter is a neutron star, the one-to-one mass ratio of the two stars (at possibly one solar mass each), which is suggested by their velocity curve, implies that the companion is a main-sequence star of about type F, and photometric evidence of an eclipsing effect is to be expected. Such evidence, however, is inconclusive; recent data also suggest that the normal star may be more massive than hitherto believed.

Other models postulate two narrow streams of evaporating gas that originate in a cloud of denser gas and are ejected by radiation pressure through two "nozzles," a formation of the cloud's inner edge, aimed in opposite directions out of the plane of the binary. According to this theory, the strange spectral lines are produced after the gas jets leave the "nozzles" and cool, and are then subjected to friction with stellar wind from the normal star, causing repeated ionization of the hydrogen atoms. In this process, each time an ion recombines with an electron, the newly formed atom emits strong radiation as the electron cascades to its ground state.

Scientists at Ohio State University proposed a model in which the gas is accelerated and spun around the binary by the strong rotating magnetic field of one of the stars; ions ejected from its magnetic poles collide on opposite sides of the binary to produce the moving spectral lines. This magnetic emission, they suggest, forms a huge ring some thirty times the diameter of the Solar System, and the orbiting gas assumes relativistic speed to complete a revolution in about 164 days. The constantly circling gas, remaining in the emitting region for several revolutions, then accounts for the brightness of the emission lines. Neither the jet nor the supposed ring has been spotted on photos, however, for SS433 is too distant: about 5.1 kiloparsecs (16,626 LY), according to recent estimates.

X-ray observations from the Orbiting Einstein Observatory, however, have confirmed the existence of two collimated, colinear jets that originate at SS433 and extend for 0.5 degree quite accurately aligned with the major axis of the radio emission of debris cloud *W 50*. These data show that the gas jets profoundly affect its evolution, and W 50 may therefore not be a supernova remnant at all, nor SS433 a residual hot core of that explosion, as formerly suggested. Recent radio observations have confirmed these X-ray findings regarding W 50 and the ejected twin-jet model of SS433; furthermore, radio maps obtained with the VLA (p. 56) have added another dimension to observations of the proper motion of the two jets, allowing a much more accurate calculation of the distance of SS433.

There is an amazing resemblance between the twin-lobed structure of SS433 and that of energetic radio galaxies, as well as quasars; astronomers are eager to explain the nature of the ejected jets, because jets in radio galaxies may operate under a similar mechanism, although on a much grander scale. Thus a solution to the mysteries of

SS433 will provide astronomers with a fund of knowledge about vastly more distant objects in the universe.

The Lamp Burns Low

Just north of Fl. #22 and only about 1½ degrees east of SS433 lies the faintest known star in the heavens, a red dwarf discovered in 1943, *VAN BIESBROECK'S STAR (Ross 652b, Wolf 1055b)* (RA 19 h 14.5 m   δ +05° 04')   m 18. The secondary component of a "midget binary," it is separated by 400 astronomical units from its primary, an ordinary ninth-magnitude red dwarf of 1/250 L ⊙. However, with a luminosity only about 1/570,000 the Sun's (700 times Jupiter's) luminosity, Van Biesbroeck's star "barely shines," much of its radiation detectable only in the infrared, owing to its temperature. A mere 19 light years distant from us, it shows the large proper motion usual for close stars. With its absolute magnitude +19.3, this very dim sun, if placed at the standard distance of 10 parsecs, would be visible only with the largest telescopes. Because of its very tiny mass, it probably cannot support thermonuclear fusion, but may be releasing energy by gravitational contraction as it gradually fades away over the centuries, destined to become a cold, dark sphere.

The Lamp Turns On and Off

About 3 degrees east of Lambda and 1 degree northwest of the spectroscopic binary #26 Aquilae lies a source of intermittent X-ray emission called an "X-ray burst source" and designated *4U1915-05* in the standard catalogue of X-ray objects. A group of astronomers at the University of California, using the most highly developed technical equipment on the Orbiting Einstein Observatory, are conducting X-ray observations of this so-called "burster." These scientists suspect the involvment of a close binary system containing a degenerate star, as indicated by a number of short dips in the X-ray counting rate that occur at intervals consistent with a period of approximately fifty minutes, the proposed orbital period of the binary. The dips, each lasting a few minutes, are irregular and are most pronounced at the lowest frequencies, apparently the effect of obscuration of the X-ray source by a gas stream passing from the degenerate companion (probably a neutron star) as this gaseous material interacts with an accretion disk encircling the companion. A search for the optical counterpart of this unusual object has also been underway; images obtained with a CCD camera at the prime focus of the 4-meter telescope at Kitt Peak Observatory reveal an extremely faint (m22) star as the most likely candidate. The absorption dips in the X-ray flux of 4U1915-05, the first ever detected in a burst source, are important evidence that bursters are accreting binary systems.

MINOR GROUPS: Sagitta and Vulpecula

To the north of Aquila lies the little feathered arrow called *SAGITTA,* formed by four stars of magnitudes 3.7 to 4.4. Many stories were told about it: Some chartists pictured

Sagitta in the Eagle's talons when Aquila was armor bearer to Jove; in contrast, it was also the arrow that slew the eagle of Jove! It bore the title "Herculea" when shot by Hercules toward the Stymphalian birds, whereas Eratosthenes said it was the arrow with which Apollo slew the Cyclops. It has also been identified as a stray arrow shot by Sagittarius the Archer, and even as the "Arrow of Cupid"! The title *Telum,* meaning "Dart," survived to Kepler's time; others were *Temo Meridianus,* the "Southern Beam," and *Virgula Jacens,* a "Falling Wand." The Hebrew *Cheits,* the Armenian *Tigris,* and the Arabic *Al Sahm* all signify *"Arrow,"* as does the title *Sagitta,* its popular name with the Romans and with us today.

An interesting quadruple system in the Arrow is *(ζ) ZETA Sagittae (Σ 2585),* (RA 19 h 46.8 m δ +19° 01′) m 5.50, just to the northeast of fourth-magnitude δ. Alvan Clark, the discoverer of the companion of Sirius, first detected duplicity in Zeta A, the primary, a close and rapid binary, its white type-A components only 0.2 second apart. Zeta B, the ninth-magnitude companion, is 8.4 seconds distant from the close pair, and an eleventh-magnitude member of the system lies much farther out, at 75.7 seconds.

Far to the west of the Arrow lies *U SAGITTAE* (RA 19 h 16.6 m δ +19° 31′), an eclipsing variable of the Algol type (p. 338), discovered by F. Schwab in 1901. Suitable for small instruments, it lies 1.7 degrees west and slightly to the south of a group of stars (#s 4, 5, and 7) in the neighboring constellation of Vulpecula. To the southeast of U lies the binary Σ 2504, a good comparison star. The components of U, a blue-white main-sequence star and its faint solar-type yellow subgiant companion, are separated by only 8.5 million miles, producing a total eclipse of the primary, during which its apparent magnitude increases from 6.4 to 9.0.

A Dwarf Nova

Certain types of novae are recurrent—and some astronomers believe that this is true of all ordinary novae—but with very long periods, possibly thousands of years or more. *WZ Sagittae* (RA 20 h 05 m δ +17° 33′), about 2.8 degrees southeast of γ (the bright star at the head of the Arrow), is a rare example of a recurrent nova with a short period. It has undergone three known outbursts: in 1913, when it changed from a sixteenth- to a seventh-magnitude star, with its luminosity increasing four thousand times; in 1946, when it reached magnitude 7.7, undergoing similar variations but fading somewhat more rapidly; and again in 1978. According to spectroscopic studies, WZ seems to have the characteristics of a very dense white dwarf with its luminosity about $1/175\ L\ \odot$ at minimum and 20 or 30 $L \odot$ at maximum, in contrast to such stars as T Coronae, which attain high luminosities at maximum, rivaling "normal" novae. WZ is also an eclipsing variable with a period of only 81.6 minutes, the shortest known, another example of a close and rapid binary; two very small components separated by only 230,000 miles— i.e., nearly in contact—which are exchanging mass. A unique feature of WZ is that its spectrum contains emission lines of hydrogen, emanating from a rotating gas ring that surrounds the white dwarf, supplied by matter ejected at high speed from the cooler and larger red-dwarf companion. The white dwarf's diameter is computed to be 11,000 miles, whereas its companion's radius is about equal to Jupiter's and its mass is about 0.04 $\mathcal{M} \odot$, one of the smallest stellar masses known.

A Planetary in Formation?

Of especial interest is the variable **FG Sagittae** (RA 20 h 09.7 m   δ +20° 11′), in the northeastern part of the constellation, about 0.07 degree southeast of the binary θ. FG is a supergiant whose spectral class has changed from B to A in a dozen years and whose surface temperature is decreasing accordingly. This strange star is covered by a round gaseous disk, presently about 1.2 light years in diameter, which the 120-inch reflector at Lick Observatory revealed in 1960: an apparent planetary nebula in formation. Spectroscopic studies indicate the ejection of great quantities of material that may form a second shell. However, R. P. Kraft has pointed out a growing production of heavier metals, as now indicated in the spectrum; he suggests that FG is changing from a normal giant to a barium (S-type) star, and that the nebula is not the normal planetary of an aged star approaching the white-dwarf stage. New developments are awaited with great interest as this star evolves before our eyes!

The Little Fox

Just north of Sagitta and south of Cygnus is a zigzag formation of about five stars of magnitude 4.5 and fainter, spanning the northern part of the Milky Way's two branches. Created as a constellation by Hevelius, who included twenty-seven stars in its boundaries, **VULPECULA** (*Vulpecula cum Ansere,* "The Little Fox with the Goose") was selected for its "appropriateness" as a neighbor to Aquila, Vultur Cadens (Cygnus), and the obsolete constellation Cerberus, the dog of Hades. Hevelius, no admirer of the fox, believed it shared qualities of greed and ferocity with the foregoing creatures.* Modern charts omit Anser, the Goose, showing the Fox without his prey.

The most noteworthy object in the Fox's territory is **M27 (NGC 6853)**, the famous **DUMBBELL NEBULA,** one of the largest planetaries, lying about 2 degrees southeast of #13 Vulpeculae, 7 degrees southeast from Albireo (Cygnus' head), and almost 3 degrees north of γ Sagittae. The exact position is RA 19 h 57.4 m   δ +22° 35′; #14 lies in its field. It was also called the "Double-Headed Shot," because of its peculiar appearance, which led Sir William Herschel to suggest that it might be "a double stratum of stars . . . one end of which is turned toward us." (Seven stars are visible with a 10-inch telescope, but M27 is mostly gaseous.) The feathery protuberances on its sides meet very faintly, forming, in Sir John Herschel's words, "a regular ellipse"; and along with its two glowing masses, which fuse in a slightly narrower central zone, they give it, roughly, the shape of a dumbbell, although, seen through smaller telescopes, it may be likened to a spool (fig. 118).

The first planetary discovered by Messier (fifteen years before Darquier discovered the Ring Nebula, in Lyra), the Dumbbell, in contrast to the latter, is not ringlike in shape; expanding at about 17 miles per second, it is among the largest of the planetaries, with a possible diameter of 2½ light years. The determination of its diameter depends on estimates of its distance, which range anywhere from 490 to 979 light years. Its 13.5-magnitude central star, a bluish dwarf that emits intense ultraviolet radiation, is one of the hottest known, at about 85,000 degrees K.; a seventeenth-magnitude

---

* We disagree!

118. M27 (NGC 6835), the Dumbbell Nebula in Vulpecula, a large planetary nebula photographed with the 100-inch telescope on Mount Wilson. (*Hale Observatories*)

yellowish companion has been detected at a separation of some 1,800 astronomical units.

CYGNUS

Several characters of Roman mythology, some attractive, one quite hateful, share a single fate: each, on his death, is changed into a swan. According to Ovid, Cygnus (or *Cycnus),* son of the sea god Poseidon, was a king of Colonae, in Troas, who defended Troy, under siege by Achilles. No weapon could affect Cycnus' body, so Achilles strangled him with the thongs of his own helmet; whereupon Poseidon changed Cygnus into a swan.

In another story, told by Hesiod, Pindar, and others, Cycnus, son of Mars (Ares), the god of war, was afflicted with peculiar tastes in architecture: he collected the heads of those strangers unfortunate enough to wander into Thessaly; it was his intention to use these gruesome trophies as "building blocks" for the construction of a temple to his father. With a stroke of his sword, Hercules, the ancients' ever present superman and righter of wrongs, lost no time in putting an end to Cycnus' macabre practice; but this deed was a kindness to the miscreant, for it enabled him to assume the form of a swan, a decided improvement over his previous state of being.

Yet another myth depicts Cycnus as the friend of Phaëthon, who tried to drive Apollo's Sun chariot across the sky, with disastrous results, until Zeus struck him down. While lamenting Phaëthon's fate, Cycnus was changed by Apollo into a swan and placed among the stars, north of the Eagle and east of the Lyre.

With the Greeks, however, this constellation was *Ornis* (a bird), often a "Hen," the meaning of its Arabian title: *Al Dajājah*. The Romans first adopted it as *Cycnus*, the "Swan," associated with the foregoing myths and others, among them its identification as *Helenae Genitor*, i.e., father of Helen and seducer of Leda, the egg-laying mother of Castor, Pollux, Clytemnestra, and Helen. Popular Roman names were *Ales* and *Avis*, or "Bird," and *Olor*, or "Swan," the latter surviving to modern times; but it also was *Myrtilus*, the bird of Venus, named after her sacred myrtle, as well as Orpheus, the tragic musician, positioned next to his lyre.

In the Alfonsine Tables, of the thirteenth century, Cygnus appeared as the hen "Galina," a name which remained in use until the eighteenth century with Bayer and others. The "Cross of Calvary," *Christi Crux*, and similar religious titles of the seventeenth century, suggested by the Swan's eight principal stars, evolved into its present-day popular name, the **NORTHERN CROSS**, familiar to all who observe the heavens.

DENEB (α) *Alpha Cygni*
(RA 20 h 39.7 m   δ +45° 06')   m 1.26

Marking the head of the Cross but the tail of the Swan, Deneb is derived from the Arabic *Al Dhanab al Dajājah*, "The Hen's Tail." An alternate name, "Arided," is apparently from *Al Ridhādh*, of uncertain meaning, which was applied to the whole constellation by the sixteenth-century French scholar J. Scaliger, and which degenerated to "El Rided." Another name is "Aridif," similar-sounding but derived from *Al Ridf*, "The Hindmost," which Bayer changed to "Arrioph"; he also used "Galina" for α alone.

A luminous white supergiant of spectral class A2 and therefore a "Sirian," or hydrogen, star, Deneb, lying at the distance of 500 parsecs, has no gravitational or optical companions. Its luminosity is about 60,000 $L \odot$, and its mass some 25 $\mathcal{M} \odot$; a comparison of Deneb's absolute magnitude, -7.2, and that of the supergiant Rigel, -7.0, to the absolute magnitude of the Sun, only $+4.8$, gives us some idea of the modest properties of our own luminary! Probable pulsation or atmospheric turbulence imparts slight variability (0.05 magnitude) to Deneb, which is approaching us at the small radial velocity of 3 miles per second.

ALBIREO (β) *Beta Cygni*
(RA 19 h 28.7 m   δ +27° 51')   m 3.09

A misunderstanding concerning the Latin words *ab ireo*, which appear in the 1515 *Almagest*, apparently led to β's proper name, Albireo, but the Arabians called it *Al Minhar al Dajājah*, "The Hen's Beak." Lying at the head of the Swan, which marks the foot of the Cross, Albireo is a famous wide double, its magnitude—3.09 orange and 5.11 blue components creating, in the words of nineteenth-century English astronomical writer Agnes Clerke, "perhaps the most lovely effect of colour in the heavens." A moderate-sized telescope best reveals their striking color contrast. At the very wide separation of 34.3 seconds, the components are 400 billion miles apart; if circled by a planet, each star would appear only as a very tiny bright disk in the planetary heavens of the other! (If these stars were as small as our Sun, at 400 billion miles their disks would be reduced to mere points.) Albireo A, a bright giant of the "Arcturian" type, is the chief contributor to the system's mean spectral class of K3, while Albireo B, its fifth-magnitude distant companion, is a class-B8 main-sequence star with an interesting hydrogen emission spectrum. The spectrum of A also indicates an old story of duplicity: here (excluding B) we really have *two* stars, the bright orange giant and a hotter (late-class-A or early-B) unseen component, the separation too small for telescopic resolution.

The Other Stars of the Cross

A bright star marks the juncture of the great upright (nearly 20° long) with the transverse: *SADR (γ) GAMMA* Cygni, (RA 20 h 20.4 m   δ +40° 06')   m 2.23, derived from *Al Sadr al Dajājah,* "The Hen's Breast"; together with ζ, ε, and δ, it forms an obsolete Arab asterism, *Al Fawāris,* or "The Riders." Its late-class-F spectrum is almost like the Sun's, but Gamma, 750 light years distant, is a supergiant, its luminosity about fifty-eight hundred times that of the Sun. Gamma has a very distant tenth-magnitude optical companion at an angular separation of 142 seconds; the companion itself is also a very close double.

On the northwestern arm of the Cross, marking the tip of the Swan's left wing, lies *(δ) DELTA Cygni* (RA 19 h 43.4 m   δ +45° 00')   m 2.87, a giant of spectral type B9 or A0. A close and difficult binary, the separation of 2.2 seconds is nearing maximum, making it a little easier to glimpse its faint bluish companion with a small telescope. On the southeastern arm of the Cross is *GIENAH (ε) EPSILON* (RA 20 h 44 m   δ +33° 47')   m 2.46, its name derived from the Arabic *Al Janāh,* "The Wing." To its north-northeast lies a fifth-magnitude optical companion, the variable T Cygni, a class-K visual double. A true physical companion of ε lies farther away, at 78 seconds due west, the fifteenth-magnitude star designated LTT 16072. Gienah, the primary, 75 light years distant from us, shines with the luminosity of forty Suns. About 5½ degrees to its southeast lies *(ζ) ZETA,* magnitude 3.20, a yellow star that marks the tip of the Swan's right wing; 390 light years distant and six hundred times as luminous as the Sun, ζ is classed as a type-G8 "bright giant." The practical Chinese, however, regarded ζ, together with (ρ) Rho, some 14 degrees to its north-northeast, and two adjacent small stars, as *Chay Foo,* a Storehouse for Carts!

Other Stars in the Swan

About 5½ degrees east of ζ, occupying the southeast corner of Cygnus, is the close visual binary *(μ) MU,* magnitude 4.45, consisting of two orange main-sequence stars, 65 light years distant, their mean separation from one another 85 astronomical units.

Just 7 degrees north of ζ, approximately 8 degrees southeast of Deneb, lies another close binary, *(τ) TAU* (RA 21 h 12.8 m   δ +37° 49′)   m 3.74; its combined spectral type is F IV, indicating that the primary is a pale yellow subgiant. At 75 light years, separation of the pair is only about 20 astronomical units, and a faint red dwarf also orbits this binary at a separation of more than 2,000 astronomical units.

In the long neck of the Swan, some 5½ degrees northeast of Albireo and less than 2½ degrees southwest of the golden binary *(η) ETA,* lies the long-period variable *(χ) CHI Cygni,* a red giant easily observed with the naked eye when it reaches fifth-magnitude (and sometimes even magnitude 3.5). After its minimum brightness, when its magnitude is larger than 12, Chi brightens up again, as its luminosity increases three thousand times on the average and sometimes even as much as ten thousand, giving it a broader range than that of Mira, in Cetus. The highly irregular light cycle averages 407 days, and χ's temperature at minimum, less than 1,900 degrees K., places it among the coolest stars known.

About 4 degrees northwest of Deneb, we find an area rich in color contrasts for the small telescope. *(o) OMICRON 1 and 2 (Fl. #31 and 32),* lying in the Swan's left wing (RA 20 h 12–13 m   δ +46–47° 35′), constitute a very wide optical double, both visible with the naked eye as separate fourth-magnitude stars 1 degree apart. Omicron 1, magnitude 3.76, an orange giant, is the principal star of a group of three (not gravitationally bound). It has two bluish companions, one at 107 seconds of arc, and the other, fifth-magnitude #30, at 338 seconds. The latter is also a long-period eclipsing variable (catalogued as V695), consisting of an orange giant separated by 1.2 billion miles from its secondary, a blue main-sequence star, of smaller dimensions, that passes behind the giant once every 10.42 years in a total eclipse lasting sixty-three days. Forming a very similar system, o 2 Cygni, to the northeast, magnitude 3.97, is also an eclipsing variable, consisting of a class-K orange giant and a bluish companion of normal dimensions that passes behind the giant every 3.15 years in a total eclipse lasting just eleven days. Also suitable for the small telescope, the variable U Cygni, a deep-red carbon star with a bluish companion, lies 45 minutes of arc east of o2.

Some 3½ degrees north-northwest of Deneb and approximately 2½ degrees northeast of the Omicron group, we find *RUCHBA (ω) OMEGA 1 and 2* (Fl. # 45 and 46), from *Al Rukbah al Dajājah,* "The Hen's Knee," of ancient Arabia, but part of the Swan's left wing with us. Omega 1 is a tertiary; the components are a close bluish-white pair, magnitudes 5.5 and 13, and a more distant tenth-magnitude companion. Omega 2, a visual double, is a red giant, suitable for smaller instruments; to the west-northwest lies Fl. #43, a class-F main-sequence star.

A Historical Star

About 5½ degrees northeast of ε and just 1⅓ degrees west-northwest of τ is famous *61 CYGNI* (RA 21 h 04.4 m   δ +38° 28′), the first star whose parallax was measured. This was done by the German mathematician Friedrich W. Bessel, who, at age twenty, distinguished himself by deriving the orbit of Halley's comet, and then abandoned a promising commercial career for a humble post as assistant in an observatory, from which he advanced, within a few years, to the directorship of the famous Königsberg

Observatory. There he made many important discoveries and deductions, including his prediction of the existence of a companion to Sirius. In 1837, Bessel began his observations of 61 Cygni by measuring its change of direction against faint background stars as the Earth moved in its orbit, and in December of 1838 he published his results: 61 Cygni was about 600,000 astronomical units distant, or close to 60 million million miles (10 light years). Thomas Henderson's successful measurement of the parallax of Alpha Centauri followed within two months, and two years later Wilhelm Struve measured the distance of Vega. On the occasion of the Royal Astronomical Society's Gold Medal Award to Bessel in 1841, Sir John Herschel declared: ". . . we have lived to see the great and hitherto impossible barrier to our excursions into the sidereal universe . . . almost simultaneously overleaped at three different points. It is the greatest and most glorious triumph which practical astronomy has ever witnessed."

Recent observations place 61 Cygni at 11.1 light years; one of the closest stars, it has the large proper motion of 5.22 seconds per annum in the general direction of σ, to its northeast. Struve in 1830 discovered 61's duplicity: the components are two class-K dwarfs, only a fraction of the Sun's mass and luminosity but visible in small telescopes because of the system's closeness to us (fourth among the naked-eye stars). A and B are separated by about 84 astronomical units (28.4″), and its period of revolution is about seven hundred years. Approaching us at 38 miles per second, the binary 61 Cygni has aroused additional interest because of the discovery, in 1942, that it also has an astrometric component, now called 61 Cygni C. An extensive series of observations by Dr. K. A. Strand, at Dearborn Observatory, plotted on a graph at intervals of a month or two, show that the path of B around A is wavy, indicating that either A or B is a double system. Apparently orbiting the A component, C is very intriguing to astronomers because of its small mass (.008 $\mathcal{M}\odot$), only eight times that of Jupiter. The unseen body 61 Cygni C has probably not achieved thermonuclear ignition and may be nonluminous, like a planet. That the major components of 61 are far enough apart not to disrupt C's orbit apparently makes the existence of such a small body possible, although the probability is small that it is a true planetary system, containing much less massive bodies in orbit around either A or B. (See fig. 119.)

Individualistic Stars

About 2 degrees south-southwest of Sadr (γ), at the junction of the Cross, we find a rather odd star: the variable *P CYGNI (Fl. #34),* which some charts designate as "NOVA 1600" or "NOVA CYGNI #1" (RA 20 h 15.9 m   δ +37° 53′)   m 4.88. In the year 1600, the star was first observed by the Dutch globe maker Caesius (Willem Blaeu), when it reached third magnitude. After six years, it gradually faded above the sixth magnitude, reappearing half a century later as it rose to magnitude 3.5, only to "vanish" three years after that and then repeat this "off-again, on-again" process within another three years. Its irregular behavior continued until 1715, when it settled down at fifth magnitude, enjoying a stable existence ever since. Astronomers no longer regard P as a nova, but they attribute its light fluctuations to an expanding shell, ejected from the star's interior at the time of its sudden increase in luminosity. This shell is indicated by peculiarities in P's spectrum, including displacement toward the violet of

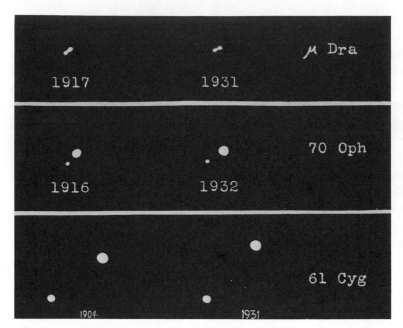

119. Orbital motion of μ Draconis, 70 Ophiuchi, and 61 Cygni during fourteen, sixteen, and twenty-seven years, respectively. (Yerkes Observatory photograph, University of Chicago)

its absorption lines and broadening of its emission lines. That P is extremely luminous is implied by its great distance, at least 3,000 light years but possibly as much as 7,000 light years.

About 1.3 degrees northeast of ∈, on the Swan's right wing (and east-northeast from T Cygni), lies **Y CYGNI** (RA 20 h 50 m   δ +34° 28′)   m 7–7.6, a gigantic eclipsing binary at the remote distance of some 9,000 light years, consisting of two luminous blue giants whose centers are separated by only 12 million miles. Some 5 degrees to the east of the binary (ζ) Xi and about 9 degrees east-southeast of Deneb, we find the brightest known "cataclysmic variable" or "dwarf nova," **SS CYGNI** (RA 21 h 40.7 m   δ +43° 21′), which puts on a display several times a year, its brightness increasing from the twelfth magnitude to the eighth within a day or two. A very close binary of the U Geminorum type (pp. 151–4), SS consists of a yellow dwarf that has a "solar" spectrum but is only half as massive as the Sun, and an underluminous but hot blue star similar to a white dwarf. With an orbital period of about $6^1/2$ hours, the components may be as close to one another as 100,000 miles and are probably exchanging mass, a factor that triggers the violent eruptions of the yellow or the blue star (astronomers do not yet agree which).

A Wolf-Rayet Binary

Less than 1 degree north-northeast of P Cygni lies an interesting eclipsing binary, **V444 CYGNI** (RA 20 h 17.7 m   δ +38° 34′). Its components, a hot blue class-O star and a much hotter "Wolf-Rayet" star (p. 23), separated by 17 million miles, revolve about their common center of mass in a little over four days, during which there are two eclipses, the primary eclipse occurring when the W star partially occults the O star for about twenty-four hours, and the secondary eclipse, lasting about twelve hours, when the reverse occurs. No more than a few hundred Wolf-Rayet stars have been observed, all at great distances, with absolute magnitudes ranging from −4 to −8. They have surface temperatures ranging from 60,000 degrees to 100,000 degrees K., which, together with their luminosities, indicate radii about twice that of the Sun. V444 Cygni is about 4,900 light years distant; the W component is about 1,450 times as luminous as the Sun and eighteen times as massive, and its diameter is estimated to be 2.3 times that of the Sun. The wide emission lines in the spectra of Wolf-Rayet stars, owing to doubly ionized helium and highly ionized atoms of carbon, nitrogen, oxygen, and silicon, indicate that each of these stars is surrounded by a shell of expanding gas; such a shell, three times the star's diameter, surrounds the W star of V444, and in addition, a much thinner, outer shell, eight times the star's diameter, encloses the inner shell.

V444 is the first known binary that contains a Wolf-Rayet star; some astronomers believe that all W stars are members of binaries but the duplicity of such stars cannot be detected because the planes of their orbits do not pass through the Earth, so eclipses cannot be observed.

A Recent Nova

In August of 1975, about 5 degrees east-northeast of Deneb, some 4 degrees north-northeast of ξ, and exactly 1 degree northwest of #63 Cygni, an exciting event occurred with the appearance of *V1500* or *NOVA CYGNI 1975* (RA 21 h 09.9 m   δ +47° 57′), which attained an apparent magnitude of 1.8, to become the brightest nova since Nova Puppis, in 1942. Initially a twenty-first-magnitude star, Nova Cygni increased its luminosity 40 million times in its 19-magnitude rise, the greatest range of any known nova. At maximum, its absolute magnitude was about −9.5, or five hundred thousand times the Sun's luminosity. The gaseous shell ejected by the star's tremendous explosion reached a velocity of 1,200 miles per second, and the star assumed a roseate hue owing to radiation of the Hα Balmer line from the hydrogen atoms. After that, Nova Cygni faded rapidly, without fluctuations, until, in January 1977, it reached the twelfth magnitude, undergoing small variations (see fig. 120).

An X-Ray Powerhouse

Not quite 0.5 degree northeast from η, in the Swan's neck, lies an exceptionally strong source of X-ray energy called *CYGNUS X-1* (RA 19 h 56 m   δ +35° 04′); it was an early discovery in the search for galactic X-ray sources, in which the department of U.S. Naval Research, NASA, and other agencies collaborated. The position of this invisible source coincides with that of HDE 226868, a massive supergiant of the ninth apparent magnitude, which is also a spectroscopic binary with a period of 5.6 days; the invisible companion of this supergiant is probably the X-ray source. This unseen component has aroused extreme interest and much speculation. X-ray pulses accompanying rapid changes in its luminosity are emitted in intervals of about 0.1 second, which means that the X-ray-emitting region in this variable source is extremely small, less than one hundred miles in diameter (about as small as or smaller than a pulsar); it must also be extremely dense, with as much as fifteen to twenty solar masses compressed within its tiny radius. This mass estimate is based on the very small period of revolution

120. V1500, Nova Cygni 1975, showing the decline in light from second magnitude at maximum luminosity (above) to fifteenth magnitude. Photographed with the 20-inch Astrograph. (*Lick Observatory*)

of the orbiting supergiant, whose mass is computed to be 20 to 30 solar masses. According to its mass, the X-ray source, if it were a normal star, would be at least as luminous as the great Rigel, but, on the contrary, X-1 is so underluminous that it cannot be seen at all! Thus its invisibility and rapid pulsation imply that it is a very small, and therefore highly condensed, object, the end product of a massive star's evolution. Since a collapsing star whose mass exceeds three solar masses cannot exist as a white dwarf or neutron star, Cygnus X-1 may be the first example of the core of a supernova that is collapsing to a black hole.

Recent studies show that a hot stream of gas is passing from the supergiant to X-1, producing intense X-ray emission in the process; if X-1 is a rotating black hole, these X-rays originate when the gas stream falls onto an accretion disk formed of particles spiraling into X-1's vortex from surrounding space. Efforts to detect gravitational radiation from the binary pulsar in Aquila (PSR 1913 + 16, p. 271) are important, because black holes are only theoretical concepts at present; the detection of gravitational waves from sources like Cygnus X-1 would be concrete evidence that such bizarre objects exist.

Also of Interest

NGC 6819, an open cluster 8 degrees due west of γ, contains 150 stars, magnitudes 11 to 15, within a diameter of 6 minutes. *NGC 7092 (M39),* about 9 degrees east-northeast of Deneb, contains twenty-five loosely scattered seventh-magnitude stars within a diameter of 30 minutes, suitable for binoculars. *NGC 6960 and 6992,* lying some 2½ to 3 degrees south of ε, near the Cygnus-Vulpecula border, together form the *BRIDAL VEIL NEBULA* (fig. 121), also called the "Lacework Nebula" because of its delicate filamentary structure, as revealed in the long-exposure photographs of large reflectors. The Veil may constitute part of an expanding gas cloud, the remnant of a supernova that exploded many thousands of years ago. Like the "interstellar bubble" in Monoceros (pp. 125–7), the Veil appears to be forming a shell around a low-density cavity that is noticeably clearer than the dusty regions that surround the nebula. The golden binary Fl. #52 (RA 20 h 44 m   δ +30° 32′)   m 4, lies in the field of NGC 6960, the Veil's western segment, but as a foreground object it is not a source of the nebula's illumination, which, in fact, astronomers are still trying to identify.

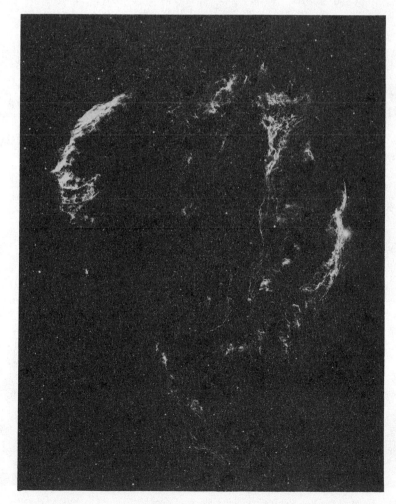

121. *NGC 6960 and 6992, the Bridal Veil Nebula, in Cygnus,* an "interstellar bubble" photographed in its entirety in red light with the 48-inch Schmidt telescope. NGC 6992 is the bright portion seen at the upper left (east). (*Hale Observatories*)

122. *NGC 6992*, northeastern edge of the Veil Nebula, showing the filamentary structure in striking detail. Photographed with the Shane 120-inch reflector. (*Lick Observatory*)

123. NGC 6960, western loop of the Veil Nebula with the star 52 Cygni a foreground object. Viewed with north to the left, this eerie scene resembles a sunset deep in space over a ghostly terrain. Photographed with the 100-inch telescope on Mount Wilson. (*Hale Observatories*)

An outstanding bright nebula lies less than 3 degrees east of Deneb and just to the west of the reddish star ξ; famous as the *NORTH AMERICA NEBULA* because its form bears a striking resemblance to our continent (fig. 124), *NGC 7000* (RA 20 h 57 m   δ +44° 08′) is 100 minutes in diameter (about one third the size of the Orion Nebula) and about 1,600 light years distant, glowing brightly in an area of the Milky Way richly jeweled with stars, many of which may contribute to its illumination, although Deneb has traditionally been credited with that function. About 1 degree to the west of the nebula, look for the smaller *PELICAN NEBULA (IC 5067)*, in the field of Fl. #s 56 and 57. To its south lies tripartite *IC 5068*. The filamentary details of all these nebulae are indescribably lovely, especially on long-exposure photographs.

A Great, Extragalactic Radio Source

In 1940, Grote Reber, pioneer radio astronomer, began to investigate mysterious radio signals, first discovered by Karl Jansky, that had been hampering transatlantic communication. Setting up, in his backyard, a 31-foot antenna he had built, Reber detected radio signals from the regions of Cassiopeia, Sagittarius, and Cygnus. The first of these cosmic "transmitters" is a supernova remnant; the second, our own Galaxy core; the third, however, is a galaxy 500 million light years distant, its radio emission 3 million times more intense than that of our Galaxy! Reber's discovery, now called *CYGNUS A* (RA 19 h 57.8 m   δ +40° 35′), some 3½ degrees west of γ and lying far behind a region of the Milky Way strewn with nebulae, is the second-strongest radio source in the heavens. In 1951, Walter Baade, using the 200-inch telescope at Palomar Mountain, first revealed it optically as an oddly shaped fuzzy double image, suggesting two colliding galaxies (fig. 125). This notion, at first the more or less accepted explanation of its appearance, lost favor when it became evident that the peculiar galaxy Cygnus A has two large lobes, at its faint, outer regions, that apparently produce its intense radio waves. If a collision produced those waves, the most intense signals would come from the colliding interface, rather than from the lobes; moreover, a collision could not account for such a tremendous outpouring of energy. Objects similar to Cygnus A have subsequently been discovered, and new theories are being formulated to account for them, suggesting one or another unusual explosive process deep within their nuclear regions (see NGC 4038, the Ring-tail, in Corvus, p. 172). These nuclei,

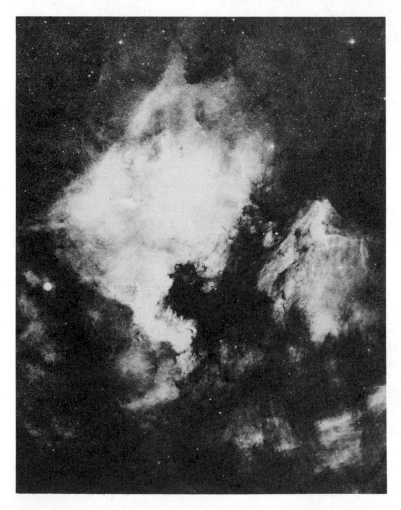

124. NGC 7000, the
North America Nebula in
Cyngus, a medium-sized
bright emission nebula,
photographed in red light
with the 48-inch Schmidt
telescope on Palomar
Mountain. (*Hale
Observatories*)

125. *Cygnus A.* This peculiar galaxy is an intense source of radio energy. Two large radio lobes at its outer regions are apparently powered by an explosive process in the nucleus or by an accretion torus surrounding a spinning black hole. Photographed with the 200-inch telescope. (*Hale Observatories*)

however, emit little detectable radiation in the optical or in the X-ray band; to explain the tremendous source of power that would have to be in the nuclei to supply energy to the radio lobes of such galaxies as Cygnus A, it has been proposed that the center of each is a spinning black hole surrounded by an ion-supported gas torus too hot and tenuous to radiate efficiently. Magnetic fields in the torus extract the rotational energy of the black hole in the form of two collimated beams or jets that accelerate fast-moving electrons in the radio lobes, while the black hole radiates underluminous waste heat. In this case, it is suggested, at an earlier epoch the hole rapidly acquired mass supplied by the debris of stars that passed so close to it that they were tidally disrupted.

Proponents of this theory see a close relationship between radio galaxies like Cygnus A and the enigmatic quasars, which may consist of black holes supported by radiative clouds without dominant magnetic fields. A similarity to Seyfert galaxies is also suggested, in which the latter are powered by somewhat less massive central black holes. Further observation of all these radio sources should lead to an explanation of the mechanism governing the most baffling objects in the universe!

Autumn

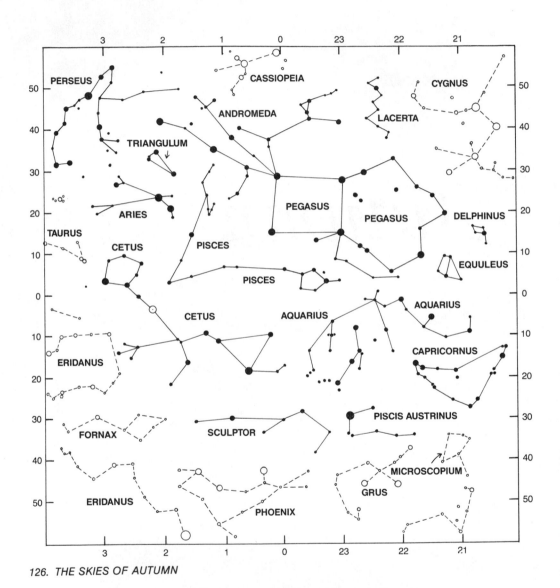

126. THE SKIES OF AUTUMN

ELEVEN

A Watery Domain

A group of water-related constellations occupy a region of the heavens known to the ancient Euphrateans as the "Sea," a vast area spanning most of the celestial sphere (as measured in hours of right ascension) and extending for some 90 degrees of declination, from about +30 degrees to −60 degrees. We have already visited the river Eridanus and Hydra the Water Serpent, major constellations of the "Sea," and as we continue our stellar cruise, we now sail eastwardly from Capricornus, the fanciful Goat-Fish. We dock in the realm of Aquarius the Water Bearer as he pours water into the mouth of Piscis Austrinus, the Southern Fish; after that, our ports of call are Cetus, the great Whale; Pisces, the Fishes, to the north; and, in a later chapter, the minor figure of Delphinus, the Dolphin (somewhat separated from its watery companions by the vast figure of Pegasus). In their fanciful "Sea," the Euphrateans also included Crater the Cup and Argo Navis the Ship, constellations which we encounter later on. All these sky pictures abound with astronomical wonders.

CAPRICORNUS (CAPRICORN)

We mentioned the struggle between the Olympian gods and the Titans, from which emerged the constellation Draco. Although the gods destroyed the Titans, their victory was short-lived; for the fearsome Typhon, son of Earth and the Titan Tartarus, rose up to avenge his forebears, as he carried a new challenge to the divine family. Typhon was bigger than the Titans; the spread of his arms could be reckoned in miles, and his hands terminated in one hundred dragons' heads; his legs were also serpentine, while his vast, winged body, taller than the mountains, bore an animal-like head which brushed the stars as fire spurted from his eyes and mouth. When the gods saw Typhon coming, they fled wildly, ultimately taking refuge in Egypt, but the monster pursued them to the banks of the Nile. To hide from him, they all assumed disguises in the forms of various animals: Zeus became a ram, Hera a snow-white cow, Apollo a crow, and Bacchus turned into a goat; although some poets tell us that this deity was not Bacchus, but the horn-headed and goat-footed Pan, who leaped into the Nile in panic, where he acquired a fish tail. After defeating Typhon, Zeus, highly amused by Pan's (or Bacchus') disguise, transferred it to the heavens as a zodiacal constellation east of Sagittarius and southeast of Aquila, allowing mankind to share his merriment.

Capricorn, the Sea Goat, originated long before the time of classical Greece, probably entering the zodiac in prehistoric days; its figure was apparently known as the "Father of Light" to the ancestors of the Babylonians about fifteen thousand years ago, when the Sun rose in its stars at the summer solstice; and in that connection, their descendants regarded the goat as sacred. Similarly, Capricornus appeared in very ancient oriental legends as the nurse of a young sun god, and subsequently as the "Southern Gate of the Sun" because of its position on the ecliptic; in ancient Chinese astronomy it was the zodiacal Ox, which later became *Mo Ki,* the "Goat-Fish." Some scholars claim that it was also known to the Assyrians as a goat-fish, identified with their tenth month, Dhabitu (December–January), and it is said that the biblical tribe of Naphtali adopted the sign of Capricorn as their emblem.

Other identities of Capricornus include an Egyptian water god bearing a goat's horns, who was thought to control the rising of the Nile, and an ibis-headed man riding on a goat, as shown on the Egyptian zodiac of Denderah. Capricorn's stars were also associated with the figure of a goat by the Persians, Turks, Syrians, and Arabians. In Greece, the followers of Plato called Capricornus the "Gate of the Gods," because they believed that the human soul, after death, ascends to heaven through its stars.

A Few of These Stars

Capricorn, a faint constellation, more nearly resembles a flying bat than a goat. On the tip of its western wing, (or horn), far to the south of Deneb and some 20 degrees south-southeast of Altair, lies *AL GIEDI (ALGEDI) (α) ALPHA 1* and *2* (RA 20 h 15 m δ − 12° 42′), from the Arabic *Al Jady,* "The Goat." The two stars, not gravitationally bound, form an optical naked-eye double because of their accidental alignment but are, in fact, widely separated (6′ 16″); α2, magnitude 3.56 and the brighter of the pair, is a class-G9 star 100 light years distant from us, whereas α1, magnitude 4.24, is some 500 light years distant. Alpha 2 has a close eleventh-magnitude bluish companion at a separation of 6.6 seconds, which the English astronomer William H. Smyth (1788–1865) said he glimpsed "in little evanescent flashes, so transient as again to recall Burns' snow-flakes on a stream." In 1862, Alvan G. Clark (who first saw the companion of Sirius) discovered that α2's faint companion (AGC 12 in his catalog) is a binary, both of its components being of magnitude 11 and their separation being only 1.2 seconds. Alpha 1, of spectral class G3, is a true triple system, with two gravitationally bound companions (m 9 and 13) at separations of 45.4 and 44.3 seconds, respectively.

About 2½ degrees south-southeast of α lies *DABIH (β) BETA,* a wide-contrast double (sometimes called Dabih Major and Minor), from *Al Sa'd al Dhābiḥ,* "The Lucky One of the Slaughterers," probably referring to a sacrificial rite performed by ancient Arabs when Capricorn rose with the Sun. Dabih Major, magnitude 3.08 and 150 light years distant, is a deep orange star, class F8 V (main sequence), and a spectroscopic triple with periods of 8½ days and about 4 years; its gravitationally bound companion, Dabih Minor, at a separation of 3 minutes 25 seconds, is a sixth-magnitude class-B9 binary, its components only 0.8 second apart.

At the opposite wing tip, marking the Sea Goat's fish tail, is **DENEB ALGEDI (δ) DELTA** (RA 21 h 44 m   δ −16° 21′) magnitude 2.82, from *Al Dhanab al Jady,* "The Goat's Tail," a Sirian-type main-sequence star only 50 light years distant. Delta is also an eclipsing binary with a magnitude variation of only 0.2; the Doppler shift of its spectrum indicates an almost circular orbit for its extremely close components. To early Arabs, δ and its neighbor γ, about 2 degrees to the west-southwest, were known as the "Bringer of Good Tidings" and also *Al Muḥibbain,* "The Two Friends."

The orange giant #41 Capricorni and the globular cluster **M30 (NGC 7099),** to its west-northwest (RA 21 h 38 m   δ −23° 25′)   m 8, form a glowing exclamation point in the night skies when they appear together on long-exposure photographs. Look for M30 about 3 degrees east-southeast of Zeta Capricorni, a yellow fourth-magnitude double which lies on the eastern rim of the "Bat's" wing. M30 is nearly 40,000 light years distant, but its slightly elongated shape shows up well in moderate-sized instruments. The cluster has a rich, compressed center from which emerge star streams that include several variables.

PISCIS AUSTRINUS

Formerly called "Piscis Australis," but Piscis Austrinus on modern charts, the **SOUTHERN FISH** lies on its back, drinking the starry water that flows from Aquarius' urn; this ancient concept is probably the origin of its occasional Latin title *Piscis Aquosus.* The Southern Fish figured in early legends as the parent of the more well-known although fainter Pisces, of the zodiac. Scholars believe Piscis Australis was associated with the Assyrian fish god Dāgōn and with several Egyptian deities, including the great human fish Oannes, Lord of the Waves and Source of All Knowledge. Other names for this constellation were the Arabic *Al Ḥūt al Janūbiyy,* or "The Large Southern Fish," and the Roman *Piscis Solitarius* and *Piscis Capricorni,* all adapted from the Greek. Rufus Avienus, Aratos' Latin versifier, called it the "Greater Fish," and it was also the "Golden Fish" of Longfellow's *Divine Comedy.* All its stars may be viewed by those of us no farther north than about 50 or 55 degrees latitude, but for these observers, it lies very low toward the southern horizon.

FOMALHAUT (α) ALPHA
(RA 22 h 54.9 m   δ −29° 53′)   m 1.17

Derived from *Fum al Ḥūt,* "The Fish's Mouth," its name has suffered innumerable variations, including "Fumahant" (Bayer), "Fomahand" (Caesius), "Phomault," "Phomaant," etc. (Riccioli), and "Fomalcuti" (Schickard). Early Arabs, however, called it *Al Difdi al Awwal,* "The First Frog," and it appears as *Ṭhalīm,* the "Ostrich" on the Borgian globe. Some three thousand years ago, when the Sun rose near Fomalhaut at the winter solstice (minimum solar declination, about December 22), the Persians called the star *Hastorang,* a "Royal Star," one of the four Guardians of Heaven, along with Regulus, Aldebaran, and Antares. About 500 B.C. it was worshiped at sunrise in the ancient Greek temple of Demeter (Ceres), and in the modern world it attained an exalted position in astronomy as well when Boguslawski, circa 1839, suggested that Fomalhaut might be the Central Sun of the Universe; an idea soon to be discarded, however, with a deepening understanding of the structure of the universe and its billions of galaxies. Years later, the universe of relativity, finite but unbounded, would offer no star such a preferred position.

The eighteenth-brightest star and only 23 light years distant, with fourteen times the Sun's luminosity, Fomalhaut, in a region where there are few other naked-eye stars, seems brighter than it is; atmospheric interference near the horizon gives it the illusion of reddishness, but its spectrum indicates a white main-sequence star (A3 V), with a color index of +0.09, comparable to Deneb's (+0.08). Fomalhaut has a faint companion, at a separation of 30 seconds, that is not gravitationally bound to it; but α's proper motion is shared by a much more distant star, the reddish dwarf *BS 8721,* lying 2 degrees, or nearly 1 light year, to its south; both stars recede from us at 4 miles per second.

The head of the Southern Fish is rounded out by three stars: *(β) BETA,* a magnitude-4.4 class-A common-proper-motion double about 6 degrees to the southwest of Fomalhaut, with a fairly large separation (30″.4), which presents no problem for the small telescope; *(γ) GAMMA,* a binary of the same type, lying some 4 degrees east-southeast of β; and nearby *(δ) DELTA,* a rather close yellow (G-type) double, just a degree to γ's northeast and about 3 degrees south of Fomalhaut. The Fish's tail is formed by Eta, Theta, Iota, and Mu, in the western part of the main figure, although the Arabians located the tail far to the south, in the present constellation of *Grus,* the "Crane." The Chinese had a more appealing identification of these four stars, as far as humans, rather than fish, are concerned: they called them *Tien Tsien,* "Heavenly Cash."

Less than 3 degrees southeast of γ and δ lies the F-type Cepheid variable *(π) PI* (RA 23 h 00.7 m   δ −35° 01′)   m 4.7 to 5.0. About 1 degree to its south-southeast, near the western border of the faint constellation Sculptor (introduced by Lacaille), lies another high-velocity star (his discovery), *LACAILLE 9352 (LFT 1758, CORDOBA 31353)* magnitude 6.49, a red dwarf only 11.9 light years distant and, with a proper motion of 6.90 seconds per annum, the star with the fourth-largest proper motion in the heavens. Most of its apparent motion is transverse to our line of sight, and at 75 miles per second, it covers about 1 degree every 520 years.

SCULPTOR

Before continuing on to Aquarius, the next promised stop on our celestial cruise, it may be well to pause briefly within the borders of Sculptor, one of fourteen modern groups dedicated to the sciences and the fine arts by the French astronomer Abbé Nicolas Louis de Lacaille (1713–62). Its complete title was *L'Atelier du Sculpteur,* "The Sculptor's Studio," which translates to *Bildhauerwerkstätte* on German charts. Sculptor lies just to the east and slightly to the south of Piscis Austrinus; it is notable for the cluster of galaxies called the "Sculptor group," or "south galactic pole group," which, at 110 kiloparsecs, is one of two aggregations of galaxies nearest our Milky Way, the other being the M81 group, of Ursa Major.

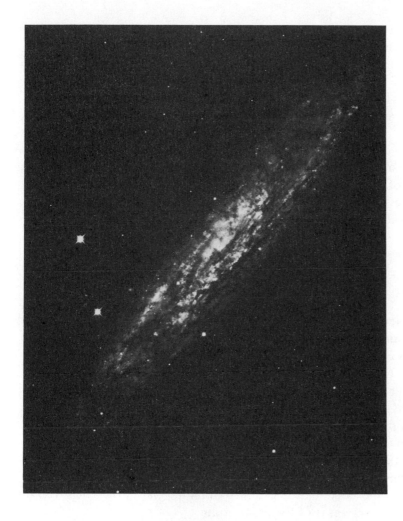

127. *NGC 253, a spiral galaxy in Sculptor,* a very large highly inclined spiral, spectacular for viewing in the southern hemisphere. Photographed with the Shane 120-inch reflector. (*Lick Observatory*)

An important member of the Sculptor group is *NGC 55* (RA 0 h 12.5 m   δ − 39° 50′), lying near the southern border of Sculptor, about 3.6 degrees north-northwest of the bright star Ankaa, in Phoenix. There is some dispute as to its type, probably an irregular galaxy or perhaps a loosely defined barred spiral. Its differential redshift arising from its rotation indicates that NGC 55 has about 46 billion solar masses, with a luminosity of some 6 billion L ⊙. Its distance is uncertain, estimates varying between 7.5 million and 9.8 million light years, but it is considered one of the nearest galaxies outside the Local Group.

Near the northern border of Sculptor, some 7½ degrees south of β Ceti, lies a spectacular spiral, the elongated, highly tilted galaxy *NGC 253* (RA 0 h 45 m   δ −25° 34′) (fig. 127), brightest member of the Sculptor group and comparable to the Andromeda galaxy, although rather low in the sky for northern observers. This galaxy, at its boundary, is rotating at about 360 miles per second and emitting gaseous material from its central regions at some 70 miles per second; fairly strong radio waves have also been detected.

About 8½ degrees south-southeast of NGC 253 and 4 degrees south of α Sculptoris, lies a very faint, diffuse dwarf elliptical galaxy known as the *SCULPTOR SYSTEM* (RA 0 h 57 m* δ −34° 00′), not suitable, however, for telescopic observation. Discovered by accident in 1938 at the Harvard Observatory as a supposed defect on a long-exposure time photograph, it soon revealed its stellar content on subsequent time-exposure photographic plates. Its brightest stars are about magnitude 18; located at a distance of about 270,000 light years as determined from a few Cepheids, it has a diameter of about 5,500 light years, and its luminosity is some 3 million $L \odot$. Another subluminous dwarf galaxy, the Fornax system, was discovered at the same time, and since then, similar systems have also been found in Leo, Draco, and Ursa Minor; all these midget galaxies are members of the Local Group. It is suspected that such dwarf galaxies may lie in large numbers beyond the Local Group but cannot be detected because of their extremely low luminosity.

AQUARIUS

Following our brief excursion to the galaxies of Sculptor, we resume our sea trip, cruising toward the boundaries of Aquarius, the Waterman, under whose control lies the celestial "Sea." Universally pictured as a man pouring water from a bucket, this figure appears on very early Babylonian stones, but the Arabians represented it by a mule carrying two water barrels, or by a water bucket alone.† The Babylonians associated Aquarius with their eleventh month, Shabatu, the "Curse of Rain" (January–February); their *Epic of Creation* associates this zodiacal sign with the Deluge. Aquarius' urn was the Babylonian *Gu,* an "Overflowing Water Jar," and also two Akkadian figures, *Ku-ur-ku,* the "Seat of the Flowing Waters," and Rammān, the god of the storm. The Egyptians called Aquarius "Monius" (derived from their *muau,* or "water"), who caused the Nile to rise as he sank below its surface to fill his giant urn. The glyph or symbol of Aquarius, water waves, is of Egyptian origin.

Aquarius was the Arabian *Al Dalw,* or "Well Bucket," and was also imaged as a water bucket by the Persians, Israelites, Syrians, and Turks. On a Roman zodiac, it was the "Peacock" or "Goose," Juno's symbol, because during her month, Gamelion (January–February), the Sun was "in Aquarius," which was also called *Junonis Astrum* when used as a diurnal sign. In ancient India and China it was the first zodiacal sign, the "Rat," whose ideograph signified water, and it was also a symbol of the Chinese emperor Tchoun Hin, in whose reign there was a great deluge; in fact, the legend of the Flood, familiar to us from the biblical account, appears to be worldwide and can be found in one form or another among all cultures. The "Noah" of Greek mythology was Deucalion, king of Phthia and son of the Titan Prometheus; he lived at the time of the so-called "Bronze Age," when, in Ovid's words, ". . . Piety lay vanquished, and the maiden Astraea,‡ last of the immortals, abandoned the blood-soaked earth." At a council of the gods, which was held in Zeus's palace on the Milky Way, man's impious behavior was dis-

---

* *(Not to be confused with the Sculptor group.)*
† *(The Muslim faith forbids reproduction of the human figure.)*
‡ *Virgo.*

cussed at length, and Zeus then announced his plan to destroy the human race by sending a vast downpour of rain upon the Earth. On the advice of Prometheus, Deucalion built a floating chest, which he filled with provisions, and then he and his wife, Pyrrha, entered this Greek version of the Ark, drifting within it upon the floodwaters for nine days and nights, until they came to Mount Parnassus, the only dry land. When he perceived that all the wicked were destroyed, Zeus tore the rain clouds apart and allowed the waters to recede. The innocent and pious Deucalion and Pyrrha, alone on the devastated Earth, consulted with the oracle Themis to learn by what means humankind might be restored, and they were advised, "Depart hence, and with veiled heads and loosened robes throw behind you as you go the bones of your great mother."* After some perplexity, Deucalion understood that her words referred to the stones of Mother Earth; thus, following the goddess's bidding, they took up stones and cast them backward over their heads; those thrown by Deucalion were turned into men, and those by Pyrrha, into women. The Earth itself spontaneously produced the lower forms of life, some in the ancient shapes and some new and strange.

Since the time of these mythical events, which supposedly occurred in the year 1500 B.C., the ancient Greeks associated the urn of Aquarius the Water Pourer with Deucalion's great Deluge, as the unending source of the waters that flooded the Earth. However, Aquarius was sometimes identified with Ganymede, Jove's Cup Bearer, the lad carried off by Aquila to form the now obsolete Antinoüs. Aquarius is an inconspicuous constellation, although it stretches across a large area of sky to the north and east of Capricornus, north of Piscis Austrinus and west of Cetus; thus a clear night is necessary for its proper observation and the identification of its stars.

Major Stars

The Water Carrier's right shoulder is marked by **SADALMELIK (α) ALPHA** Aquarii (RA 22 h 03 m δ −00° 34′) m 2.93, from *Al Sa'd al Malik*, "The Lucky One of the King" (or "Kingdom"), lying only 1/2 degree south of the celestial equator and a few degrees from the border of Pegasus, to the north. About 1,100 light years distant, α is a supergiant with a diameter about eighty times the Sun's and a luminosity of 6,000 $L \odot$; its spectral class (G2) and its surface temperature are similar to the Sun's.

Some 10 degrees to the southwest of α lies **SADALSUUD (β) BETA** (RA 21 h 28.9 m δ −05° 48′) m 2.86, from Al Sa'd al Su'ud, "The Luckiest of the Lucky," which later became the astrologers' *Fortuna Fortunarum*, a name signifying its heliacal rising at the start of the spring rainy season. The Euphrateans called Sadalsuud the "Star of Mighty Destiny." At 1,030 light years, β's spectral class, G0 Ib, is nearly the same as that of α, as is its luminosity, 5,800 $L \odot$. Two very faint optical companions lie in its field. Look for the bright globular cluster **M2 (NGC 7089),** about 5 degrees to the north and slightly to the east, suitable for small telescopes.

About 12 degrees west-southwest of β and far out toward the western border of Aquarius is **AL BALI (ε) EPSILON** (RA 20 h 45 m δ −09° 41′) m 3.77, which Bayer called *Mantellum*, because it marks the "towel" held in the Water Pourer's hand. Its Arabic name is derived from *Al Sa'd al Bula'*, "The Good Fortune of the Swallower," a

---

* *Ovid*, Metamorphoses, *Book 1, lines 149–50.*

title that included nearby Mu, and Nu some 6 degrees to the east-southeast. Epsilon, about 108 light years distant, is a white, Sirius-type, main-sequence star, with a radial velocity of 9.94 miles per second in approach.

Four stars, Gamma, Zeta, Eta and Pi, form the inverted water jar, the Roman *Urna*, which also looks like a tilted *Y*. Also marking the right elbow, about 4 degrees east-southeast of α, is **SADACHBIA (γ) GAMMA** (RA 22 h 19 m   δ −01° 38′)   m 3.84, derived from *Al Sa'd al Aħbiyah,* said to mean "The Lucky Star of Hidden Things," which may refer to its rising ("emerging from the sun's rays") at sunset during the spring thaw, when reptiles end their hibernation; but it is also said to signify the raising of the nomads' tents in the springtime. In a strange forecast of "things to come," the Capuchin friar De Rheita, of Cologne, thought he saw five new satellites of Jupiter when it entered Gamma's field in 1643, but they proved to be several faint stars just south of the urn. Gamma, a class-B9 giant 95 light years distant, with a luminosity twenty times the Sun's, is a spectroscopic binary with a period of about two months. About 2.5 degrees northeast of Gamma and almost on the celestial equator is *(ζ) ZETA,* magnitude 3.66, the central star of the *"Y."* This class-F2 subgiant is also a binary; its magnitude-4.4 and magnitude-4.6 white components, separated by 1.7 seconds, or about 100 astronomical units, are suitable for medium-sized instruments and are a good test for the quality of your 3-inch refractor. At 75 light years, ζ is receding at some 15 miles per second. An unseen red dwarf is apparently orbiting the secondary component at a separation of only 9 astronomical units.

Some 17° southeast of Gamma lies **SKAT (δ) DELTA** (RA 22 h 52 m   δ −16° 05′)   m 3.28, formerly known as "Scheat," the name now applied to β Pegasi. Delta Aquarii's proper name is derived from *Al Şāk,* "The Shinbone," which it indicates; the Euphrateans identified this star with their king Hasisadra, hero of their version of the ancient Deluge. Delta is a Sirian-type main-sequence star 85 light years distant. *Fl. #77,* a naked-eye companion immediately to its south, is not gravitationally bound. About 8½ degrees north and slightly to the west of δ lies *(λ) LAMBDA,* magnitude 3.8, which the Greeks called "The Water" and "The Outpouring," because of its position at the head of the stream of stars cascading from the Water Jar toward the open mouth of the Southern Fish.

A Few Variables

About 10 degrees north-northeast of δ and 6 degrees east of λ and lying just north of the optical triple ψ, is the irregular variable **SITULA (χ) CHI** (RA 23 h 14 m   δ −08° 00′)   m 5 to m 5.3; *Situla,* Latin for "Water Jar," is a red giant, and its ruddy color (as well as the story of Ganymede) may have influenced the Alexandrian Theon the Younger when he called it the "Outpouring of Wine"! Somewhat over 10 degrees to the southeast of χ we find the variable **R AQUARII** (RA 23 h 41 m   δ −15° 34′), lying just south of the very wide optical double ω *1* and *2,* and, like these stars, part of the waters "falling" from the jar. This pulsating red giant fluctuates between periods of stability lasting a few years and sudden peaks or drops in its light curve, with its magnitude ranging from 5.9 to 11.4. An emission star of class M7, with about 100 times the Sun's diameter and 230 times its luminosity, R seems to reveal by its spectral lines the presence of a hot but underluminous O- or B-type companion at a separation of only 1 astronomical unit. These dissimilar partners appear to take turns in displaying their variability, the red component becoming relatively quiet and placid when the blue star is actively fluctuating. The R Aquarii system is surrounded by an expanding lens-shaped gaseous nebula, and the binary itself is embedded in a variable nebulous disk. It is suspected that one of these two "symbiotic" stars may have suffered a nova-like eruption six hundred or seven hundred years ago, probably caused by an exchange of material between them, as in the case of erratic variables and novae.

Directing our attention to the northern part of Aquarius again, in the region of the Water Jug or *Y*, we find the faint variable *CY AQUARII* (RA 22 h 35 m δ +01° 17′) m 10.5–11.5, a pulsating B8 to A3 star with the second-shortest known period, 88 minutes. (Fourth magnitude η lies less than 1° to its south-southwest.) The term *dwarf cepheid* has been coined for CY Aquarii, which has a smaller radius and is denser and less luminous than RR Lyrae stars, with which it was formerly grouped.

Some 8½ degrees north-northwest of ε, in the northwest region of Aquarius near its border with Aquila, and in the field of #71 Aquilae (7½° east of Theta Aquilae) lies the irregular variable *AE AQUARII* (RA 20 h 37.6 m δ −01° 03′) m 10.7–11.5, an erratic

128. NGC 7293, the Helical Nebula in Aquarius, largest and closest planetary nebula. The central star and fine details including radiative threads are clearly shown. Photographed in red light with the 200-inch telescope. (*Hale Observatories*)

star with an amazingly complex pattern of light changes of small and large amplitude. Classed as an SS Cygni star, AE is also a very close binary, with a separation of a few hundred thousand miles and a period as short as ten hours; its components are a normal K-type dwarf, just leaving the main sequence as it approaches the subgiant phase, and a hot, dense star approaching the white-dwarf stage. Again, an interchange of gaseous matter between them probably causes the eruptive activity on the surface of the degenerate star, resulting in the wild fluctuations of the system's light curve.

A Great Planetary Nebula

About 8 degrees southwest of δ, just west of υ Aquarii and about 11 degrees north-west of Fomalhaut, lies the greatest and closest of the planetaries, *NGC 7293, the HELICAL NEBULA* (RA 22 h 27 m   δ −21° 06′), diameter twelve by sixteen minutes, about 450 light years distant. The thirteenth-magnitude central star is plainly visible in long-exposure photographs, as are numerous fine "threads" in the nebula radiating outward from the direction of the central star, an indication of the envelope's expansion (fig. 128); refer to the Ring Nebula, in Lyra, (p. 227).

Less than 6 degrees east-southeast of ε and just west of ν, on the Water Carrier's left (western) arm, is the bright planetary *NGC 7009, the SATURN NEBULA* (RA 21 h 01.4 m   δ −11° 34′), so named because of rays curving around both sides of the main gaseous disk, somewhat similar to Saturn's rings (fig. 129). Noteworthy is the nebula's deep green color, owing to doubly ionized oxygen excited by radiation from the extremely hot central star (55,000° K.).

129. *NGC 7009, the Saturn Nebula in Aquarius,* photographed in red light with the 120-inch reflector. (*Lick Observatory*)

CETUS

But see! as a swift ship with its sharp beak
plows the waves, driven by stout rowers' sweating
arms, so does the monster come, rolling back the
water from either side as his breast surges through.
Ovid—*Metamorphoses* IV, 706

Measured eastward from the flowing waters of Aquarius' urn to the banks of the River
Eridanus, Cetus' great body spans nearly three hours of right ascension, and its head
pokes some 20 degrees northwardly between Pisces and Taurus. Known to the Greeks
and Romans as the Sea Monster that Poseidon sent to devour Andromeda, Cetus the
Whale, despite its vast size, remains but an ineffectual threat, confined to its assigned
location among the stars and unable to reach its intended victim. It may have originated
in much earlier times, however, as the Euphratean dragon Tiamat, sharing that fero-
cious identity with Draco, Hydra, and Serpens. Before they adopted the Greek constellations, the Arabians separated Cetus' stars into three groups: a "Hand," "Hen Ostriches," and a "Necklace," but afterward they knew it as *Al Ḳeṭus*, which became the
more recent *Elketos* and Elkaitus. On medieval charts and globes it was pictured very
differently from a whale, showing ears, forelegs, and a long tail; this concept gave rise
to the *Belua* ("Beast") of the 1515 *Almagest* and the Alfonsine Tables, and to Lalande's
Chien de Mer, or "Sea Dog." A modern title is the "Easy Chair," probably the most apt
description of its stellar form.

Major Stars in the Whale

MENKAR (α) ALPHA, Fl. #92 (RA 2 h 59.7 m   δ +3° 54′)   m 2.52, sometimes called
"Menkab," is derived from *Al Minhar,* meaning "The Nose," although this reddish star
supposedly marks Cetus' open jaws. A class-M2 giant, Menkar is 150 light years distant,
with the luminosity of 175 Suns. A fifth-magnitude blue star to its north, **Fl. #93,** is not
gravitationally bound, but with #92 it forms a wide optical double and presents an
agreeable color contrast for the small telescope.

At the opposite end of the Whale lies **DENEB KAITOS (β) BETA** (RA 0 h 41 m   δ
−18° 16′)   m 2, also called **DIPHDA,** which is sometimes spelled "Dipda" or "Difda."
The name "Deneb Kaitos" is derived from *Al Dhanab al Ḳaiṭos al Janūbīyy,* "The Tail of
the Whale Toward the South" (its southern branch). "Diphda" is from *Al Ḍifdi 'al Thānī,*
the Second Frog (the first being Fomalhaut). The Chinese, however, saw β as *Too Sze
Kung,* "Superintendant of Earthworks"! About 60 light years distant from us, β is a K-
type orange giant with luminosity of forty Suns. Ptolemy registered it as only of the
third magnitude, but Agnes Clerke explains that β brightened up during the eighteenth
century. Some 10 degrees to β's northeast is *(η) ETA,* the Northern Tail, a magnitude-
3.5 class-K3 giant.

Returning to the stellar circle forming Cetus' head, we visit *(γ) GAMMA* (RA 2 h 40.7
m   δ +3° 2′)   m 3.5, marking the juncture of the head with the neck, 5 degrees west
and slightly to the south of α. Seventy light years distant, γ is an interesting contrast

double; its components are a class-A main-sequence star and a normal F-type dwarf separated by some 60 astronomical units; their period of revolution is so long that the observed change in their relative position is only about 3 degrees per century. *LTT 10888,* a faint red dwarf 14 minutes to the northwest, slowly circles these stars at a distance of some 18,000 astronomical units, thus forming a gigantic trinary system with them. About 3 degrees south-southwest of γ is the fourth-magnitude short-period variable *(δ) DELTA Ceti,* a class-B2 giant of the spectrum-variable, or β Canis Majoris, type.

Centrally situated in the Whale's body is *BATEN KAITOS (ζ) ZETA, Fl. #55* (RA 1 h 48 m δ − 10° 02′) m 3.9, from *Al Batn al Kaitos,* "The Whale's Belly"; about one-half degree to its southwest is the binary *(χ) CHI,* which forms a very wide double with Zeta. Writers of the seventeenth-century "biblical school," who renamed all the constellations according to the Scriptures, saw this "belly" as the one occupied by Jonah!

About 6 degrees south-southwest of Zeta lies *(τ) TAU, Fl. #52* (RA 01 h 41.7 m δ −16° 12′) m 3.5, 11.8 light years distant and estimated to be the seventh-nearest star. Very similar in type to our Sun, Tau Ceti is a class-G8 main-sequence star, its diameter 90 percent the Sun's and its luminosity some 45 percent $L \odot$, with an absolute magnitude of +5.7, as compared with the Sun's +4.7. There is much interest in solar-type stars as the possible possessors of planetary systems similar to our Sun's. This raises the fascinating question of life beyond the Solar System, a concept that is vigorously contested by those who believe that we on the Earth are truly alone. In the words of British astronomer Fred Hoyle, with ironic reference to his colleagues, "Most of us are emotionally pre-Copernican and like to think that the Earth is at least the centre of life."† However, the notion that life may exist only on our tiny outpost at this one location in the vastness of space seems to impute a lopsidedness, even a grotesqueness, to the unimaginably huge, dynamic Universe with its billions of evolving galaxies. We therefore believe that some proof of extraterrestrial life will one day arise, shaking to its roots the last pre-Copernican orthodoxy, that of exclusively terrestrial life!

About 2½ degrees southwest of Tau Ceti lies the red dwarf binary *L 726-8,* containing the flare star *UV CETI* (RA 1 h 36.4 m δ −18° 13′) m 7–12. Only 9 light years distant, it is probably the sixth-nearest star, with the very large proper motion usual for such close objects. The components of this system are among the least massive stars known, and their luminosities are correspondingly small. UV Ceti B, the fainter member, also called Luyten's flare star, is the standard example of its type, with sudden eruptions that may occur in less than a minute and lower its magnitude by as much as one or two steps, followed by a return to normal luminosity (0.00003 $L \odot$) within a few minutes. The exact spectral type is dM6e, indicating an emission spectrum, but during an eruption a bright continuous spectrum is superimposed upon it. Radio waves have also been detected following each flare-up.

A "Wonderful" Star

About 6 degrees southwest of δ and almost on a direct line between γ and ζ is a famous star, the classical long-period variable *MIRA THE WONDERFUL (o) OMICRON*

---

† The Observatory, *October 1980, meeting of the Royal Astronomical Society.*

Ceti (RA 2 h 16.8 m   δ −3° 12′)   m 3–9.5; other names are *Stella Mira* and *Collum Ceti.*
Its variability was accidentally (and unknowingly) discovered in the year 1596 by David
Fabricius, an amateur astronomer and disciple of Brahe. Fabricius noticed that a third-
magnitude red star that marks the Whale's neck was no longer visible; the temperamen-
tal star reappeared sometime before the year 1603, however, when Bayer designated it
o in his *Uranometria,* where he introduced the system of applying Greek and Roman
letters to the stars. In 1638–39, some thirty years after the invention of the telescope,
the Dutch astronomer Phocylides Holwarda observed Omicron Ceti's light cycle, a
period fixed with fair accuracy at 333 days by Bouillaud in 1667. This was the first
established record of a variable star, preceding John Goodricke's observations of Algol,
in Perseus, and Delta Cephei, by nearly a century and a half. The excitement among
astronomers generated by the phenomenon of o Ceti's light changes and the detection
of stellar periodicity was detailed by Hevelius in 1662 in his *Historiola Mirae Stellae* ("A
Short History of the Wonderful Star"), and ever since then o has been known as "Mira
the Wonderful."

On November 6, 1779, Sir William Herschel observed that Mira had become almost
as bright as Aldebaran, thus temporarily joining the ranks of first-magnitude stars.
Normally, however, its amazing light variations range from an occasional second magnitude at maximum luminosity all the way to ninth or even tenth magnitude at minimum, within an average period of 331 days; the interval between successive maxima
ranges anywhere from 320 to 370 days, with an average period of 330 days. Its spectral
class varies from M6e at maximum to M9 at minimum, and its red color gradually
intensifies as it reaches minimum luminosity, a characteristic of all pulsating red giants.
Mira's temperature at minimum, about 1,900 degrees K., places it among the coolest of
stars; one of the ten largest, with a mean diameter of some 350 million miles, or about
four hundred times the Sun's diameter, it is extremely tenuous, with an average density
no more than 0.0000002 that of the Sun. Mira's distance, about 220 light years, gives it
an estimated luminosity of 250 $L \odot$ at maximum and slightly less than the Sun's
luminosity at minimum!

Mira's spectrum is typical for variables with periods ranging from 150 to 450 days; it
shows strong, dark bands of titanium oxide with the appearance of bright emission
lines of hydrogen at maximum, which originate deep below the photosphere, as we
have previously discussed (p. 167). Mira is also much more luminous at maximum than
its minimum luminosity leads us to expect, according to the principles of the *Stefan-
Boltzmann law* of black-body radiation, which states that if the absolute temperature of a
black body‡ is doubled, each square centimeter of its surface emits 2^4 times, or 16
times, as much energy per second as previously, and if it is tripled, 3^4, or 81, times as
much as previously, etc. According to this principle, each square centimeter of Mira's
surface at maximum luminosity, when its temperature is 2,600 degrees K., should
radiate 3½ times more rapidly than at minimum, when its temperature is 1,900 degrees K. If, then, the surface temperature alone changed during a cycle, Mira would be
3½ times brighter at maximum; however, it is five magnitudes, or a hundred times,
brighter (both visually and photographically) at maximum. There is thus a twenty-five-
fold discrepancy between the observed luminosity and that calculated from the Stefan-

‡ *I.e., a perfect absorber, or surface that completely absorbs all wavelengths of radiation falling on it.*

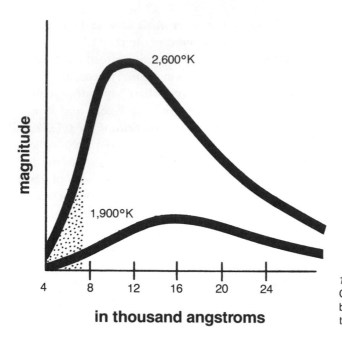

130. The energy distribution of Mira Ceti at maximum and minimum brightness. The shaded area is in the visible part of the spectrum.

Boltzmann law. Part of this is due to Mira's larger radius at maximum luminosity, about a 20-percent increase, which gives it an area that is larger by a factor of 3/2. But this accounts for only a small fraction of the discrepancy and we must look elsewhere to explain it all. A large part of the remaining discrepancy is due to the change in shape of the energy distribution curve, showing a shift in Mira's radiation from the invisible infrared wavelengths (at minimum) over to the visual and photographic part of the spectrum at its maximum luminosity and correspondingly higher temperature of 2,600 degrees K.

At 2,600 degrees K. much more energy is in the visible and photographic wavelengths (the shaded part of figure 130) than there is at 1,900 degrees K., so a proper comparison can be made only if all the infrared radiation at minimum is added to the visible radiation. If the total radiation (most of it in the infrared) is taken into account at both maximum and minimum, the luminosity during a single cycle of Mira changes by about one magnitude, and the discrepancy between the observed and calculated luminosity is reduced considerably. The remaining discrepancy may be due to the intensification of the absorption bands and lines when the star is faint, along with the possible formation of an obscuration cloud composed of substances able to exist at lower temperatures, factors tending to reduce Mira's minimum observed luminosity still further.

The Companion of Mira

In 1918, A. H. Joy discovered a tenth-magnitude B-type spectrum associated with Mira's characteristic M-supergiant spectrum. This indicated a companion, which was finally detected visually as a hot bluish subdwarf by A. G. Aiken in 1923 with the 36-inch refractor at Lick Observatory. Subsequent observations showed it to be intrinsically variable, with a period of about fourteen years and a luminosity range from about the tenth to the twelfth magnitudes. Mira and its companion have the same proper motion, with a radial velocity of recession, of about 38½ miles per second. With a mean separation of some 50 astronomical units, the two components complete one revolution about their common center of mass in two hundred years. This indicates that the Mira system has a total mass of about 3.7 $\mathcal{M}\odot$, much smaller than earlier estimates. The companion is thirty-three hundred times as dense as the Sun, although its diameter is one-eleventh the Sun's; its rather variable spectrum contains strong bright lines of hydrogen, helium, and ionized calcium. Other long-period variables of the Mira type belonging to various late spectral classes are the pulsating giants R Aquilae, S Cephei, R Hydrae and Chi Cygni.

A Galaxy Group

A small group of spirals lie clustered to the east of δ. Chief among them is *M77 (NGC 1068, 3C 71)* (RA 2 h 40 m δ −0° 14′) m 10, about 1 degree southeast of δ. About two by two and a half minutes in area, its unusual, triple-armed spiral pattern shows up best in larger telescopes (fig. 131). A point of interest concerning M77 is its very large redshift, indicating a recession speed of 620 miles per second; along with the "Sombrero," in Virgo, it is one of the first galaxies whose extremely large rate of recession, measured by Slipher in 1913, led to the theory of the expanding universe. M77, estimated at some 60 million light years, with its extremely condensed, bright nucleus and faint outer portions, is classified as an Sb-type Seyfert galaxy. Moderately strong radio emission appears to originate in the central area, presumably the site of vast explosions from which high-velocity and very massive gas clouds have been ejected, as determined by measurements of the Doppler shift in the spectrum of M77.

In the northwestern part of the Whale, almost on the border with Pisces, lies the irregular dwarf galaxy *IC 1613* (RA 1 h 02.5 m δ −01° 52′) m 11.5, similar to the Magellanic Clouds and also a member of the Local Group of galaxies, but very dim and not suitable for smaller instruments. IC 1613 is of interest chiefly because of its many Cepheids, whose period-luminosity relationships have been used to determine its distance, about 1.8 million light years. A gas- and dust-filled region in the northeastern part of the galaxy, containing a huge association of blue giants, is another possible "baby-star cradle": like the Orion nebula, a birthplace of new stars.

ARIES

The "Sea" also relates to the story of Aries, a golden ram that saved the life of Phrixus, eldest son of Athamus, king of Boeotia, in Thessaly. Ino, Phrixus' stepmother, in bitter conflict with Queen Nephele, Phrixus' mother, intrigued against Nephele's children by persuading peasant women to parch the wheat and cause a crop failure. Then, under Ino's influence, messengers returning from the Oracle at Delphi gave the king a false report, instructing him to sacrifice Phrixus to Zeus and thus restore fertility to the soil.

131. M77 (NGC 1068, 3C
71), an emission-line
spiral nebula in Cetus with
an unusual spiral pattern.
Photographed with the
120-inch reflector. (*Lick
Observatory*)

At the insistence of his subjects, the sorrowing Athamus was about to carry out this advice when Hermes sent a dazzling golden ram to Nephele, who placed Phrixus and his sister Helle upon its back. Before the awestruck populace could react, the ram bore the children swiftly away; but as it flew over the Thracian straits, Helle lost her hold and fell into the sea, which was ever afterward known as the "Hellespont." Phrixus arrived safely on the eastern shore of the Black Sea and was well received by Aeëtes, king of Colchis. In thanks to Zeus (who, conveniently, was also the god of escape), Phrixus sacrificed the ram and gave its fleece to Aeëtes, who nailed it to an oak in a sacred grove of Ares (Mars). Guarded by a sleepless dragon, it was eventually claimed by Jason and the Argonauts, who returned it to Thessaly; but the ultimate resting place of Ares' Golden Fleece was in the zodiac to the north of Cetus' head and west of the Pleiades.

Known to the Romans as *Portitor Phrixi*, the "Bearer of Phrixus," the ram is also associated with the flight of the Olympian gods from Typhon as the disguise assumed by Zeus (Jupiter), hence Aries' Roman titles of *Jupiter Ammon* and *Jovis Sidus.* Aries, though, may have originated, long before the days of Greek mythology, as the heavenly representation of the Euphratean *Tammuz Dum-uzi,* the "Only Son of Life"; its entrance into the early zodiac as a Babylonian sign apparently occurred some three thousand years ago, when the Sun first rose "in its stars" at the vernal equinox. That point on the celestial equator from which right ascension is measured, still referred to as the First Point of Aries, now lies in the neighboring constellation of Pisces, owing to the westward precession of the Earth's axis; but the Sun was in the stars of Aries, or the Hebraic *Teli,* during the Hebrew month of Nisan (March–April), when the Israelite slaves were released from Egypt to follow Moses into the wilderness; forty years later, a ram's horn rallied their forces in battle and seven of them brought down the walls of Jericho! Known as a Ram or Sheep to the Arabians, Syrians, Persians, and Turks, Aries became the leader of all the signs of the zodiac, although as the ancient Chinese *Kung,* which progressed in opposition to the Sun's apparent path, it was the eleventh sign and pictured as a "Dog." In seventeenth-century Europe, astronomers and writers of the "biblical school" identified Aries as Abraham's Ram, sacrificed in place of Isaac, a story with elements similar to that of Athamus and Phrixus, insofar as the impending tragedy is averted when a ram appears; in a similar vein, the Dutch writer Caesius designated Aries as the Lamb of Calvary, sacrificed in place of all humanity.

Principal Stars

The Ram's forehead is marked by **HAMAL (α) ALPHA** (RA 2 h 4.3 m δ +23° 14') m 2, a golden giant some 75 light years distant, its name derived from *Al Ḥamal,* "The Sheep," which is the Arabian title for the entire constellation. Another name for α is "Arietis," "of Aries", the Roman name for the whole figure. References to Hamal are found on cuneiform inscriptions of the ancient Euphrateans, which have been interpreted either as *Dil-kar,* the "Proclaimer of the Dawn," or *Dil-gan,* the "Messenger of Light"; other Euphratean names were *Lu-Lim,* the "Ram's Eye," and *Si-mal,* the "Horn Star." According to the English orientalist Robert Brown, Jr., it was associated with the first of ten mythical kings of Akkad, each represented by one of the ten chief ecliptic stars, Hamal, Alcyone, Aldebaran, Pollux, Regulus, Spica, Antares, Algedi, Deneb Algedi, and Scheat, the duration of the kings' reigns proportionately coinciding with the apparent distances separating these stars. Many Greek temples dating from 1580 B.C. to 360 B.C., particularly those dedicated to Zeus and Athene, were oriented to Hamal because of the fleeing god's transformation into a ram, his celestial symbol. In recent centuries, Hamal was an important navigational star because of its position about 11 degrees north of the ecliptic.

Some 4 degrees southwest of α, marking one of Aries' horns, is **SHERATAN (β) BETA** (RA 1 h 51.9 m δ +20° 34') m 2.65, from *Al Sharatain,* "The Signs," a designation including γ, to the south, and referring to the new year, because β marked the vernal equinox when these stars were named, a century or so before the present era. They also shared the title *Al Nāṭiḥ,* "The Butting One," which duplicates the Arabic name of β Tauri. As the twenty-seventh lunar station of India, β and γ Arietis were known as "Horsemen," whereas the Persians, Sogdians, and Copts saw them as a "Protecting

Pair." A Sirius-type main-sequence star, sometimes described as "pearly white," β is also a spectroscopic binary with an extremely eccentric orbit and a separation of 15–20 million miles.

About 1½ degrees to the south and slightly to the west of β is *MESARTHIM (γ) GAMMA* (RA 1 h 50.8 m   δ +19° 03')   m 3.90, also "Mesartim," possibly from the Hebrew *Mesharetim,* or "Ministers," a reference to the lunar station it marked with β. The three stars α, β, and γ also formed one of the Arabic "tripods," three stones on which the desert wanderer placed his kettle. At 160 light years, Mesarthim is a fine double, suitable for the small telescope; its duplicity was discovered by the English scientist Robert Hooke while he was observing the comet of 1664. The components are 7.8 seconds apart, with a combined luminosity of 50 Suns; they are nearly matched in spectral type, one of them a B9 main-sequence star and the other a class-A0 magnetic variable of the Canes type (Cor Caroli), but with strong silicon lines in its spectrum. A faint close double, not gravitationally bound to the bright pair, lies 3 minutes 41 seconds distant.

Near the borders of Taurus, about 10 degrees southwest of the Pleiades, is *BOTEIN (δ) DELTA,* a 4.5-magnitude star marking the Ram's tail, although its name derives from the Arabic *Al Baṭn,* "The Belly"; *(ε) EPSILON,* magnitude 4.6, a close A-type double, lies some 3½ degrees to δ's northwest, marking the base of the tail. Some 6½ degrees to ε's northwest we find *#30 Arietis,* (RA 2 h 34 m   δ +24° 26')   m 6.67, 190 light years distant, an excellent double for the small telescope; its components are normal dwarfs of types F5 and F6, magnitudes 6.5 and 7.5, respectively, and the brighter "twin" is also a spectroscopic binary. Two degrees southwest of δ lies *#53,* a class-B variable of the β Canis Majoris type and one of three "runaway" stars, whose high space velocities apparently originated a few million years ago when they were ejected from the association of young, hot stars within the Orion nebula. Fl. #53 speeds along at about 35 miles per second, resulting in its large proper motion of 0.025 second per annum.

The small triangle of stars some 8 degrees to the northeast of α, Fl. #s 35, 39, and 41, which more or less indicates the Ram's back, has been pictured as a separate constellation by Flamsteed and others, named MUSCA BOREALIS, or the "Northern Fly." The brightest member of this group, *Fl. #41,* is a magnitude-4.5 class-B8 star forming an optical triple with two very faint (m-11) companions at separations of 24.6 and 31.3 seconds. *Fl. #33,* just 2 degrees to its west, is also an A-type double; the magnitude 6 and 8½ components lie 28.6 seconds apart.

TRIANGULUM

Immediately to the north-northwest of Aries lies the small, faint group of stars forming a Delta (Δ) or a triangle. The Greeks called it *Deltoton,* but it was also associated with the Nile as *Aegyptus* and *Nilus.* Its three-sidedness suggested such names as *Trigonum* and *Tricuspis,* and the triangular shape of the island of Sicily, which Ceres begged Jove to reproduce in the sky, gave it the names *Sicilia* and Trinacria. It was also the Hebraic *Shalish,* a biblical three-stringed instrument.

About 7 degrees north-northwest of Hamal lies the star that marks a vertex of the triangle, **CAPUT TRIANGULI (α) ALPHA** (RA 1 h 50 m   δ +29° 20′)   m 3.49, meaning the "Head of the Triangle," the Arabians' *Rās al Muthallath.* An orange subgiant 65 light years distant, with a luminosity of some thirteen Suns, α is also a spectroscopic binary with a period of only 1.7 days.

At the northern corner of the triangle, some 6 degrees northeast of α, lies *(β) BETA,* at magnitude 2.99 the brightest star in this small figure. A white "Sirian" giant, 140 light years distant, it has a luminosity of 90 L ⊙. Like α, β is a spectroscopic binary, but its components, at a mean separation of 6 million miles, revolve around their common center of mass about once a month.

Just outside the triangle, about 3½ degrees east-northeast of α, lies a fine double, *(ι) IOTA,* magnitude 5.05, its colorful components often described as "yellow and blue," although their spectral types, G5 and F6, indicate that they are a "golden yellow" pair. Some 200 light years distant, separated from one another by 3.8 seconds, or 250 astronomical units, each of these stars is also a spectroscopic binary.

An interesting object in Triangulum is **M33, the PINWHEEL GALAXY** (RA 1 h 31 m   δ +30° 24′), lying almost on the border of Pisces, about 3½ degrees west-north-west of α Trianguli and 7 degrees southeast of β Andromedae. A member of the Local Group of galaxies and about 2.4 million light years distant from us, M33 is the nearest Sc galaxy; it was discovered in 1764, but its faintness made it virtually impossible for earlier observers to resolve its stars. To view the "Pinwheel," use very low-power wide-angle glasses on a clear night, in a location free of interfering lights. In the long-exposure photographs of our great observatories it is revealed as a large double-armed spiral, irregular in structure, with a curdled appearance caused by bright clumps of stars and dark nebulous masses of gas and dust lying in its broad and fuzzy arms. It is difficult to see from what points of the nucleus these irregular and partly formed arms originate (fig. 132).

A large number of class-B supergiants give the galaxy a somewhat bluish color, but it also contains a rich variety of other stellar objects, including luminous O and B stars, at least three thousand red supergiants, many Cepheids and irregular variables, and both open and globular star clusters. Between 1960 and 1972, five novae were observed. The visible diameter, of 60 minutes, is some 42,000 light years across, but outlying areas may increase this figure by nearly a third. With a luminosity of about 3 billion L ⊙ and an absolute magnitude of − 19, estimates place its mass at about 8 billion times the Sun's, and recent radio studies show that more than one fifth of it (1.6 billion ℳ ⊙) consists of neutral hydrogen. The Pinwheel rotates clockwise, probably completing a "full turn" once every 200 million years!

On the northeastern edge of the galaxy, out near the end of one of its spiral arms, lies **NGC 604,** a great emission nebula and star association some 1,000 light years in diameter, considered to be one of the largest HII regions within any galaxy; it is thirty times larger than the Orion nebula.

The First Quasar Discovered

We told the story of quasi-stellar radio source 3C 273, in Virgo, and its baffling spectrum. The first of these remarkable objects optically identified was *3C 48,* in

132. *M33 (NGC 598), the Pinwheel Galaxy, in Triangulum,* a large but faint double-armed spiral, one of the Local Group of galaxies; contains a rich variety of stellar types. Central region, photographed with the 200-inch telescope on Palomar. (*Hale Observatories*)

Triangulum (RA 1 h 35 m   δ +32° 54′)   m 16.2, lying about 5 degrees northwest of α and some 5 degrees west of the white double star ε Trianguli. The riddle of the broad emission lines in the spectrum of 3C 48 was solved when they were identified as the ordinary hydrogen lines redshifted by 37 percent. Its optical spectrum, containing a considerable ultraviolet excess, is rich in synchrotron radiation (p. 141), indicating the presence of magnetic fields and relativistic electrons. From its speed of recession, some 60,000 miles per second, utilizing Hubble's law, astronomers estimate the distance of 3C 48 at about 6 billion light years. During the past twenty years, thousands of quasars have been discovered; as many as one million are now believed to exist.

PISCES

When the gods fled to Egypt and assumed various animal-like disguises to escape the monster Typhon, Aphrodite (Venus) and her son Eros (Cupid) transformed themselves into two fishes and swam away in the Nile. The images that concealed them were placed in the zodiac, where they are known as Pisces, a lengthy double constellation occupying the regions south of Andromeda and Pegasus and northwest of Cetus. The "Fishes" acquired the title "Leaders of the Celestial Host," after precession brought the vernal equinox to a point south of the star Omega in the southwestern fish, about 2 degrees west of the line formed by the "Three Guides," β Cassiopeiae, α Andromedae, and γ Pegasi (fig. 31, p. 66). In very ancient times, Pisces was the Babylonian Nūnu, or "Fish"; it symbolized the Syrian goddess Derke as well as Atargatis, who was important to many ancient nations and depicted as a huge fish with a woman's head. Pisces was also the Hebrew *Dagaïm,* or "Two Fishes" (some say the sign of the tribes Simeon and Levi), and the Syrian god *Dāgōn*—names of similar origin. With the Arabians, it was *Al Hūt,* "The Fish," but also part of much more extended stellar figures, the lunar mansions of ancient India and Arabia. Such names as *Venus et Cupido, Venus cum Adone,* etc., refer of course to its Greek myth; but Pisces was also significant to the early Christians and is said to represent the "Miracle of the Loaves and Fishes," by which Jesus fed the multitudes.

At the juncture of the two Fishes and marking the constellation's southeastern corner, and, we note, some 7 degrees northwest of Mira the Wonderful, is *AL RESCHA (AL RISCHA) (α) ALPHA* (RA 1 h 59 m   δ +02° 31′)   m 3.96, 130 light years distant from us; its name is derived from *Al Risha,* "The Cord," because it represents the knot in a flaxen thread formed by several stars uniting the Fishes. Alpha Piscium was first observed as a binary by Sir William Herschel in 1779, when its components, then 3.1 seconds apart, were near their widest separation; approaching each other in an elongated orbit, they are now about 1.8 seconds apart and no longer suitable for small instruments. Their period of revolution, about 720 years, will bring them closest together in the year 2060. Both components are A-type stars, but the spectrum of the fainter star has strong metallic lines; together, they have the luminosity of about thirty-five Suns.

Near the westernmost border of Pisces, marking the mouth of the Western Fish, lies *FUM AL SAMAKAH (β) BETA,* from the "Fish's Mouth" (RA 23 h 02 m   δ +3° 49′)   m 4.5. With Gamma, Theta, Iota, and Omega, it forms a jagged line that the Chinese called *Peih Leih,* or "Lightning"! Beta is a blue-white (class-B5) star with a peculiar spectrum that indicates no radial velocity and only a slight transverse motion toward the east; thus, relative to our line of sight, β Piscium seems to be almost standing still. Its distance has not yet been determined.

To the east of β, a ring of six bright stars called the *Circlet of Pisces* indicates the Fish's head. The easternmost of these stars are Iota and Lambda; lying between them and somewhat to their east is the striking red variable *#19 (TX) PISCIUM* (RA 23 h 44 m   δ +3° 13′)   m 5.5 to 6, a cool, gigantic N-type (carbon) star, noted for its excellent color and easily viewed with field glasses or the small telescope. No parallax has been detected for it, but its estimated distance is some 1,000 light years.

On the long line of stars that constitute the body of the Western Fish, midway between α and the spectroscopic binary ω, and about 12 degrees southeast of third-magnitude γ Pegasi (one of the "Three Guides" already referred to), we find *(δ) DELTA* (RA 0 h 49 m   δ +7° 35′)   m 4.43, a class-K5 orange giant, 88 parsecs, or 288 light years, distant; in its field is sixth-magnitude *Fl. #62,* a deep yellow giant 326 light years distant. About 2 degrees to its south is the noted white dwarf *VAN MAANEN'S STAR (WOLF 28, LFT 76)* (RA 0 h 47 m   δ +5° 9')   m 12.4, only 13.8 light years distant. Its spectral class, late F or early G, is similar to the Sun's, but its luminosity is only about .00017 $L \odot$. With an Earth-like diameter, about 7,800 miles, and a mass equal to the Sun's, its density is about twenty tons per cubic inch, or ten times that of the companion to Sirius; its surface gravity is more than fifty thousand times the Earth's. Because it is nearby, Van Maanen's star has the large proper motion of 2.98 seconds per year.

About 6½ degrees to Delta's east is the fine double star *(ζ) ZETA* (RA 1 h 11 m   δ +7° 19')   m 4.8, the first star of the Hindu lunar mansion Revati, which means "rich"; ζ was prominent in the Hindu astronomy of A.D. 572, when the vernal equinox lay near it. Its components, 23.5 seconds apart, are easily viewed in a small telescope, and their spectral types, A5 and dF6, present an attractive color contrast. Their Doppler shifts indicate that each is a spectroscopic binary. A very faint (m-12) companion, apparently an underluminous dwarf, lies only 1 second from Zeta 2; these factors make its detection extremely difficult.

An interesting but difficult double in the Northern Fish, almost 15 degrees northwest of α, is *(η) ETA* (RA 1 h 29 m   δ +15° 05')   m 3.7, which may have belonged to an early Babylonian constellation, *Kullat Nūnu,* the "Cord of the Fish," an indication of Pisces' great age. The primary component of η is a class-G8 yellow giant about 44 parsecs, or 143.44 light years distant, and the eleventh-magnitude secondary, at the small separation of 1 second, shares its slight transverse motion toward the east.

Just 1½ degrees east-northeast of η lies the large face-on spiral galaxy *M74, NGC 628* (RA 1 h 34 m   δ +15° 32'), its dimensions nine by nine minutes. Although M74 was discovered in 1780, its spiral character was undetected until 1848, and it has remained a difficult object because of its faintness. Under optimum conditions it may be viewed with a low-power eyepiece. Long-exposure photographs reveal an excellent example of the type-Sc "classical" spiral, its arms originating from two well-defined points in the nucleus (fig. 133). Nearly 30 million light years distant, M74 has a diameter of some 80,000 light years, and its luminosity is about 13 billion times that of the Sun, with a mass of 40 billion $\mathcal{M} \odot$. These figures indicate that M74 is underluminous for its estimated size and distance, but its redshift of 426 miles per second implies that it may be farther away than hitherto realized. Compare this large galaxy to the much less well-defined Pinwheel galaxy, in Triangulum (p. 310).

A Fast Departure

Near the Pisces-Pegasus border, about 2½ degrees east-northeast of Gamma Pegasi (the star marking the southeastern corner of the Great Square of Pegasus), lies the quasar *3C 9* (RA 0 h 18 m   δ +15° 24')   m 18.21, noteworthy because of its tremendous redshift, 2.012, which corresponds to a speed of recession equal to about 0.8 the speed of light! (Refer to p. 192 for 3C 273, in Virgo, and the discovery of quasars.)

About 1 degree to the west of 3C 9 lies another of the very distant galaxy clusters recently under observation. The very faint and extremely blue cluster *0014.4-1551,* which occupies an area of about 1.5 square minutes, was photographed in ultraviolet, blue, and red light with the Kitt Peak Mayall 4-meter telescope. Its extreme blueness is expected for a galaxy cluster with a very high redshift, which implies that it is young and is receding at a very high velocity. Not a radio or X-ray source, this cluster of galaxies lies at the extreme limits of the observable universe, billions of light years distant. As new techniques are developed for understanding the composition of such remote galaxies, astronomers will learn more about the large-scale structure of the universe and its rate of expansion, as well as the origin of galaxies.

A New Discovery

In the far northern part of Pisces, approximately 15 degrees east-southeast of α Andromedae (the bright star occupying the northeastern corner of the Great Square of Pegasus), is the elliptical galaxy *NGC 326* (RA 0 h 57 m   δ +26° 44′), associated with the radio source *4C 26.03;* the latter consists of two "inversion-symmetrical" cosmic gas jets that depart from the center of NGC 326 in opposite directions, one bending upward and the other downward, and terminating in extended radio lobes. This type of bending may result from precession of the galaxy's spin axis. (For a detailed account of cosmic jets, see radio source 3C 449 in Lacerta, p. 324.)

133. M74 (NGC 628), spiral galaxy in Pisces; a beautiful example of a "classical" face-on spiral with unwound multiple arms. Note the well-defined dust lanes and small, compact nucleus. Photographed with the 200-inch reflector. (*Hale Observatories*)

TWELVE

A Royal Rescue

A chain of mythic events links several autumnal groups. The hero Perseus inadvertently acts as midwife when he lops off the Gorgon Medusa's head and the winged horse Pegasus springs from her blood. Perseus then slays the sea monster Cetus and rescues the princess Andromeda, daughter of Cepheus and Cassiopeia. Auriga the Charioteer stands nearby; we like to imagine that he is ready to whisk Perseus and Andromeda off on their honeymoon! Most of these constellations lie far to the north and are therefore visible for the greater part of the year to observers at latitudes of 40 degrees or more. They contain many important stellar objects and new discoveries for us to explore, beginning with a visit to mythology's famous thoroughbred, Pegasus.

PEGASUS

At the command of Polydectes, king of Seriphos, Perseus set out to fetch the head of Medusa, which was covered with writhing serpents and turned to stone any who beheld it. Fixing his gaze upon her image reflected in his shield, he beheaded her while she slept, and in that instant a snowy white horse which she had conceived by Poseidon (Neptune) sprang to life. In the earliest tales, the horse Pegasus had no wings, but as his story grew, so did his wings. Pegasus' first deed was to cause the Muses' fountain to flow at Pirene, on Mount Helicon, by striking the ground with his hoofs; there beside the stream, Prince Bellerophon of Corinth harnessed the spirited horse with a golden bridle, the gift of Pallas Athene, daughter of Zeus and a lover of horses. Bellerophon, a descendant of Deucalion and Pyrrha (p. 298), had been recommended to Iobates, king of Lycia, who sought a hero capable of saving his land from the Chimaera, a destructive three-headed beast with the fore part of a lion, tail of a dragon, and a goat's head that belched fire. Flying through the air upon his newly acquired horse, Bellerophon easily dispatched the Chimaera; after other successful exploits, Bellerophon ordered Pegasus to take him to Mount Olympus, but Jupiter in his wrath caused an insect to sting the winged steed, which threw its impious rider to the Earth. Succeeding where Bellerophon had failed, Pegasus became a resident of Mount Olympus as the "Thundering

Horse of Jove," carrying the god's lightning; according to a later tradition, however, Pegasus was Perseus' mount, which the hero was riding when he sighted the maiden Andromeda.

Like many of the constellations, Pegasus has a preclassical origin; a winged horse is represented on tablets and vases of the early Etruscans, Euphrateans, and Hittites, as well as on the coins of Corinth (its mythical founder was Sisyphus, grandfather of Bellerophon), Carthage, and other ancient cities. The Romans called Pegasus *Equus Ales,* the "Winged Horse," *Cornipes,* "Horn-Footed," and *Sonipes,* "Noisy-Footed." Aratos' translator Germanicus apparently originated the name "Pegasus" from *Pegai,* the "Springs of the Ocean," referring to its birth, or perhaps from *pegos,* or "strong," but a seventeenth-century scholar, the Reverend Samuel Bochart, traced it to the Phoenician *pag* or *pega,* and *sūs,* the "bridled horse," an emblem used as a figurehead on ships. The name *alatus,* or "winged," appears in the Alfonsine Tables and became the seventeenth-century *Pegasus Equus alatus.*

Pegasus shared many daring adventures with Bellerophon, but the Winged Horse's crowning achievement was its permanent transfer to the heavens, north of Aquarius and the western fish of Pisces and southeast from the Northern Cross, where its famed quadrangle dominates the autumn skies and provides us with excellent orientation.

The Great Square

Pegasus appears nearly inverted, its feet in the air; four bright stars enclosing the Horse's body constitute the quadrangle called the **GREAT SQUARE OF PEGASUS.** Its southwest corner is occupied by **MARKAB (α) ALPHA** (RA 23 h 2 m δ +14° 56′), from *marchab,* meaning "saddle," "ship," or "vehicle"; but it was also called *Matn al Faras,* the "Horse's Withers," or "Shoulder." A white A-type giant, Markab, with 95 $L \odot$, is more than four times as luminous as Sirius, the model A-type star, but Markab's greater distance, 110 light years, reduces its apparent brightness to magnitude 2.5. The interesting S-shaped spiral galaxy **NGC 7479** lies just 2.9 degrees to the south.

At the northwest corner of the Square lies **SCHEAT (β) BETA,** whose puzzling name seems to derive from *al sā ʿid,* "the upper part of the arm," or, more closely, from *seat* or *scheat alpheraz,* "the horse's shinbone or ankle." Arabians called it *Mankib al Faras,* the "Horse's Shoulder," but this location is more correctly applied to α. The Arabians sometimes combined α and β with the Water Bucket of Aquarius, either as its crossbar or as its fore spout. Scheat itself is a red giant, 210 light years distant, that varies irregularly like Betelgeuse; it ranges from magnitude 2.1 to about 3. Its diameter expands from 145 to 160 times that of the Sun, and its luminosity varies from 240 to 500 $L \odot$; yet it equals only five solar masses, having the extremely low density characteristic of red giants.

At the southeast corner of the Great Square is **ALGENIB (γ) GAMMA,** magnitude 2.84, its name derived either from *al janāḥ,* "the wing," which it marks, or from *al janb,* "the side" (i.e., of Pisces, where Tycho Brahe erroneously placed it). Algenib is one of the "Three Guides" that lie near the zero hour circle. A hot blue subgiant, class B2, γ is about 570 light years distant, and its great luminosity is some 1,900 $L \odot$. A short-period variable of the Beta Canis Majoris type, it exhibits tiny light changes that take place within about 3½ hours.

The Great Square is completed by *(δ) DELTA,* at the northeast corner, but modern astronomy has given this star to Andromeda so that the Princess might have a head; as Alpheratz (α Andromedae), we shall discuss it a few pages farther along.

The Winged Horse never knew defeat; thus, appropriately, the nose of this winner is marked by a golden supergiant, ***ENIF (ε) EPSILON*** (RA 21 h 41.7 m   δ +9° 39′)   m 2.31, from *al anf,* Arabic for "the nose"; it was also the Arabian *Fum al Faras* and Flamsteed's *Os Pegasi,* both meaning the "Horse's Mouth." Also known in desert lands as *Al Jahfalah,* "The Lip," Enif lies 780 light years distant from us and belongs to spectral class K2, with a luminosity approximately equal to that of 5,800 Suns; its absolute magnitude is −4.6 and its mass equals nearly ten solar masses. A magnitude-8.5 blue optical companion, at 2 minutes 23 seconds, has been observed from the time of Herschel, and more recently, a very faint (m-11.5) optical companion at 1 minute 21.8 seconds has also been discovered.

Look 4 degrees northwest of Enif for the 8-minute compact globular ***M15 (NGC 7078),*** one of the brightest globular clusters in the northern skies (m 6.5); owing to its great condensation, a 6-inch reflector is necessary to resolve even a few of its outermost stars. However, it is an impressive object in time-exposure photographs made with large telescopes. Approximately 40,000 light years distant, M15 has a luminosity estimated at about 200,000 *L* ⊙, and it is now known as an X-ray source.

Almost directly east of ε and some 7 degrees southwest of α is ***HOMAM (ζ) ZETA*** (RA 22 h 39 m   δ +10° 34′)   m 3.46, probably derived from *Sa'd al Humām,* the "Lucky Star of the Hero"; according to R. H. Allen, many titles signifying "good luck" were given to morning stars of the early spring. A white main-sequence star of late spectral-class B, Homam lies some 210 light years distant from us, with a luminosity of about 145 *L* ⊙. An eleventh-magnitude optical companion at 1 minute 2 seconds was discovered by S. W. Burnham in 1879. Just about 2 degrees northeast of ζ is *(ξ) XI (#46),* magnitude 4.5, a class-F7 main-sequence star with a faint red dwarf gravitationally bound to it.

More or less indicating the top of Pegasus' foreleg is another pair of stars, *(λ) LAMBDA* (m 4) and *(μ) MU* (m 3.5), lying a little over a degree apart, and roughly 13 degrees north of ζ and ξ. Together, Lambda and Mu were the Arabic *Sa'd al Bāri,* the "Good Luck of the Excelling One." The Persian astronomical writer Kazwini called them *Sa'd al Nazi,* the "Good Luck of the Camel Striving to Get to Pasture." Mu, a yellowish class-G8 giant 100 light years distant, shines with the light of thirty Suns. To the north and slightly to the west of Lambda and Mu is the pair ***MATAR (η) ETA*** (m 3) and 5th magnitude *(o) OMICRON.* "Matar" derives from *Al Sa'd al Matar,* "The Fortunate Rain," a title including o. Eta is a bright giant, class G8, with about 630 *L* ⊙. A faint F-type companion lies very close to it; physically bound, these components complete one revolution about their common center of mass every 818 days.

To the east of Lambda and Mu and within the Great Square lies a fourth stellar pair, *(τ) TAU* and *(υ) UPSILON,* each of magnitude 4.6, designated as *Sa'd al Na'amah* by Al Sufi, but according to the nineteenth-century English astronomer E. B. Knobel, *Al Na'āim,* "The Cross Bars" over a well; yet another name was *Al Karab,* "The Bucket Rope." The star Tau alone was called *Sagma,* or *Salm,* a "Leathern Bucket." The three latter pairs, λ-μ, η-o, and υ-τ, formed the Chinese asterism *Li Kung.*

Near the western border of Pegasus, one of a zigzag row of faint stars stretching north from ϵ, lies *(κ)* **KAPPA** (RA 21 h 42 m   δ +25° 25′)   m 4.27 (about 15° north of ϵ), marking Pegasus' right foreleg. About 100 light years distant from us and a class-F5 subgiant with a luminosity of some sixteen Suns, it has a faint optical companion first recorded in 1776 by Sir William Herschel. The primary is also a close and rapid binary, whose components, magnitudes 4.8 and 5.2, are an average 0.2 second apart, with a period of revolution of about twelve years, each star with a mass of approximately one and a half solar masses. To add to this merriment, the magnitude 4.8 component is itself dual, i.e., a contact binary whose components take only about six days to complete one revolution.

An Exception in Science

In the northeast corner of the Great Square, about 1.8 degrees south-southwest of Alpheratz (the star common to Andromeda and Pegasus), lies *85 PEGASI* (RA 23 h 59.5 m   δ +26° 49′)   m 5.75, an enigmatic binary system, 40 light years distant, discovered by S. W. Burnham in 1878 and under continuous observation since then. It consists of two gravitationally bound components only 0.8 second apart and two faint optical components at 1 minute 15 seconds and 1 minute 49 seconds. The physically bound stars, about 9.5 astronomical units apart, have a 26.27-year period of revolution and are of apparent magnitudes 6 and 11. The primary component, a class-G3 main-sequence star similar to the Sun, has a luminosity about 70 percent that of the Sun, but the observed orbit of this strange pair indicates two stars of equal mass, approximately 3/4 $\mathcal{M}$ ⊙. The eleventh-magnitude component is apparently a very tiny, underluminous subdwarf[*]; therefore it seems to violate the mass-luminosity relationship, according to which the stars should be equally luminous. So far, this discrepancy has not been satisfactorily explained, but it may be caused by a continuous exchange of mass between the two components, with the luminosity of the subdwarf, which is acquiring mass, not yet adjusted to the new mass.

Variable Stars in the Horse

About 3.8 degrees west-northwest from γ, at the southeast corner of the Great Square, lies *U PEGASI* (RA 23 h 55.4 m   δ +15° 40′)   m 9.2–9.9, a dwarf eclipsing binary, with its two spectral-class-dF3 components nearly in contact. Each, with a diameter about 60 percent of the Sun's, contains about one solar mass (1.10 $\mathcal{M}$ ⊙ and 0.88 $\mathcal{M}$ ⊙). The orbital period has been slowly decreasing since the binary was discovered, in 1894, and its light curve also seems to be undergoing some changes. Sudden flares detected in 1951 point to the usual exchange of matter between close components of a binary system, affecting the masses and the evolution of these dwarfs, so that U Pegasi may be a precursor of the erratic dwarf novae typified by SS Cygni.

About 3.3 degrees northeast of Enif, at the horse's nose, is the erratic variable *AG PEGASI* (RA 21 h 48.6 m   δ +12° 23′)   m 6.4–8.2. Peculiarities in its spectrum, that have developed rapidly since 1920, indicate two unlike, or "symbiotic," components, possibly a Wolf-Rayet star and a red giant. Strong titanium-oxide bands have been

---

[*] *A star midway in its evolution between a red giant and a white dwarf.*

noted, as well as certain complex spectral features associated with one or more ex-panding gaseous shells. Flare activity on the surface of the Wolf-Rayet star seems to be creating a planetary nebula in its early stages. The distance of this unusual system is too large for determination of its parallax.

Galaxies

About 4.3 degrees north-northwest of η lies a tilted spiral with fine, delicate arms, *NGC 7331* (RA 22 h 34.8 m δ +34° 10′) m 10 (fig. 134), which is often compared to our own Galaxy and to the great Andromeda galaxy. Much smaller than the latter, however, and much farther away, it lies nearly 50 million light years distant and, in accordance with its redshift, is receding at about 656 miles per second. With the faint outer portions included, its mass may be as large as 140 billion $\mathcal{M}\odot$, with a luminosity of about 50 billion $L\odot$.

Just one-half degree south-southwest of NGC 7331 is an intriguing group of "discordant musicians," galaxies *NGC 7317, 7318 A* **and** *B, 7319*, and *7320,* known as *STEPHAN'S QUINTET* (fig. 135), an object of much speculation and controversy because the redshift (480 miles per second) of #7320, the largest of these galaxies, differs widely from that of the other four (3,600 miles per second). The most plausible theory is that this galaxy cluster is exploding and therefore individual members are separating at various space velocities. Many astronomers believe that NGC 7320, which shows a much greater stellar resolution in photographs than the others, is a dwarf galaxy lying closer to the Earth, although some long-exposure photographs show apparent filaments of light connecting all five cluster members; laboratory experiments, however, have demonstrated that illusory bridges of light may appear between bright objects on such photographs.

Theories have been presented by such prestigious scientists as Sir Fred Hoyle, and

134. *NGC 7331*, tilted spiral galaxy in Pegasus, similar in type to our own. Photographed with 200-inch reflector. (*Hale Observatories*)

Geoffrey Burbidge of Kitt Peak Observatory, which challenge the accepted explanation of redshifts, i.e., the expansion of the universe, and impute a significance to the Doppler effect not related to this expansion to explain such phenomena as the discordant redshifts of Stephan's Quintet (as well as the apparent supcrluminosity of quasars based upon the assumption of cosmological distances for these objects); these new ideas, however, which apparently need to be more firmly grounded despite some interesting observational evidence, have not won widespread support in the scientific community.

An Abnormal Galaxy

Roughly 5 degrees north-northwest of γ and just west of the fifth-magnitude star (χ) Chi, lies a most unusual object, the Seyfert galaxy *MARKARIAN 335* (RA 0 h 5.18 m δ +19° 33′) m 14.09, a focal point of today's scientific spotlight. Until now, the outstanding feature of Seyfert galaxies has been their extremely luminous nuclei; recently, however, radiation from regions outside the nucleus of MK 335 has shown strong emission lines; these apparently originate in a large photoionized gas nebula with an inferred mass of about 200 million $\mathcal{M} \odot$ and a very large internal kinetic energy (random energy of its electrons), a unique combination of factors. The hydrogen atoms in this nebula are ionized by extremely energetic photons originating in the Seyfert nucleus. The source of the gas is uncertain, because some of the spectral lines are emitted equally in all directions and others are distorted by blending or intervening absorption, thus providing no indication of the direction in which the gas clouds are moving. If these clouds originated in the nucleus, then the age of the gas, only 1 million years, indicates that their expulsion from the nucleus lasted for a very short time in comparison with the expected life span of the galaxy, about 100 million years. That these clouds have not been slowed by drag in their long passage out of the nucleus, and that the nebula is strongly asymmetrical, raise further questions.

Although Seyfert galaxies are known as spirals, the distribution of MK 335's surface

135. *NGC 7317–20, Stephan's Quintet,* a small cluster of galaxies with differing redshifts. Three of the galaxies are barred spirals, two of which are intertwined and gravitationally connected. The largest galaxy of the group is an irregular spiral, and the smallest is an elliptical galaxy. Photographed with the 120-inch reflector. (*Lick Observatory*)

brightness is more like that of an elliptical galaxy; however, unlike ellipticals with line-emitting nuclei, MK 335 is definitely not a bright radio source. The many contradictions of this strange galaxy, one of nature's more challenging puzzles, call for continued study and observation.

EQUULEUS

Four faint stars, the brightest only of fourth magnitude, form a tiny trapezium to the west of ε Pegasi. Known as *Equuleus,* the "Foal" or "Little Horse," and the "Horse's Head," which it resembles, the origin of this small constellation is uncertain; it may have been created by Hipparchos from stars originally belonging to Delphinus, and some mythologists claim that it represents Celeris, the brother of Pegasus. Among its various Latin and Arabic titles are *Equus Primus* and *Al Faras al Awwal, "the First Horse,"* because it rises before Pegasus, as well as *Equi Sectio* and *Al Kiṭah al Faras,* both of which mean "Part of the Horse"; the latter is the origin of the name *KITALPHA (α) ALPHA,* a fourth-magnitude G-type giant, about 150 light years distant, marking the southern corner of the trapezium.

Directly to α's north, at the northeast corner, is *(δ) DELTA* (RA 21 h 12 m   δ +9° 48′)   m 4.49, poetically described as "topaz yellow," which more or less agrees with its spectral class of F7; a main-sequence star and a close visual binary, its fifth-magnitude components at a mean separation of 0.26 second, or some 4.6 astronomical units, complete a revolution about their common center of mass in about 5.7 years. A tenth-magnitude star not gravitationally bound to the close components lies at 60 seconds and is moving steadily away from them; thus δ is sometimes referred to as a triple star.

About 4½ degrees west-southwest of α and virtually the only naked-eye star outside the trapezium, indicating the neck of the Little Horse, is fifth-magnitude *(ε) EPSILON (#1),* a true triple system, about 200 light years distant. It consists of two class-F5 subgiants, of magnitudes 5½ and 6, their mean separation only about 0.66 second (40 AU), and a seventh-magnitude normal dwarf, also of class F, which orbits the close pair at a separation of 10.9 seconds. The subgiants, in a nearly edge-on orbit, complete a revolution every 101 years.

DELPHINUS (JOB'S COFFIN)

Another group of fourth-magnitude stars forms a small trapezium or kite-like figure lying roughly 10 degrees to the northwest of Equuleus and about 17 degrees west-northwest of ε Pegasi. This minor constellation is also one of those occupying the heavenly "Sea" and is related to Amphitrite, a nereid, or sea nymph, persuaded by a dolphin to marry Poseidon (Neptune); the dolphin was rewarded with a place in the skies as *Persuasor Amphitrites.* Another Greek myth tells how it saved the life of the poet Arion, whose songs charmed all the creatures of the sea; he was returning from Tarentum to Corinth with a valuable horde of jewelry when his crew mutinied and seized his fortune. Arion, pleading for his life, asked to be delivered live into the ocean. His sailors complied, and just when it seemed that he was facing quite a rugged swim, an admiring dolphin offered him a lift! From this noble deed arose the title *Vector Arionis,* and also *Musicum Signum,* apparently a reference to Arion and his lyre. Similarly, the Hindus saw a porpoise here called *Zizumara,* and the Arabians adopted the Greek dolphin as *Dulfim,* a marine animal whose function it was to save drowning sailors. Ptolemy credited Delphinus with influence over human character, along with other constellations not necessarily of the zodiac. The biblical school saw it as the "Leviathan" or "Whale" of Psalm 104, and possibly invented the title "Job's Coffin," which is still popular.

A Hidden Name

The northwest star in the trapezium, fourth-magnitude *SUALOCIN (α) ALPHA,* and its neighbor to the southwest, the double star *ROTANEV (β) BETA,* aroused great curiosity because of their peculiar names, which first appeared in the Palermo Catalogue of 1814. The Reverend Thomas W. Webb, English astronomer, reversed their spelling: "Nicolaus Venator," the Latin form for "Niccolò Cacciatore," Giuseppe Piazzi's assistant at the Palermo Observatory! Alpha is a class-B9 main-sequence star, magnitude 3.77; Beta, magnitude 3.78, an F5 subgiant some 125 light years distant, consists of two components at a mean separation of 0.48 second, with a total luminosity of about 36 $L \odot$, which require 26½ years to complete one revolution. A larger telescope is usually necessary to resolve them.

At the northeast tip of the "kite" is the beautiful contrast double *(γ) GAMMA* (RA 20 h 44.4 m δ +15° 57') m 3.91, 100 light years distant, a golden K-type giant and a lemon-yellow or greenish main-sequence star, class F8, 10 seconds apart and recommended for small instruments. In the field is the faint double *Σ 2725.*

(ζ) ZETA, at the base of the quadrangle, (just west of β) is a magnitude-4.62 type-A main-sequence star, and *(η) ETA,* to the south, nearly a magnitude fainter, is of similar type. At the tip of the dorsal fin and southernmost of Delphinus' brighter stars is *(ε) EPSILON,* magnitude 3.98, called "Deneb," from *Al Dhanab al Dulfim,* "The Dolphin's Tail," but this name confuses it with brilliant Deneb in Cygnus. Delphinus was important to early Christians as the "Cross of Jesus," and Epsilon was known in Arabia as *Al Amud al Salib,* "The Pillar of the Cross." Its spectral type, B6 III, classes it as a hot, bluewhite giant; this color gives us no clue to its Chinese title, *Pae Chaou,* the "Rotten Melon"!

A Long-Range Variable

Some 3 degrees north of γ is the variable *V DELPHINI* (RA 20 h 45.5 m δ +19° 09'), with its magnitude ranging from 8 to 16 and higher; a red giant of the Mira class, 5,000 light years distant, it has a period of about 534 days and a calculated absolute magnitude of −2 at maximum luminosity. Just to its southwest is the irregular red variable *U,* with the modest range of magnitude 5.7 to 7.6. West of U is another M-type star, the semiregular *EU,* varying only from magnitude 6 to 6.9.

A Modern Nova

Somewhat over a degree to the west of V Delphini is the nova *HR DELPHINI*, or *NOVA DELPHINI 1967* (RA 20 h 40 m   δ +18° 59'), magnitude range 3.5 to 12, discovered by the amateur astronomer G. E. D. Alcock on July 8, 1967, as it became a naked-eye object of magnitude 5.6. Older photographic plates indicate that a slightly variable twelfth-magnitude bluish star had occupied that position. Nova Delphini, which flared up in about a month, reached magnitude 3.5 in mid-December, and, after many increasingly erratic fluctuations, faded slowly, reaching the eighth magnitude by March, but taking several years to return to its original faintness. At the time of its maximum luminosity, its spectrum indicated four sets of expanding gaseous shells.

LACERTA

North of Pegasus' forelegs and immediately south of Cepheus' head is the faint constellation of Lacerta the Lizard, formed by Hevelius in 1687 from stars between Cygnus and Andromeda. Perhaps he was inspired by the early Chinese, who had combined stars of Lacerta and the eastern part of Cygnus to form their "Flying Serpent." Hevelius also called Lacerta *Stellio,* the "Stellion," a Mediterranean newt with starlike dorsal spots.

The brightest star of Lacerta is *(α) ALPHA* (RA 22 h 29 m   δ +50° 01')   m 3.8, about 98 light years distant, a type-A "Sirian" main-sequence star and also an optical double, its very faint companion, at 36.3 seconds, being only of twelfth magnitude. About 2 degrees to α's north-northwest, marking the Lizard's head, is *(β) BETA,* magnitude 4.4, a deep yellow class G9 giant, about 215 light years distant. A few degrees to β's north, near the border of Cepheus, the noted *NOVA LACERTAE 1936 (CP LACERTAE)* was first seen on June 18, 1936, as a naked-eye star that reached magnitude 2.2 two days later. Before its sudden outburst, CP, some 5,400 light years distant, had been a fifteenth-magnitude star. During maximum luminosity, i.e., 300,000 *L* ☉, it shed its outer layers at a speed of about 2,400 miles per second, which W. M. Smart called "the greatest linear velocity of a shell thrown out by a nova" (until T Coronae Borealis in 1946), and it then quickly faded, reaching the sixth magnitude in six weeks. After that, Nova Lacertae continued to show expanding nebulosity for many years.

A Mirror-Symmetrical Cosmic Jet

About 2⅓ degrees west-northwest of the fifth-magnitude double star Fl. #10, which lies within the Lizard's tail, in the southern part of Lacerta, are an elliptical galaxy 100 million light years distant, *A2229-39* (RA 22 h 29 m   δ +39° 6'), and the weak radio source *3C 449,* a most remarkable object identified with it. In 1979, radio astronomers working with the VLA (Very Large Array) in New Mexico discovered two narrow streams of plasma (ionized gas) emerging from the center of this galaxy in opposite directions. Their observations yielded an amazing picture: a northern jet which extends for about 100,000 light years and bends abruptly toward the east and then back again toward the north, and a southern jet, bending toward the east and then back toward the south; thus, two opposing jets which seem to be "mirror images" of each other! Each jet terminates in a large radio lobe similar to the lobes in such extragalactic radio sources as Cygnus A, although the latter, as is typical for a powerful double source, has no large-scale jets.

Traveling at speeds greater than sound, the gas in the cosmic jets of 3C 449 transmit mass and energy, and magnetic flux, from the galactic center of their origin out to the extended radio lobes, where there is a tremendous buildup of energy. Hot spots of accumulated energy at the end of each jet decelerate the gas, generating a shock wave that converts the ordered kinetic energy of the gas into high-speed electrons accelerated by the energy of a newly created magnetic field. After remaining in the jet's hot spot for 10,000 to 1 million years, each gas molecule returns to the galaxy, carrying with it the relativistic electrons and the lines of magnetic force; the backward flow of this gas inflates the large lobe of radio emission at the end of each jet. The bends in these jets, scientists theorize, may reveal orbital motion of the galaxy around a close companion galaxy. We must also mention that the zigzag pattern of the jets is, in a sense, an apparent one, because their individual pellets of gas move always in straight lines (much like drops of water emerging from a garden hose that is being swung back and forth).

Since the discovery, in 1953, of the double structure, i.e., the lobes, characteristic of extragalactic radio sources detected at frequencies of less than one gigahertz (1 billion cps), an explanation for this phenomenon has been sought; an understanding of the dynamic processes within cosmic radio jets may clarify the mystery.

An example of a very well-studied one-sided jet is associated with the elliptical galaxy *NGC 6251* (RA 16 h 35 m δ +82° 36'), northwest of ϵ Ursae Minoris, the star at the bend of the handle of the Little Dipper. This galaxy is 300 million light years distant, and the jet itself, with an angular width of only 3 degrees, is more than 100,000 light years long and the straightest jet known. Its narrow end, lying at the center of galaxy NGC 6251, coincides with a small, pointlike source of radio waves; in 1978, radio maps revealed this point as a small jet, only 3 light years long, colinear with the large jet and containing a point source of its own at one end. One wonders what a surprise supersensitive instruments of the future would reveal if the point source of the small jet were mapped: perhaps another, still tinier jet, also with a point source of its own, which proves to be an extremely minute jet with a point source of its own, etc., ad infinitum!

ANDROMEDA

. . . And Perseus saw her there, Andromeda,
Bound by the arms to the rough rocks; her hair,
Stirred in a gentle breeze, and her warm tears flowing
Proved her not marble, as he thought, but woman. . . .†

† *Ovid*, Metamorphoses, *IV, 673–76.*

136. *Perseus Delivers Andromeda,* by
Ingres.

The ancient myths seem unfair in condemning Andromeda for her mother's boastfulness. When Cassiopeia offended the Nereids, the land of Philistia was threatened with a sea monster, as we have told, and the princess Andromeda became its intended offering. It is said that the story of Andromeda's rescue by Perseus originated in the Babylonian *Epic of Creation,* which describes the defeat of the dragon Tiamat, often related to Draco. The Greek tragic poets Euripides and Sophocles wrote about Andromeda, but she was also known to the Romans as Andromeda *Mulier Catenata,* the "Woman Chained," and also as *Virgo Devota;* the Arabs called her *Al Armalah,* "The Widow"!

The constellation of Andromeda is joined to the Great Square of Pegasus by a common star, Alpheratz, and if we include some fainter stars southeast of Alpheratz, the Chained Woman extends for some 26 degrees to the northeast, occupying a region south of Cassiopeia and west of Perseus. Pisces and Triangulum lie to the south, and the great Northern Cross, farther over to the west, also provides orientation.

Andromeda's hair or head is marked by ***ALPHERATZ (α) ALPHA Andromedae,*** formerly δ Pegasi, the second-magnitude star occupying the northeast corner of the Great Square. Its name is derived from the Arabic *Al Surrat al Faras,* "The Horse's Navel," but since Ptolemy's time the Arabians called it "The Head of the Woman in Chains." With β

Cassopeiae and γ Pegasi, it marks the equinoctial colure, as we have mentioned. A late-B or early-A-type star, about 120 light years distant, α has a luminosity about 160 times the Sun's, and its spectrum shows unusually strong lines of manganese. It is also a spectroscopic binary; its close components, some 20 million miles apart, complete one revolution about their common center of mass every 96.697 days. A physically unrelated optical companion of magnitude 11 lies at 81.5 seconds from the close pair.

Along the more prominent (eastern) branch, or "left leg" of Andromeda, we find an attractive double star, the orange giant *(δ) DELTA* (RA 00 h 36.6 m   δ +30° 35')   m 3.25, which is 160 light years distant and has a luminosity of about a hundred Suns. Its companion, at 28.7 seconds, is a twelfth-magnitude red dwarf with only 1/40 the Sun's luminosity, too faint for small instruments. Farther to the northeast, also marking the left leg, lies the second-magnitude star *MIRACH (β) BETA,* its name indirectly derived from the Arabic *mi'zar,* or "girdle"; about 75 light years distant, β is a red giant with a luminosity of about 75 *L* ☉. It has an extremely dim companion at 28 seconds, a fourteenth-magnitude dwarf with only 1/800 the Sun's luminosity.

Marking Andromeda's left foot, on the "eastern branch," is *ALMACH (γ) GAMMA* (RA 2 h 4 m   δ +42° 20')   m 2.18, a beautiful contrast double, especially suited for small instruments. The primary is a golden giant, about 650 *L* ☉, and its fifth-magnitude companion, at 10 seconds, is of a warm blue tint. Almach is really a quadruple system: in 1842, the blue star was revealed as a close double with a period of 61 years; and the brighter of its components is now known as a spectroscopic binary consisting of two identical B9 stars that take 2.67 days to complete one revolution. Northwest of

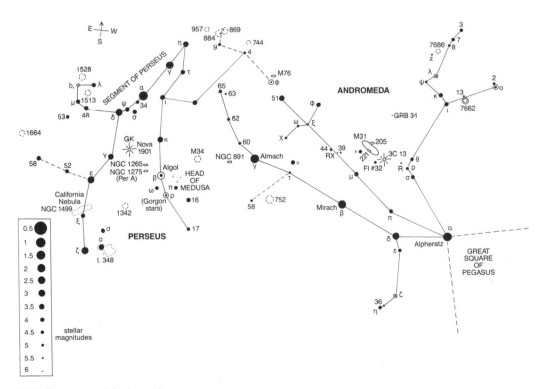

137. *Perseus and Andromeda.*

Almach, toward the end of Andromeda's right leg, lies fifth-magnitude *ADHIL (ξ) XI,* derived from *Al Dhail,* "The Train" (of a garment). Immediately to the east is *(ω) OMEGA,* magnitude 4.84, an F4 subgiant and long-period binary, possibly a member of the Hyades moving group.

Andromeda's "right arm" stretches toward the northwest from Sigma to Iota, to the binary Kappa, and to the fourth-magnitude *(λ) LAMBDA,* a yellow subgiant of late spectral type G, some 80 light years distant. A spectroscopic binary, λ's peculiar spectrum shows bright emission lines of calcium. Its slight variability (m 3.7–4.1) is unrelated to the revolutionary period of the extremely close components, about 20½ days.

Some 7 degrees northeast of α, marking the princess' breasts, is the bluish-white double star *(π) PI,* magnitude 4.43, its components 36 seconds apart. The primary is also a spectroscopic binary, with a period of 144 days. Andromeda's left arm is indicated by fourth-magnitude *(ζ) ZETA,* about 100 light years distant, a slightly variable contact binary consisting of two K-type stars, egg-shaped from mutual tidal action, whose ovoid shapes create slight light changes as they circle one another, presenting larger and smaller surface areas to our line of sight. A faint red dwarf shares their proper motion at a separation of 96 seconds, creating a triple system.

About 2 degrees southeast of Zeta is the sixth-magnitude binary *Fl. #36,* a K-type object of a deep yellow hue. The components, with a mean separation of 1.0 second, complete a revolution in some 165 years; close to periastron (0.6 second) for the past fifty years, they are just beginning to draw apart again as this book goes to press. Immediately southeast of #36, lying very near the Andromeda-Pisces border, is *(η) ETA,* a magnitude-4.5 late solar-type giant; some 114 light years distant, it marks the Princess' left hand.

Unusual Variables

In the far western region of Andromeda, near the border of Lacerta, lies the variable *(o) OMICRON* (RA 22 h 59.6 m   δ +42° 03')   m 3.63. At times, its spectrum is that of a class-B6 star of normal characteristics, but at rather irregularly separated years, it indicates a surrounding gaseous shell. Omicron's equatorial speed of rotation at its surface, 215 miles per second, exceeds that of most known stars. Estimates of its period vary between three-fourths of a day and one and one-half days, and its light changes resemble those of various types of objects at different times (e.g., RR Lyrae, the cluster variable, as opposed to β Lyrae, the eclipsing binary), suggesting that Omicron is a close binary one of whose components undergoes short-period pulsations and some shell activity. The American astronomer Arne Slettebak, who analyzed the rotational broadening of spectral lines for stars from classes B0 to G0, observed the stratification of Omicron's shell produced by the ejection of such elements as magnesium and silicon from the star's interior into various levels above its surface.

About 2⅓ degrees north of Lambda, in the northwestern part of Andromeda, lies the semiregular variable *Z ANDROMEDAE* (RA 23 h 31 m   δ +48° 32')   m 8 to 11.5, a spectroscopic binary whose close components are a cool red giant and a hot bluish B-type subdwarf. Sudden increases in luminosity tend to occur every fifteen or twenty years, when wide, bright lines appear in the spectrum, indicating the presence of a gaseous shell. The flares may be caused by matter falling upon the surface of the hot blue star, as in the case of recurrent novae. Both components have a light cycle and radial velocities of some seven hundred days.

About 6 degrees north of β and less than 3 degrees northeast of *(μ) MU* (a 4th-magnitude star on Andromeda's right leg), lies the irregular variable *RX AN-DROMEDAE* (RA 01 h 2 m   δ +41° 02′)   m 10.3 to 13.6, one of the so-called "cataclysmic variables" or "dwarf novae." Apparently similar in spectral type to Z Camelopardi (p. 77), RX Andromedae is a close binary whose tiny components, probably a red dwarf and a very hot, dense blue dwarf, circle one another in only five hours, with the inevitable exchange of material, which results in small-scale explosions every few weeks.

At Andromeda's right shoulder, near the triangle formed by Theta, Rho, and Sigma and 4 degrees southwest of the famous galaxy M31, lies the long-period variable *R ANDROMEDAE* (RA 0 h 21.4 m   δ +38° 17′)   m 5.3 to 15.1. Its light cycle averages 409 days, and when at maximum luminosity it may be seen with binoculars or small telescopes. R is a pulsating Mira-type red giant but is classed as an S6 emission star because bands of zirconium oxide (as well as titanium oxide) are present.

An Attractive Planetary

Near the Chained Woman's right hand, about 2 degrees west and slightly north of Iota, lies the bluish, slightly elliptical bright planetary nebula *NGC 7662* (RA 23 h 23.4 m   δ +42° 12′)   m 8.5. The fifth-magnitude spectroscopic binary *#13* Andromedae lies just one-half degree to its northeast. Almost starlike in small telescopes, the planetary appears ring-shaped with greater magnification, because of its darker center. In very large telescopes, the fourteenth-magnitude central star is visible, and although it may appear yellowish in comparison with the deeper tint of the nebula, this object is an extremely hot bluish dwarf, with a surface temperature of about 75,000 degrees K.

A Large Open Star Cluster

About 5 degrees south-southwest of γ is *NGC 752* (RA 1 h 54.8 m   δ +37° 26′), 1,300 light years distant, a large open cluster especially suitable for high-quality binoculars. Types F to K predominate among its brightest stars, and most of its members are well-evolved subgiants of class F, indicating that the cluster is at least 1.5 billion years old.

An Edge-On Spiral Galaxy

Approximately 3 degrees east of γ lies a fine example of an edge-on spiral galaxy, *NGC 891* (RA 2 h 19.3 m   δ +42° 07′)   m 12, suitable for small telescopes. The equatorial dark lane is particularly prominent in the central regions, which in wide-angle photographs appear similar to our Milky Way.

The Great Galaxy in Andromeda

Just to the west-northwest of the spectroscopic binary (ν) Nu, more or less centrally located in the figure of the Chained Woman, and about 7 degrees northwest of β, lies one of the most spectacular and famous of all celestial objects, *M31 (NGC 224)*, (RA 0 h 40 m   δ +41° 00′)   m 3.5, still known under its popular name, the "Great Nebula in Andromeda," but more correctly, as the **GREAT ANDROMEDA GALAXY** (fig. 138). Visible to the naked eye as a tiny patch of light, it may have been known as early as A.D. 905, and in the same century, Al Sufi, of Persia, called it the "Little Cloud"; it was first examined telescopically on December 15, 1612, by the German astronomer Simon Marius, who described it as resembling "the diluted light from the flame of a candle seen through horn" (alabaster). Immanuel Kant, in 1755, proposed that there is an analogy between the numerous faint nebular patches (like M31) then under observation and our Milky Way, and Sir William Herschel also regarded them as galaxies; however, the instruments of his time could not resolve them into individual stars, and he abandoned the idea in 1791. Toward the end of the nineteenth century, M31 was still described as cloudlike, and astronomers erroneously believed that it was a vast planetary system in formation. The Mount Wilson 60-inch and 100-inch telescopes did resolve some of the brightest stars, but controversy over the nature of spiral nebulae continued until 1923, when Hubble discovered Cepheid variables in M31; M33, in Triangulum (the Pinwheel); and NGC 6822, in Sagittarius (Barnard's Galaxy), establishing such objects as galaxies.

The Andromeda galaxy is the nearest spiral, with an angular size of 160 by 40 minutes; it has a structure similar to our own Galaxy's, oriented in such a way that its detailed features can be analyzed. Its inclination of 15 degrees gives it an oval appearance, but it is actually a round, flat Sb (intermediate-type) spiral, typical of those containing both population-I and population-II stars, in which the dust, gas, and population-I stars are confined to the spiral arms. The nucleus is composed of population-II stars, which pervade the flattened disk and are found far out beyond the spiral structure (see fig. 139). From analyzing the type-I Cepheids in its arms, astronomers know that its distance is about 675 kiloparsecs, or 2.2 million light years. Its apparent optical diameter is about 2.4 degrees as revealed by a good photograph, but the latest studies with the *microdensitometer,* a very sensitive electrical instrument that picks up images on photographic plates not visible to the naked eye, increase the diameter to 4.5 degrees, or 180,000 light years. With a luminosity of about 11 billion Suns, the absolute magnitude is −20.3, so that it is about twenty-five magnitudes more luminous than the Sun. This means that it contains the equivalent of at least 11 billion stars as luminous as the Sun. However, since stars less luminous than the Sun are difficult to detect and probably contribute little to the visible radiation, it is probable that the total number of stars is far greater; estimates place this figure at 300 billion or more, and thus its computed mass is about 400 billion $\mathcal{M}\odot$.

Walter Baade, of the Mount Wilson and Palomar Observatories, studied the Andromeda galaxy in great detail, observing seven spiral arms, the outermost and innermost of them 21 and 0.3 kiloparsecs, respectively, from the center. These arms contain the same stellar content and interstellar material as our own Galaxy. The outermost arm of M31 is defined by scattered groups of blue supergiants, with no obvious signs of

138. M31 (NGC 224), the Andromeda Galaxy. Our famous neighboring galaxy, the largest of the Local Group, shown with its satellites NGC 205 and 221. North is to the top. Photographed with the 36-inch Crosley reflector. (*Lick Observatory*)

gas and dust. Blue supergiants dominate the fifth and fourth spiral arms, 12 to 9 kiloparsecs from the center, with the dust becoming quite conspicuous. In the third and second arms, HII regions become visible, and in ordinary telescopic plates the population-I giants are much less resolvable into individual stars. We see them only as bright unresolved clumps of stars in the innermost spiral arm, but the HII regions stand out clearly. As Baade remarked, "The association of dust and gas and population-I stars is most strikingly revealed by the HII regions of the Andromeda Nebula, which are strung out like pearls along the spiral arms. That these HII regions and with them their exciting O and B stars of high luminosity are deeply embedded in the dust of the spiral arms is shown by their strong reddening."

A comparison of the slight reddening of the globular clusters with the stronger obscuration of the HII regions proves that the dust in M31 is concentrated in the spiral arms. The innermost arm is richly studded with the aforementioned (unresolved) clumps of population-I supergiants, which gradually diminish in number toward the nucleus and abruptly disappear. This arm, however, continues as a lane of dust into the nucleus itself, which is so condensed that individual stars have only recently been resolved with the introduction of red-sensitive photographic plates used in the largest telescopes. With a diameter of about 50 light years, the core may contain more than 10 million stars, with an average separation of only a few hundred astronomical units. These population-II red and yellow giants form an elongated system and must be identified with the disk of the Andromeda galaxy as well as the large halo surrounding it. Intermixed with the population-II giants, there are undoubtedly ordinary giants, similar to those found in open clusters, which have evolved from population-I stars. Long-exposure red-sensitive plates show that the spaces between the spiral arms are filled with dense sheets of population-II giants, whose numbers slowly decrease as we go outward. According to this analysis the spiral arms are thin layers of gas, dust, and O and B population-I supergiants, embedded in the disk.

139. Nucleus of the Andromeda Galaxy. An important photograph showing population-I stars in the spiral arms and population-II stars resolved in the nucleus of the galaxy, like millions of sand grains on some vast beach. Taken with the 200-inch telescope. (*Hale Observatories*)

Radio Emission of the Andromeda Galaxy

We spoke of the 21-centimeter radiation (p. 264) emitted by the neutral hydrogen in our own Galaxy and the information this spectral line gives us about the structure of our Galaxy. By studying the 21-centimeter line of the Andromeda galaxy, we find much the same situation. The 21-centimeter line can be detected along the major axis in a number of regions, and these correspond to the spiral arms. This radiation can be traced out to almost 3 degrees on either side of the nucleus; from it one can calculate the rotation rate of these outer portions of the galaxy. It is interesting that the sharp maximum in the 21-centimeter line at about 8.7 kiloparsecs from the center of M31 corresponds to a similar maximum in our own Galaxy at 6.5 kiloparsecs from its center. This is just where Baade placed the fourth spiral arm in Andromeda, where the population-I stars, gas, and dust are at their maximum concentrations. We can also detect a radio halo around M31 similar to that surrounding our Galaxy; one hundred radio sources have been identified in the M31 halo, distributed out to great distances from the plane of the galaxy and forming an almost spherical system which extends more than 10 kiloparsecs at right angles to the plane. The actual size of the Andromeda galaxy, as determined from radio waves, is considerably larger than its optical size; according to Shklovsky, the part that contains radio sources has a diameter of about 100 kiloparsecs.

Rotation

An analysis of the radio spectrum of M31 verifies that it is rotating like our own Galaxy. The optical analysis is fairly simple, since we can measure the Doppler shift of the spectral lines in the radiation emitted by stars on either side of the nucleus of M31. Out to an angular distance of 65 to 70 minutes on either side of the nucleus, the angular velocity appears to be uniform, as though the nucleus were rotating like a solid disk; it completes one rotation every 11 million years. From about 70 minutes out to 155 minutes on either side, the rotational velocity decreases with increasing distance; the period of rotation for these outer regions ranges from 90 to 200 million years. This is in accordance with Kepler's third law of planetary motion (p. 31), which is applicable to all celestial bodies in gravitational orbits and thus to the various groups of stars in M31 as they revolve in circular orbits at different distances from M31's center.

The Motion of the Spiral Arms

An important question in connection with the rotation of galaxies is the direction of rotation of their spiral arms: whether the arms are trailing (i.e., winding up) or advancing (unwinding). If, like M31, the galaxy is tilted, we take as the side closer to us the one marked by the dark lanes of obscuring matter. However, in such cases the arms cannot be clearly traced out, so we are left with some uncertainty as to whether they are trailing or not. If the angle of tilt is sufficiently large to enable us to trace out the spiral arms unambiguously, the lane of obscuring material no longer stands out sufficiently to tell us which is the closer side. According to G. de Vaucouleurs, who analyzed most of the available data, the balance of evidence favors the winding up of the arms and agrees

140. Central region of the Andromeda Galaxy, showing bright unresolved nucleus joined on opposite points by emerging spiral arms, which consist of massed star clouds crossed by many dark lanes of dust and gas. Red population-II stars form the nucleus. Photographed with the 100-inch telescope on Mount Wilson. (*Hale Observatories*)

with the rotation of our own Galaxy as determined from radio data. Such measurements lead to the same conclusion regarding the Andromeda galaxy.

Novae in M31

In addition to the Cepheid variables, clusters, supergiants, and planetary nebulae used in determining the distances of such galaxies as M31, novae have also been used. Even before M31 was resolved into individual stars, H. D. Curtis applied the photometric method to its novae by comparing their average apparent magnitudes at maximum luminosity with the known mean absolute magnitudes at maximum of similar novae in our Milky Way. Over a hundred normal novae have been observed in M31, the brightest of which was the nova of 1925, which had an absolute magnitude of −9.3, comparable in luminosity with the famed Nova Aquilae. The appearance of a supernova 16 seconds southeast of the nucleus of M31, in the year 1885, caused great excitement. The first supernova ever observed in an external galaxy, this brilliant "new star," undetected in photographs taken a day before its discovery, may have reached naked-eye visibility at an apparent magnitude of 5.4 to 6; after six months, it faded from sight, and was last seen, as a sixteenth apparent magnitude object, by Asaph Hall on Feb. 1, 1886, using the 26-inch refractor at Washington, D.C. At the time of its outburst, spectroscopes had only recently been turned toward the stars, but available data show that the supernova, now called *S Andromedae*, had a spectrum similar to that of galactic novae near maximum luminosity. At peak brilliance, S may have reached an absolute magnitude of -18.2, which implies a luminosity of some 1.6 billion Suns!

The Companion Galaxies of M31

A familiar sight in long-exposure photographs of the Andromeda galaxy are the two tiny companions flanking the great galaxy. The smaller of these, *M32 (NGC 221)*, is a ninth-magnitude object lying 24 minutes south of the bright disk of M31. With a luminosity of 70 million $L \odot$, it has a diameter of 2,400 light years; the apparent dimensions are 3.6 by 3.1 minutes. About 35 minutes northwest of M31's central portion lies **NGC 205 (HV 18)**, (fig. 141), the larger of the companions but only of eleventh magnitude, measuring 8.0 by 3.0 minutes. Its diameter is about 5,400 light

141. NGC 205, larger companion of the Andromeda Galaxy. An elliptical galaxy shown with resolution of the stars in its outer regions. Taken in red light with the 200-inch Palomar reflector. *(Hale Observatories)*

years, but its luminosity is equivalent to only 21 million Suns. Both galaxies are ellipticals, containing mostly population-II stars. Two fainter galaxies, NGC 185 and NGC 147, also physically bound to M31, lie about 7 degrees to its north, within the borders of neighboring Cassiopeia. All four galaxies are believed to be the same distance from us as M31 and are the closest elliptical galaxies to the Solar System.

The Most Distant Galaxies Known

In the region of the Andromeda galaxy, astronomers have detected two faint galaxies estimated at a distance of 10 billion light years, the most remote objects ever recorded. Using the Kitt Peak 4-meter reflector, the team of H. Spinrad, J. Stauffer, and H. Butcher‡ obtained videocamera frames and direct (prime-focus) photographic imagery of *3C 13,* a radio source identified with the galaxy *0031-39* (RA 0 h 31 m 33 s,   δ +39° 07′ 43″)   m 21, the brighter central object in a small chain of galaxies, situated about 2 degrees south in declination and 2 degrees west in right ascension from M31, and about 2½ degrees west of the fifth-magnitude yellow giant Fl. #32. The second galaxy photographed is radio source *3C 427.1,* about 2 degrees to the south of 3C 13; the redshifts, difficult to measure because there are no definite emission lines in their spectra, are 1.050 for 3C 13 and 1.175 for 3C 427.1.

Astronomers were surprised that these very distant and presumably young galaxies are only 0.7 magnitude bluer than local elliptical galaxies, which implies they were actually formed much earlier than 10 billion years ago, perhaps 6 billion years farther into the past. (The first epoch of star formation in most giant elliptical galaxies is estimated at 15 to 18 billion years ago.) A comparison of such remote objects as 3C 13 and 3C 427.1 with closer galaxies will enable astronomers to learn more about galactic evolution and the rate of expansion of the universe.

PERSEUS

> . . . he flew the whole world over,
> Saw the cold Bears, three times, and saw the Crab
> With curving claws, three times, whirled often eastward,
> Whirled often to the west. . . .
> (Ovid, *Metamorphoses,* IV, 626)

Acrisius, king of Argos, in the Peloponnesus, hid his extremely beautiful daughter Danaë from the sight of men, in fear of an oracle's prediction that her son would cause his death. But destiny could not be cheated, and the god Zeus, smitten by Danaë's loveliness, transformed himself into a shower of gold dust, thus gaining access to her underground prison; the result of this union was a son named Perseus.* Apprehensive, King Acrisius set mother and son adrift on the sea in a wooden chest, but a fisherman

‡ Astrophysical Journal, *March 1981.*

* *I.e., "Per Zeus": (fathered) by Zeus.*

rescued the outcasts at Seriphos, an island in the Aegean Sea, and brought them to Polydectes, his king, with whom they remained until Perseus reached manhood. Polydectes, forcing unwanted attentions upon Danaë, tried to get rid of Perseus by sending him on the dangerous mission, which we have described, wherein he slew the Gorgon Medusa. After this deed, Perseus, blown in every direction by fierce winds, flew a great distance in his winged shoes until he reached mighty Atlas' realm, at the Earth's western edge; there Perseus requested rest and hospitality but was refused, whereupon he raised Medusa's head, showing her face to Atlas, who instantly became a huge mountain that grew until all the heavens rested upon it. With dawn, the winds ceased and Perseus continued on the flight that resulted in his rescue of Andromeda. On his return to Seriphos, he showed Medusa's head to Polydectes and his court and they were turned to stone; but then, as a participant in a quoit game in Thessaly presided over by his grandfather King Acrisius, Perseus accidentally struck him in the foot with a badly flung disk, causing the old man's death and thus unwittingly fulfilling the oracle's prophecy.

Perseus the champion is of ancient origin, his name possibly derived from the Hebrew *parash,* a "horseman," or from the Babylonian *parsondas.* According to Herodotus, the land of Persia received its name from Perseus and his son Perses. Other titles were the poetic "Pinnipes," referring to Perseus' winged sandals, and "Victor Gorgonei monstri," the slayer of Medusa. Despite the usual obstacles faced by heroes of mythology, Perseus won the hand of Princess Andromeda, as we know, but his ultimate reward was an eternal place by her side as a very great constellation, extending for 28 degrees all the way from the southern border of Cassiopeia to a region just 7 degrees north of the Pleiades.

Major Stars of Perseus

The Champion's brighter stars appear to form two north-to-south branches, joined at their northernmost tip by the star Eta. The eastern branch, which represents Perseus' right side, commences with a slightly curved branch of stars including η, γ, α, σ, ψ, and δ. Containing some of the constellation's brighter stars, this figure is often called the *SEGMENT OF PERSEUS.* Somewhat toward its center lies *MIRFAK (α) ALPHA* (RA 3 h 20.7 m   δ +49° 41′)   m 1.80, sometimes "Marfak" or "Mirzac," from the Arabic *Marfiḳ al Thurayya,* the "Elbow Nearest the Many Little Ones" (the Pleiades); α has also been designated as "Algenib" (from *Al Janb,* "The Side"), which is the modern name for Gamma Pegasi. Perseus' helmet lies within the circumpolar zone, and Mirfak, almost on the 50-degree circle of declination, never sets in latitudes as far north as that of New York City, but just touches the horizon. About 190 parsecs, or 619.4 light years, distant, Mirfak belongs to spectral class F5 Ib, which indicates a golden supergiant, although in some telescopes it appears white and has even been called "brilliant lilac"! Its color index is +0.48, and its luminosity is over 4,000 L_6, with an absolute magnitude of at least −4.4. Mirfak apparently belongs to a large scattered cluster of stars in its field moving to the southeast, more or less toward El Nath (β Tauri), at the rate of about 10 miles per second.

Moving along to the southeast on Perseus' right side, about 3½ degrees from α, we find the third-magnitude bluish giant *(δ) DELTA,* a class-B5 star, some 590 light years distant, that may have a luminosity greater than 1,700 Suns, although, according to revised estimates, it is only 326 light years distant, a figure that would reduce its computed luminosity. About 8 degrees south-southeast of δ and an equal distance east-southeast of famed β Persei (Algol), forming a bright triangle with them, lies an interesting double for the smaller telescope, *(ε) EPSILON* (RA 3 h 54.5 m   δ +39° 52')   m 2.88, with a computed distance of about 680 light years and consisting of a B-type main-sequence star and a class-B8 faint companion at 9 seconds.

About 4 degrees south of ε is Burritt's *MENKIB (ξ) XI,* or "Shoulder," indicating that Perseus may have been pictured with his head to the south in earlier days, but this magnitude-4.04, class O7 emission star now marks the left ankle. Another 4 degrees south and slightly to the west (and about 8° north of the Pleiades) lies *ATIK (ζ) ZETA* (RA 3 h 51 m   δ +31° 44')   m 2.85, about a thousand light years distant, a B-type supergiant marking Perseus' right foot. Derived from *Al 'Ātik,* the name refers to its earlier position between the shoulders when the figure was reversed, but according to R. H. Allen, "Atik" was formerly applied to nearby (o) Omicron, a B-type giant and variable double just to its west. Zeta has a ninth-magnitude common-proper-motion companion, at 12.9 seconds, that is accessible to small instruments; two faint optical companions lie farther out.

Zeta, Omicron, Xi, and several other O and B stars in the vicinity form the young, expanding group called "II Persei," which covers an area some 100 light years in diameter. Just north of Xi is the gaseous nebula *NGC 1499,* the *CALIFORNIA NEBULA* (fig. 142), with angular dimensions of 145 by 40 seconds but a disappointing object except on long-exposure photographs; apparently illuminated by Xi, it is also a member of II Persei.

Algol, the "Demon Star"

We told the story of young John Goodricke's important observations of variable stars, in particular of δ Cephei. The first of these fluctuating stars to capture his attention was *ALGOL (β) BETA Persei* (RA 3 h 4.9 m   δ +40° 46')   m 2.1 to m 3.4, which marks Perseus' left leg (the western branch of its stars). Algol has been known as the Arabian *Rā's al Ghūl,* "The Demon's Head" *(al Ghūl* is, literally, "Mischief Maker") and as the Hebrew *Rōsh ha Sāṭān,* or "Satan's Head," a title that probably originated with Ptolemy, who associated this bright star and a group of its fainter neighbors with the Gorgon Medusa's head, which Perseus carries. A tradition arose among astronomers attributing these early "Satanic" titles (and its popular names "Demon Star" and "Blinking Demon") to the ancients' supposed observation of its variability, but according to R. H. Allen there is no evidence to support such a notion. The Chinese title for β, *Tseih She,* or "Piled-up Corpses," is often cited as further "proof" of this idea but actually bears no more relationship to the star's variability than does the name *Pae Chaou,* the "Rotten Melon," that given to ε Delphini, a decidedly nonvariable blue-white giant; but the titles do have a common association: putrefaction! In fact, the negative and demoniac qualities many ancient and medieval peoples attributed to Algol seem to stem from its association with Medusa's head. Thus, even in the world of science, do myths arise!

The first astronomer to note β's variability was Geminiano Montanari, of Bologna,

142. *NGC 1499, the California Nebula,* a bright gaseous nebula in the vicinity of the Zeta Persei stellar association. Its delicate filamentary structure is evident in this photograph, taken with the Crossley reflector. (*Lick Observatory*)

who made his observations prior to 1672; years later, they were confirmed by Maraldi and then by Palitzsch.† On November 12, 1782, John Goodricke began his studies of Algol, and throughout the winter he recorded its light-changes; in May 1783, he communicated his findings to the Royal Astronomical Society, including in his paper a remarkably accurate deduction of Algol's period of variation (2 da. 20 hr. 49 min.), and suggesting that the phenomenon is the periodic eclipse of Algol by a large and dark stellar body revolving about Algol in the same period as that of the light-changes. At first, Sir William Herschel opposed this theory, because he could not resolve the companion star in his telescope, although he later changed his mind when he discovered the existence of binaries. In 1889, H. C. Vogel, at Potsdam, verified that Algol is a binary system, by analyzing the variable Doppler effect of Algol's spectral lines to obtain the components' radial velocities of approach and recession, an early milestone in the achievements of astronomical spectroscopy.

Current data show that Algol is 100 light years distant; the primary, a Sirius-type

† *A Saxon farmer who observed the first predicted reappearance of Halley's comet, on Christmas night 1758.*

(class-B8) main-sequence star, has a radius of about 1.33 million miles (3 times that of the Sun) and a luminosity equal to a hundred Suns; its mass is about four $\mathcal{M}\odot$ at most. The darker and larger secondary, a K-type star, has a radius of about 1.5 million miles (about 3½ times the Sun's), but its mass, not yet definitely determined, is approximately the same as the Sun's; thus the companion is somewhat more diffuse than the primary and seems to have evolved to the subgiant stage.

The Algol system is normally of apparent magnitude 2.1, but about every 2.87 days it fades to magnitude 3.4; the light period, which is measured from minimum to minimum, is exactly 2 days 20 hours 48 minutes and 56 seconds; an eclipse lasts about 10 hours, during which some 79 percent of the primary is occulted by the larger companion. Midway between these eclipses, a slight secondary minimum occurs, when the companion is partially covered by the primary. By studying the light curve, astronomers can determine the orbital features as well as the geometries of the components; according to this analysis, the primary of Algol lies about 1 million miles from the system's center of mass; estimates give approximately 6.5 million miles for the mean separation of the two stars. The inclination of the plane of their orbit to our line of sight is approximately 82 degrees.

Two special brightness effects are also produced by Algol's eclipses: the first of these, called the "reflection effect" produces a brightness increase just before the primary is eclipsed, when light from its invisible hemisphere strikes the larger, cooler companion and is reflected to us, and the same phenomenon occurs immediately after the primary minimum. As the system approaches the secondary eclipse, when the companion is seen beyond the brighter primary, the effect is even more pronounced, because a larger surface area of the companion is turned toward us, as it reflects additional light from the primary. The second of these lighting effects is called "limb darkening," the result of heavier atmospheric obscuration at the star's rim as compared to its center, so that an eclipse, beginning at the star's limb, reduces the light that reaches us from the star more gradually at first, and then much more rapidly as the star's center is covered.

In addition to the main pair, a third component in the Algol system has been detected spectroscopically; an F-type main-sequence star, its radius only 0.75 million miles, it revolves around the binary with a period of 1.86 years, about 50 million miles from its center of mass. Owing to apparent astrometric variations of Algol, the existence of a fourth star, with a period of 188 years, was announced some years ago, but this finding has been thrown into doubt by more recent studies.

An object of research and investigation, the "Demon Star" continues to add to our knowledge of the heavens: recently flares were discovered in the radio emission from the Algol system: sudden intense bursts of radio waves from the surface of the subgiant superimposed on the steady radio signal.

A Historic Nova

For three hundred years after the excitement caused by Kepler's star in 1604, only novae of relatively low apparent brightness were detected in our Galaxy; however, on February 21, 1901, as if to announce a new century of startling scientific achievement, a "new" third-magnitude star appeared about midway between Algol and δ Persei. Photographs made at the Harvard Observatory two days previously showed that it was

143. *Nova Persei 1901,* showing illumination of expanding nebular shell. The asymmetry of the outburst and the irregular structure of the shell are apparent in this photograph, taken in November 1949 with the 200-inch reflector. (*Hale Observatories*)

144. *NGC 869 and NGC 884,* the double star-cluster in Perseus, a famous spectacle of the northern skies. These two open clusters are noteworthy for their many luminous young supergiants. Photographed with the 20-inch astrograph. (*Lick Observatory*)

then of the thirteenth magnitude, indicating that, when discovered, its luminosity had increased ten thousand times! First observed by the Scottish amateur astronomer T. S. Anderson, *GK,* or *NOVA PERSEI 1901,* reached a maximum luminosity of 200,000 $L \odot$ on February 23, rivaling Capella and Vega in its apparent magnitude of 0.2, but then fading quickly to second magnitude in a week, and two weeks later to fourth magnitude. It then began to oscillate between the fourth and the sixth magnitudes for several weeks, "disappearing" and "reappearing" in four-day cycles, after which it continued fading, reaching the thirteenth magnitude again about thirty years after its initial nova-like outburst.

A few months after Nova Persei's appearance, an interesting phenomenon was discovered from photographs that revealed a faint nebulosity surrounding the star and apparently expanding at the phenomenal rate of 2 seconds per day; at the star's computed distance of 1,300 light years, this would equal 4.6 times the speed of light if this were a true speed! A careful analysis, however, indicates that this "rapid expansion" of the nebula is illusory, an optical effect created by light originating in the rapidly expanding radius of the nova and illuminating increasing areas of an already existing dark nebula. The true nebular shell of GK Persei did not become visible until 1916; its rather peculiar fanlike appearance suggests that matter was ejected mainly from one hemisphere of the star. GK Persei, varying irregularly between the eleventh and the fourteenth magnitudes, is now recognized as a close binary with a period of revolution of 1.904 days; the components are a very dense class-B (blue-white) subdwarf and a K2 subgiant, their respective masses about 1.29 $\mathcal{M} \odot$ and 0.56 $\mathcal{M} \odot$.

Clusters in Perseus

Near the border of Andromeda, about 5 degrees west-northwest of Algol, lies an attractive open cluster, *M34 (NGC 1039)* (RA 2 h 38.8 m   δ +42° 34')   m 6, its 30-minute field just visible to the naked eye and excellent for binoculars and small telescopes with low-power wide-angle eyepieces. About 1,500 light years distant, it contains approximately eighty stars, including the white double *h1123,* in the center; the cluster's estimated age is slightly over 100 million years.

The Champion apparently carries a very long sword, for, according to R. H. Allen, υ (Fl. #48), on Perseus' right side, a fourth-magnitude star, marks its tip at the eastern end of the Segment of Perseus; whereas the sword handle of Perseus is traditionally shown in the far-northwestern part of the constellation. Here we find one of the most beautiful objects for the smaller telescope, the spectacular *DOUBLE STAR CLUSTER NGC 869* and *NGC 884* (RA 2 h 17 m   δ +56° 54') (fig. 144), each cluster about 50 minutes in diameter. The Arabians called them *Mi'ṣam al Thurayya,* the "Wrist Nearest the Pleiades"; Hipparchus was apparently the first to record the clusters, designating them as a "cloudy spot"; like other objects of this type, they were regarded as nebulae until Galileo's time. Modern telescopes reveal two great swarms of giant stars, occupying a neighboring spiral arm as it branches out from the galactic center, therefore embedded in a very rich part of the Milky Way. The clusters, about one-half degree apart and between 7,000 and 8,000 light years distant, contain many B- and A-type supergiants, some with luminosities of nearly 60,000 $L \odot$; NGC 884 also contains red supergiants of spectral classes M0 to M5. Studies of the clusters have indicated that they may not be physically bound, because, according to the findings of various astronomers (between 1943 and 1954), NGC 884, at an estimated distance of 2,500 parsecs, is about 11.5 million years old, while NGC 869, lying 2,150 parsecs distant, is only 6.4 million years old. The total mass of the clusters is about 5,000 $\mathcal{M}\odot$, and their total luminosity is about 200,000 $L\odot$ Each cluster is surrounded by a much more extensive association of stars sharing the characteristics of the cluster members. The mass-luminosity ratio indicates that the stars in the clusters are primarily supergiants.

A Peculiar Radio Galaxy

More than a hundred extragalactic radio sources have been identified as individual galaxies. These can be divided into two groups: normal spirals, like our Galaxy, which, in addition to the 21-centimeter hydrogen line, emit a continuous spectrum of radio waves from their disks, nuclei, and coronas, and peculiar radio galaxies, which have unusual and complex structures and emit extremely large quantities of radio waves. In some cases this energy is emitted from a single, concentrated source, whereas in others the emission occurs from a concentrated core and an extended halo or huge jets. An example of the latter type of radio galaxy is the controversial *NGC 1275, PERSEUS A (3C 84)* (RA 3 h 16.4 m δ +41° 20') m 12.7 (fig. 145), about 2 degrees east-northeast of Algol. Classed variously as an elliptical and as a Seyfert galaxy, with its main body either an S0 galaxy (the type midway between spirals and ellipticals) or a very compact spiral, Perseus A has a core about one light-year in diameter. Huge patches of gas and dust along its northern edge, some of them with large redshifts, are interpreted by some astronomers as a loose spiral galaxy passing through (i.e., colliding with) the main elliptical galaxy, a hypothesis, however, that cannot adequately explain the vast quantities of radio energy such galaxies emit. Violent events are evidently occurring in the cores of Seyfert galaxies, because their brightnesses increase markedly as we pass from their outer regions to their small, luminous nuclei. Their strong, broad emission lines indicate concentrated regions of hot gases in the cores, expanding with speeds up to thousands of kilometers per second.

Perseus A emits tremendous quantities of energy in the infrared and X-ray regions. Various theories have been advanced to account for this output of energy, including the presence of an exceptionally hot interstellar gas cloud, much larger than the galaxy itself, in the area of the X-ray source. The high luminosity of Perseus A in the short wavelengths has also been attributed to the *inverse Compton effect*, a process in which very energetic electrons transfer part of their energy to photons which collide with them and become X rays; in the standard *Compton effect*, the electron gains energy and the photon loses energy in a collision and thus reddens. The X-ray source of Perseus A covers an area 3 million light years in diameter, extending long tentacles into space on all sides of the galaxy, as though they were produced by a tremendous central explosion.

Radio Trails

Another member of the rich cluster of galaxies including Perseus A is the elliptical galaxy *NGC 1265*, situated about 25 minutes north of Perseus A. Out of this galaxy, stream two so-called "radio trails," two comparatively low-power jets ejected from the

145. NGC 1275, Perseus A, a peculiar elliptical or Seyfert galaxy in the Perseus cluster of galaxies and a strong source of radio emission; the bright patches at its northern edge (right), which extend outward like a tentacle, may be part of a colliding spiral. Photographed with the 200-inch telescope. (Hale Observatories)

galaxy's center and then distorted sideways into a long curve by the relatively dense, hot, ionized intergalactic gas moving through the galaxy cluster at speeds of several thousand kilometers per second.

AURIGA

Near the bent Bull a Seat the Driver claims,
Whose skill conferr'd his Honour and his Names.
His Art great Jove admir'd, when first he drove
His rattling Carr, and fix't the Youth above.

Manilius

Hephaestus, the blacksmith and god of fire whose apprentice aided Orion in his quest for the sunrise, tried to rescue his own mother, Hera, when Zeus hung her by the feet from Mount Olympus as a punishment for sending a storm against Hercules, seaborne after his Trojan victory. Enraged by his son's intervention, Zeus threw him from the heavens; Hephaestus landed on the island of Lemnos, laming both legs in the fall. There he set up his forge, and, as smithy of the gods, increased his fame. When Athena came to Lemnos to fashion some arms, Hephaestus, recently deserted by Aphrodite, fell in love with Athena and clumsily tried to force himself upon her; as the goddess fled, she brushed his seed from her leg with a wool cloth and threw it upon the earth, from which a son was born, named Erichthoneus, after his origin. Secretly brought up by Athena as an immortal, Erichthoneus became the fourth king of Athens; however (contrary to genetic laws), he inherited his father's lameness and needed artificial transportation. A talented lad, Erichthoneus solved this problem by inventing the four-horse chariot; in happy contrast with Hephaestus' fall, Erichthoneus and his chariot were transferred to the Milky Way by an admiring Zeus as the great constellation Auriga, extending from the Bull's northern horn to the feet of Camelopardalis, and bounded on the west by Perseus.

Auriga is also identified as Myrtilus, the son of Hermes and charioteer to Oenomaus, king of Pisa, whom he betrayed. Oenomaus had a much sought-after daughter named

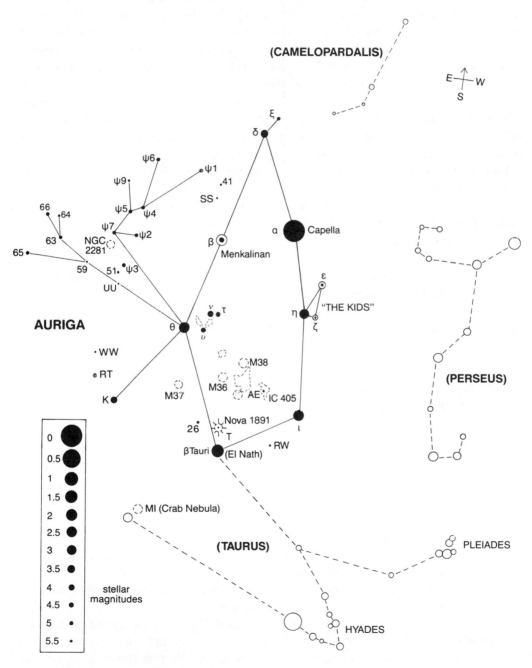

146. Auriga and surroundings.

Hippodamia; an oracle warned the king that he would be slain by the man that married her. Fearing this prophecy, Oenomaus set up a snare for her prospective suitors. Each of them was permitted to flee by chariot with Hippodamia as far as the Isthmus of Corinth; she would then belong to him who escaped Oenomaus, who, fully equipped with arms and horses by the god of war, would be in hot pursuit. At least twelve suitors died trying to outrace the king; Hippodamia remained single until the beauteous youth Pelops, son of Tantalus, came to Pisa to press his suit and she fell ardently in love with him. Answering Pelops' plea for help, the sea god Poseidon gave him a golden chariot with winged steeds that never tired. Hippodamia helped by enlisting the aid of Myrtilus, the charioteer; because he desired the lovely princess for himself, Myrtilus inserted waxen linchpins in place of bronze in the wheels of Oenomaus' chariot. This act of sabotage resulted in Pelops' victory when Oenomaus became entangled in the reins and was dragged on the ground by his own horses. Dying, the king cursed Myrtilus, who then accompanied the triumphant couple on their wedding journey; the curse was fulfilled when the faithless charioteer was slain by Pelops and hurled into the sea for trying to rape Hippodamia during a momentary absence of her husband.

Usually shown as a young man carrying a whip, Auriga holds a goat and two kids against his shoulder and wrist; this traditional way of picturing the Charioteer is so ancient that an archaic Euphratean sculpture shows Auriga much as it is today, with a goat carried on the left arm. A Graeco-Babylonian constellation called *Rukubi,* the "Chariot," occupied much the same part of the skies; even the Chinese, whose ancient constellations differ so greatly from ours, pictured some of the stars of Auriga as their *Woo Chay,* the "Five Chariots." Some Greek titles are Heniochus, the "Rein Holder" (Erichthoneus) and Hippolytus, identified with the biblical Joseph. The poetic Latin title was *Aurigator,* and various Roman titles meaning "Charioteer" and "Rein Holder" appear in medieval charts; the Teutonic peasantry called Auriga the "Ploughman with His Oxen."

CAPELLA (α) *ALPHA*
(RA 5 h 13 m δ +45° 57') m 0.1

Capella is the diminutive of *capra,* or "goat," with which, on occasion, the constellation Capricorn was also identified; as the Romans' "Little She-Goat," α Aurigae was known as a "rainy sign" from classical times. The sixth-brightest star in the heavens, it was treated by some writers as a constellation unto itself. Its mythological name "Amalthea" refers to the goat that nourished the infant Jupiter (Zeus), and it was therefore called *Jovis Nutrix.* Capella was also identified as *Cornu Copiae,* the "Horn of Plenty," one of the goat's horns which the infant deity broke off and gave to his nurses, having endowed it with the power of filling up with whatever its possessors might wish. Seen from the deserts of Arabia, Capella was *Al Hāḍī,* rising with the Pleiades and riding before them like the singer urging on a procession of camels.

Several Egyptian temples dedicated to the god Ptah were oriented to Capella's setting; it was the Hindu *Brahma Ridaya,* the "Heart of Brahma," the Akkadian *Dil-gan I-ku,* meaning "Messenger of Light," *Dil-gan Babili,* the "Patron Star of Babylon," and in Assyria prior to 1,730 B.C., Capella was known as *Î-ku,* the "Leader," because its position relative to the moon at vernal equinox determined the beginning of a new year. In very ancient times, spring began when the Sun entered Taurus, and because of Auriga's proximity to the Bull, Capella was called the "Star of Marduk," or the "Spring Sun," identified with Taurus and worshiped as the son of Ia.

The most northern of all first-magnitude stars, Capella is a deep yellow giant, spectral class G8. At a distance of 45 light years and with a luminosity 160 times that of the Sun and an absolute magnitude of −0.6, Capella is receding with a radial velocity of 18.5 miles per second, just as the Hyades cluster and its associated Taurus cluster, are moving.

Capella is a spectroscopic binary, the components only 70 million miles apart, less than an astronomical unit, therefore closer to one another than the Earth is to the Sun. The primary, or G8, star, with 90 $L \odot$, overwhelms the spectrum of the secondary, a class G0 star, which is nevertheless only somewhat less luminous, at 70 $L \odot$; but this apparent contradiction may be explained by the tremendous broadening of the G0 star's spectral lines owing to turbulence in its atmosphere. A faint red dwarf is also a component of the Capella system, but it lies at least one thousand astronomical units from the close pair. As usual, the story doesn't end here, for the distant dwarf is itself a close binary, two red dwarfs in fact, at an apparent separation of 2.7 seconds, their spectral types dM1 and dM5; together, they are only 1 percent as luminous as the Sun!

Other Major Stars

About 6 degrees east of α and slightly to its south, is **MENKALINAN (β) BETA,** derived from *Al Mankib dhi'l 'Inān*, "The Shoulder of the Rein Holder." About twice the distance of Capella, second-magnitude β has a luminosity of 110 $L \odot$ and an absolute magnitude of −0.3. First detected as a spectroscopic binary by Pickering in 1889, β was subsequently found to consist of two nearly equal A2 subgiants that eclipse one another twice during their period of 3.96 days. The diameter of each star is about 2.6 times that of the Sun, and their masses are 2.35 $\mathcal{M} \odot$ and 2.25 $\mathcal{M} \odot$. These twins keep close company, at a separation of only 7.5 million miles! An extremely faint star lying 350 astronomical units distant from the main pair, its luminosity only about 1/630 that of the Sun, shows some indication that it may be orbiting them.

The Kids

Tempt not the winds forewarned of dangers nigh,
When the Kids glitter in the western sky. . . .
                              Callimachus

On modern charts, three stars, ε, ζ, and η (Epsilon, Zeta, and Eta) constitute the famous "Kids," which accompany the Goat Star; they lie to the southwest of Capella in the form of an acute right triangle. Together with Capella, ζ and η were the Arabic *Al 'Ināz*, "The Goats." In older charts, only ζ and η are shown as the Kids, but in charts that appeared toward the end of the nineteenth century, they were joined by *(ε) EPSILON* (RA 4 h 58.4 m δ +43° 45′) m 3.0 to m 3.8, known by the Arabians as *Al*

Ma'az, "The He-Goat," in earlier times, before it discovered the secret of youth and became a kid. Its variability was suspected as far back as 1821 and confirmed by Schmidt in 1843 and Heis in 1847. Since then, Epsilon Aurigae has become one of the most observed and controversial of all known stars. Some 3,300 to 4,500 light years distant, too far for a reliable parallax, ε is an eclipsing binary with one of the longest periods of revolution known, 27.06 years. Only one star is visible, an F-type bright supergiant with a luminosity at least sixty thousand times that of the Sun and an absolute magnitude variously estimated at −7.1 and −8.5; its diameter is approximately one hundred eighty times the Sun's. When eclipsed by its mysterious companion, it remains for a year at minimum luminosity, which is preceded and followed by six months of partial fading.

The invisible companion seems to be one of the largest stars known, with a radius possibly two thousand times the Sun's, or nearly 900 million miles, and its structure has been the focal point of countless investigations and unending speculation. Calculations indicate that it is an extremely diffuse, cool, and therefore underluminous object, radiating chiefly in the infrared part of the spectrum. Because a newly evolving proto-star begins its life as an immense, cool globe of gas and dust, before the process of gravitational contraction raises its temperature and increases its luminosity, astronomers at first believed the invisible component of ε Aurigae to be a star near its birth. Modern theory of the evolution of stars leads to the opposite conclusion, that the invisible component is a very highly evolved star in the last stages of its evolution as a greatly expanded and rarefied red supergiant. A peculiar transparency of the companion and its apparent tendency to absorb equally all wavelengths of the visible star led to the suggestion that only the companion's outer layers, ionized by radiation from the visible star, are causing the eclipses.

A newer theory suggests that the eclipsing body is not a star, but a vast shell of gas, dust, and solid particles surrounding a tiny, relatively faint central star. A variation of this idea proposes that the companion, a tiny star, is surrounded by a flattened ring of rotating gases, edge on to our line of sight and periodically transiting the visible F star. However, these theories are probably not the last word concerning the Epsilon Aurigae system, and thus the controversy is likely to continue.

The "Western Kid" (so-called by tradition), which lies about 2.75 degrees south of ε, is **SADATONI (ζ) ZETA** (RA 4 h 59 m δ +41° 00′) m 3.8 to m 3.9. Its title is apparently derived from *Al Said al Thani,* "The Second Arm" (of the Rein Holder). Like Epsilon, Zeta Aurigae is an eclipsing variable of great interest; the components, separated by about 500 million miles, are an immense K-type cool star 8.3 times the Sun's mass and a hot, blue, and much smaller star 6.8 times the mass of the Sun. Uncertainty over the luminosity of the primary has led to varying estimates of the distance of this system, ranging from 160 parsecs to as much as 370 parsecs, or about 1,200 light years. The smaller star is eclipsed by the giant every 2 years 8 months, and the duration of the total eclipse is 39 days. The components complete one revolution about their common center of mass every 972 days, or roughly 32 months. A few weeks before an eclipse begins, as the light of the small hot star begins to penetrate the giant's atmosphere, conditions are favorable for spectroscopic studies, which reveal peculiarities in the structure of the giant's atmosphere reminiscent of the intricate pattern of numerous small prominences extending out from the photosphere of the Sun. As in the case of

Omicron Andromedae (p. 326), chemical stratification of the K star's atmosphere has been observed, with neutral metals indicated in the lower levels, ionized metals in the higher layers, and hydrogen and ionized calcium distributed throughout the atmosphere. A further analogy with the Sun is found in the lines of ionized metals and in the lines of those metals requiring greater excitation energies, which are also present in the upper part of the Sun's atmosphere; this means that the temperature at that level is higher than on the photosphere. The Sun's chromosphere rises to about 5,000 miles, but the atmosphere of the giant component of ζ Aurigae extends to a height of 20 million miles or more above the surface.

The eastern kid, only 45 minutes from Zeta, is *(η) ETA,* magnitude 3.18, a hot, blue main-sequence star, type B3; its exact distance is not known, estimates yielding such widely differing figures as 197 and 370 light years. In the days of classical Greece, the rising of the Kids (ζ and η) in early October ushered in the stormy season and signaled the closing of navigation; the passing of this dreaded time was celebrated with a festival called *Natalis Navigationis.*

About 10 degrees east-southeast of the Kids and about 7 degrees south of β, is *(θ) THETA,* which now marks the Charioteer's right heel since Gamma was recognized as a star of Taurus. About 110 light years distant, θ, with a luminosity eighty-five times the Sun's, is notable for the pronounced lines of silicon in its spectrum. At 3.5 seconds, a small solar-type companion of magnitude 7.5 shares θ's proper motion, and a faint optical companion lies farther out, at about 52 seconds; both are suitable for small instruments.

As we have indicated, the former (γ) GAMMA Aurigae, lying nearly 10 degrees to the southwest of θ, is now β Tauri (El Nath); Arabian astronomers first saw this star as the Heel of the Rein Holder, but in later times they relegated it to the Bull's northern horn. It was also the Romans' *Aurigae Manus,* because the Driver was pictured bent over, holding his heel in his hand; however, the Abbé Hell, an eighteenth-century Austrian astronomer, applied this title to θ. Gamma and Iota together were the Arabic "Goat's Attendants," but the title *Al Ka'b dhi'l 'Inān,* "The Heel of the Rein Holder," was later given to *(ι) IOTA,* a 2.5-magnitude K3 giant about 7 degrees northwest of El Nath and now the southernmost major star of Auriga.

A rather inconspicuous yellow star about 10 degrees northeast of Capella marks the head of the Charioteer, *(δ) DELTA,* a m 3.72 K-type giant, known to the Hindus as *Prajapati,* the "Lord of Created Beings." With nearby ξ (Xi) and several other, fainter stars, δ was part of a Chinese asterism whose title, *Pa Kuh,* the "Eight Cereals," seems to remind us of some advertisements for total-nutrition breakfast foods.

The Charioteer is not complete without his whip; its many *flagelli,* or lashes, are represented by several fifth-magnitude stars lettered *(ψ) PSI* and numbered from 1 to 10. Scattered to the east of the main figure, they include ψ *1* (#46 Aurigae), an irregular variable and supergiant of spectral class M with a magnitude range of 5 to 6. Look for it about 6 degrees northeast of β.

A few degrees southwest of ψ 1 is another variable, the cataclysmic dwarf *SS AURIGAE* (RA 06 h 09.6 m   δ +47° 46')   m 10 to m 14.8, easily located about 3½ degrees northeast of β and just south of binary #41. The star remains quiet when at minimum luminosity for approximately 55 days, then flares up suddenly with its luminosity increasing sixty times in about 24 hours. Like SS Cygni, the classical example of this type of eruptive star, SS Aurigae is a close binary whose tiny components, almost in contact, orbit one another in a period of only 4 hours 20 minutes, exchanging material in the process.

Another 20 degrees to the south, about 8 degrees southeast of θ, we find the variable **RT** (#48) (RA 06 h 25.4 m   δ +30° 32′), a bright Cepheid with a period of about 3.75 days and a magnitude variation from 4.9 to 5.9, during which this pulsating supergiant (2,300 L ☉) changes from spectral type F5 to G0. The constellation of Auriga is rich in variable stars, and within the figure of the Charioteer, approximately 3.5 degrees east of Iota, is the hot, blue-white O-type irregular and erratic variable *AE AURIGAE* (RA 05 h 13 m   δ +34° 15′)   m 5.4 to m 6.1. About 1,600 light years distant, AE is the illuminating star of the diffuse nebula *IC 405,* although photographic evidence and comparisons of their radial velocities suggest that the association of star and nebula is a recent one and that dust clouds near the star will be swept away by its intense radiation, which will greatly alter the surrounding nebula. AE Aurigae is one of three "runaway" stars that may have started their lives in the Great Nebula of Orion among the vast association of O- and B-type stars, separating from it some 2.7 million years ago.

At the Charioteer's feet, more or less between Iota Aurigae and El Nath, is the erratic variable *RW AURIGAE* (RA 5 h 04.6 m   δ +30° 20′)   m 9 to m 12. Its swift and sudden light changes may increase its brightness twenty-five times, and its peculiar emission-line spectrum indicates atmospheric turbulence, possibly interacting with surrounding particles of gas and dust, although no nebula is visible. A few stars with similar light curves, often associated with nebulae, are sometimes called RW Aurigae-type variables; they tend to have luminosities similar to that of the Sun and may be related to the important T Tauri stars, protostars in the Hayashi convective stage found in abundance in the Orion nebula.

The Nova of 1891

On January 23, 1892, Dr. T. D. Anderson, of Edinburgh, the discoverer of Nova Persei, turned a spyglass skyward and noticed a fifth-magnitude yellowish star about 2 degrees northeast of El Nath. Photographic plates subsequently showed that it had already been visible for several weeks and had reached its maximum luminosity with an apparent magnitude of 4.4 on December 20, but had gone unnoticed. It remained quite bright until March, when it began to fade rapidly, becoming invisible toward the end of April at magnitude 15, but in August it made a modest comeback, reaching magnitude 9.5 and remaining at that level for three years. It then resumed its fading, reaching magnitude 15.5 about the year 1925 and remaining at minimum luminosity since then. The character of the spectrum of *T AURIGAE (Nova 1891)* excited great interest in the astronomical world, especially since it was one of the first stars analyzed spectroscopically. On the nineteenth of August, 1892, E. Barnard, at Lick Observatory, rediscovered the nova as an apparent planetary nebula, which was the nova's expanding shell or disk, which increased from 3 seconds at the time it was first observed to some 12 seconds about fifty years later.

Nova Aurigae is known as a "slow nova," because it rose rapidly to maximum luminosity but underwent a relatively gradual decline in brightness lasting many years. At maximum, its luminosity was probably twenty-five thousand times that of the Sun, and its absolute magnitude may have reached −6.2. Its present photographic magnitude is about 14.8, and this bluish stellar remnant, at a computed distance of 4,100 light years, is now classed as an O-type (emission) star. Recent investigations have revealed that T Aurigae, like Nova Persei and Nova Aquilae, is a close and rapid binary. With a period of revolution of about five hours, the dwarf components also eclipse one another, the primary occultation lasting for about forty minutes.

Three Clusters

Three bright galactic clusters lie in or near the lower part of the Charioteer's figure. The first of these, *M36, NGC 1960* (RA 5 h 32.9 m δ +34° 07′), about 5 degrees southwest of θ, was discovered in 1749; a compact, bright object 12 minutes in diameter, it contains sixty stars, many of which are B-type main-sequence stars, subgiants and giants, whose broad spectral lines show rapid rotation; it has a total luminosity of about 5,000 $L \odot$.

About 4 degrees southeast of M36 (just outside the main figure) lies the largest and brightest of the three open clusters, *M37, NGC 2099* (RA 5 h 49 m δ +32° 33′), a beautiful object containing about 150 luminous stars in a 25-minute field. Most of them are class-A main-sequence stars, but twelve red giants are also present; plotted on the H-R diagram, the graph of M37 indicates that it is much older than M36, perhaps more than 200 million years of age.

About 2.3 degrees northwest of M36 lies the large, irregularly shaped cluster *M38, NGC 1912* (RA 5 h 25 m δ +35° 48′), containing a hundred stars in a 20-minute field. Among them are giants of classes B5 and G, and A-type main-sequence stars. M36 and M38 may be viewed together in a small wide-angle telescope. The galactic equator passes by M38, less than a degree to the west, and this region of the heavens is therefore crammed with stellar treasures. In his *Celestial Objects,* of 1893, the English astronomer Rev. T. W. Webb called it a "glorious neighborhood." Here, close to the circumpolar zone, where we began our journey, we leave the northern heavens, enriched by three aspects of human creativity: mythic lore, observation, and scientific discovery.

VI

The
Southern Skies

INTRODUCTION

Little was known about the southern heavens before Columbus' voyages, and Columbus himself sailed no farther south than the Caribbean, but his discovery of the lands to the west, or the "New World" (as it was later called), inspired numerous naval expeditions to hitherto unknown regions early in the sixteenth century. The Italian navigator Amerigo Vespucci (after whom the American continents are named), the famed Portuguese explorer Ferdinand Magellan, Vasco de Gama, who sailed around the African coast, and many others, observed the stellar regions surrounding the south celestial pole. This information was incorporated by various astronomers, including Johann Bayer, whose *Uranometria,* of 1603, was notable for its inclusion of twelve new southern asterisms.

In 1676, twenty-year-old Edmund Halley, resolved to become a practical astronomer, interrupted his studies at Queen's College, Oxford, and traveled south to St. Helena. There he spent the next three years preparing his Catalogue of Southern Stars, published in 1679, in which 341 stars and several new constellations are included. Hailed as the "southern Tycho" on his return to England, Halley received many important academic honors and pursued a distinguished career as Astronomer Royal, holding this title until his death.

Many years later, the French astronomer Abbé Nicolas Louis de Lacaille surveyed the southern stars from the Cape of Good Hope using only a 1/2-inch glass, and published his *Mémoires,* of 1752 and *Coelum Australe Stelliferum,* of 1763, in which he introduced fourteen new constellations named after the instruments and technology of the time and the fine arts. As a result of his work, Lacaille was called the "true Columbus of the Southern Sky."

About twenty years later, Sir William Herschel began extensive telescopic observations of double stars in the northern hemisphere; after Herschel's death, in 1822, his son Sir John Herschel continued this project, eventually transferring it to the southern hemisphere and setting up an 18-inch reflector at Feldhausen, near Cape Town, South Africa. There, between 1834 and 1838, he discovered 1,202 double and multiple stars, gathered data on 70,000 stars and recorded 1,708 clusters and nebulae. On his return, at age 46, Sir John devoted his life to analysing this material, which he later published in his extensive *Results of Astronomical Observations Made at the Cape of Good Hope,* his *Outlines of Astronomy,* and the authoritative General Catalogue of Nebulae (1864).

THIRTEEN

Mythical and Modern Creations—Centaurus to Grus

The southern skies offer a wealth of material adequate for a lengthy exploration, but here we restrict our discussion to some important highlights. Our first constellation is visible to southern observers during June, early winter in the south temperate zone, where the seasons are exactly the reverse of those for observers at corresponding northern latitudes. One of the largest sky pictures, more than 60 degrees long, this figure is the southern counterpart of Sagittarius the Archer, and their stories have often been confused.

CENTAURUS

The fourth of Hercules' labors was to bring back alive the destructive Erymanthian boar; en route to its lair, he journeyed through Pholoë, in the southern Peloponnesus, where he was hospitably received by the centaur Pholus. After dining on roast meat, Hercules asked for wine; despite the warnings of his host, he opened a bottle belonging to all the centaurs, who soon smelled it and stormed Pholus' cave. Using firebrands and arrows, Hercules repulsed their attack and pursued the centaurs to Malea; there they cowered about the immortal Chiron, who was the son of Cronus (Saturn) and known as the wisest of all centaurs and the instructor of many great heroes. Hercules shot a poisoned arrow at one of the centaurs, which accidentally struck Chiron in the knee. Recognizing Chiron as his own former teacher, Hercules ran to him in great distress, drew out the shaft and applied one of Chiron's ointments to it, but the wound would not heal. Although Chiron longed to die, as the son of a god he was compelled to remain alive until Prometheus, in an act of kindness, took over Chiron's immortality, thus enabling the Centaur to die and then be raised to the heavens.

After Chiron's death, the other centaurs fled in many directions, but the hospitable Pholus, while examining one of Hercules' poisoned arrows, accidentally dropped it on

147. Hercules' Combat with the Centaurs. Our mythical superhero, busy everywhere in the heavens, plays a key role in the story of Centaurus. (*The Bettmann Archive*)

his own foot and was instantly killed. Because of Hercules' gratitude, Pholus, too, was transferred to the skies, and thus two kindly centaurs, Chiron and Pholus, are identified with the constellation Centaurus, the southern Centaur.

The universal Greek title "Kentauros" also refers to the offspring of Ixion's phantom cloud, as we have told (p. 255), who fathered the race of centaurs. Centaurus was sometimes called "Philyrides," after Chiron's mother, the ocean-nymph Philyra. Chiron taught a diversity of subjects, including astronomy, and he is credited by mythology with the invention of the constellations! Chiron's skills in medicine, which he transmitted to the famed healer Aesculapius (Ophiuchus), are reflected in the names of the medieval medicinal plant *Centaurea,* the centaury, and an earlier plant, the *Chironeion.* Other Latin names for this constellation are *Minotaurus,* because of its occasional association with Minos' Bull; *Semi Vir,* the "Half Man"; and *Acris Venator,* the "Fierce Hunter." Caesius, of the biblical school, saw Centaurus as Nebuchadnezzar, overcome with madness, lying under the dew of heaven and eating grass like the oxen.

The Centaur is pictured with his back toward us and his head turned to the east; in his right hand he holds a spear aimed toward adjacent Lupus the Wolf. The minor constellation of Circinus, the Compasses, lies at Centaurus' feet, which are marked by α and β Centauri; to their west is the famous Southern Cross. Immediately north of Centaurus stretches the long form of Hydra; the immense Argo Navis, with its three subdivisions, Carina, Puppis, and Vela, lies farther to the west of the Centaur's body.

RIGIL KENTAURUS (α) ALPHA CENTAURI
(RA 14 h 36 m   δ −60° 38′)   m −0.3

Marking the eastern hoof, its name derived from the Arabic *Al Rijl al Kentaurus,* "The Centaur's Foot," this famous star is the closest in the heavens to our Sun, and the third-brightest. It was worshiped by the ancient Egyptians, and a large number of temples dating from 3800 B.C. were oriented to its rising at the autumnal equinox. In southern China, Alpha Centauri was the determinant of the ancient stellar formation called *Nan Mun,* or the "South Gate." Lying in the Milky Way and being the southernmost major star of Centaurus, α cannot be viewed by observers farther north than the 29th parallel.

The Parallax of Alpha

Scottish-born Thomas Henderson held the humble position of solicitor's clerk early in life, but like his German counterpart Friedrich Bessel, he soon abandoned a commercial career to attain the heights of achievement in astronomy. In 1831 he was appointed director of the observatory at the Cape of Good Hope, and while there, Henderson made a series of observations of Alpha Centauri. A few years later, he returned to Scotland as his country's first Astronomer Royal, and from the data he had recorded at the Cape, he deduced Alpha's parallax. In February of 1839, after considerable delay, Henderson published his observations; had he done so sooner, he would have been credited as the first astronomer to measure the distance of a star successfully, a distinction Bessel had won only two months earlier with his publication of the parallax of 61 Cygni. Henderson's findings give Alpha Centauri a distance of about 200,000 astronomical units, and although modern measurements reveal that its distance, 4.34 light years, is 30 percent greater, it is nevertheless our Sun's nearest neighbor, and its parallax (0″.751) is the largest of all first-magnitude stars, as is its proper motion (3″.68 per annum, in a northwesterly direction).

The Alpha Centauri System

Alpha Centauri has been known as a visual binary since 1689, when its components were first resolved by Father Richaud at Pondicherry, India. The primary, a class-G2 main-sequence star, magnitude −0.04, is very similar to the Sun spectroscopically; it also has a similar mass (1.10 $\mathcal{M} \odot$), diameter (1.07 times that of the Sun), and luminosity (1.5 $L \odot$). The secondary component of α Centauri is a K1 star, and with a diameter 1.22 times the Sun's, it is slightly larger than α 1, but the K star's mass, about .85 $\mathcal{M} \odot$, is smaller, and its luminosity is only 0.4 $L \odot$. The first observations gave the K star's apparent magnitude, incorrectly, as 3.5 and even 4.0; the correct figure is 1.17. Lacaille first accurately calculated the period of revolution of the pair, now computed at about eighty years. With a radial velocity of about 23⅓ kilometers, or about 14½ miles, per second, the components are orbiting their common center of mass in an eccentric and elongated ellipse, with an apparent separation ranging from 2 seconds to 22 seconds, or about 11 to 121 astronomical units; they are presently very nearly at their widest separation, which will occur in the year 2000.

Proxima Centauri

In 1915, the slightly variable radial velocity of Alpha Centauri led R. T. Innes to the discovery of a tenth-magnitude distant companion, now called **PROXIMA CENTAURI,** because its slightly larger parallax, 0.762 second, indicates that it is somewhat closer than the bright pair. Situated 1 degree 51 minutes south and 9.9 minutes west of Alpha, Proxima is a class-M5 red dwarf, with an emission spectrum that shows unpredictable sudden increases in luminosity of up to a full magnitude, changes that occur on an average of about once every six months (but at irregular intervals), owing to flares originating on the surface of this star. In the number of its recorded outbursts, Proxima exceeds all other flare stars.

Tiny Proxima, only one thirteen-thousandth as luminous as the Sun, is one of the least massive stars, with 5 percent to 15 percent of the solar mass; its diameter, only about 40,000 miles, is equivalent to that of slightly less than five Earths, or about 5 percent of the Sun's diameter. Although it lies 10,000 astronomical units, or almost 1 trillion miles, from the bright components of Alpha Centauri, it is believed to be gravitationally bound to the system; owing to its great distance from the central pair, however, it has an extremely long period of revolution, perhaps as much as a half million years, in agreement with Kepler's third law.

Other Major Stars

About 4½ degrees west of Alpha Centauri lies the brilliant *HADAR (β) BETA,* magnitude 0.66, somewhat overshadowed by the fame of its spectacular neighbor. *Ḥaḍar* and *Wazn,* meaning "Ground" and "Weight," were Arabic titles for both α and β, apparently because of their closeness to the horizon when viewed from the 30th parallel, in Cairo, a thousand years ago. Together, these stars form an impressive naked-eye double and are often referred to as the *SOUTHERN POINTERS,* because (from α to β) they point toward the Southern Cross. The Bushmen of South Africa saw α and β as "Two Men That Once Were Lions," while for Australian natives they were "Two Brothers."

Hadar, a bright B-type giant, is 490 light years distant and has a luminosity of 10,000 Suns; it is also a close binary, but the fourth-magnitude secondary, only 1.3 seconds from the primary, is not recommended for small telescopes. About 7 degrees north-west of Hadar is second-magnitude *(ε) EPSILON,* a blue-white star at the top of the Centaur's left foreleg; on the right foreleg, some 9 degrees west-northwest of ε, is the binary *(γ) GAMMA* (RA 12 h 38 m δ −48° 41′) m 2.17, 110 light years distant, among the multiple stars catalogued by John Herschel at Feldhausen. The components, nearly identical A-type giants at a separation ranging from 0.2 second to about 1.7 seconds, are only about 10 astronomical units apart at their closest approach, which occurs once in eighty-five years.

In the upper, human part of the Centaur's figure, we find *MENKENT (θ) THETA* (RA 14 h 0.4 m δ −36° 7′) m 2.04, marking his left shoulder. A golden K-type star classed as either a subgiant or a normal giant, Theta is approximately 45 to 55 light years distant, estimates varying; based on the latter figure, its luminosity is about 40 L ⊙. The Centaur's right shoulder is represented by *(ι) IOTA,* magnitude 2.76, approximately 8 degrees to the west of θ. Iota is a class-A2 main-sequence star about 52 light years distant with the large proper motion of .35 second toward the southwest. Lying about 7 to 10 degrees southeast of Theta and delineating the end of the Centaur's spear, are *(η) ETA,* a second-magnitude binary, and *(κ) KAPPA,* magnitude 3.15, also a binary; both η and κ are B2 main-sequence stars, but η, with 1,300 L ⊙, is also classed as a giant. We complete our description of Centaurus' figure with yet another B2 star, *(ζ) ZETA*, magnitude 2.5, some 8 degrees to the southwest of η. Zeta, classed as a subgiant, is also a spectroscopic binary, with a period of eight days. According to R. H. Allen, ζ was probably Arabian astronomer Al Tizini's *Al Nā'ir al Baṭn al Kentaurus,* "The Bright One in the Centaur's Belly."

Variables

Two variable stars that reach naked-eye visibility are *T CENTAURI* (RA 13 h 39 m δ −33° 21'), in the northern part of Centaurus, about 5 degrees northwest of θ, and *R CENTAURI* (RA 14 h 13 m δ −59° 41'), in the south, about 1½ degrees east-northeast of β. T, a semiregular variable, ranges in apparent magnitude from 5.5 to 10 in a period of 91 days, and its emission spectrum changes from a deep orange class K7 to a roseate M3 as it approaches minimum luminosity. The long-period Mira-type red variable R ranges from magnitude 5.4 to 12 in a period of about 547 days, and its unusual light curve, reminiscent of a roller coaster, shows double maxima and minima, with shallower dips in luminosity alternating with deeper ones!

An Outstanding Globular

Within the Centaur's torso, about 4 degrees due west from ζ, lies *(ω) OMEGA CENTAURI, NGC 5139* (RA 13 h 24 m δ −47° 13') m 4, a spectacular globular cluster, the finest of its type in the entire heavens (fig. 148). It carries the Greek letter ω, which Bayer gave it because he thought this object was a hazy star. Sir John Herschel resolved it in his reflector at Feldhausen, calling ω "a noble globular cluster, beyond all comparison the richest and largest in the heavens." With an apparent diameter of nearly 30 minutes (about 159 LY), Omega, at a distance variously estimated as from 15,000 light years to 22,000, has a possible total mass of a half million Suns and is believed to contain more than one million stars. Many of these are highly evolved giants of the later spectral classes; also present is a very large number of RR Lyrae stars, as well as six Cepheids and several long-period and irregular variables. In the highly condensed central core (diameter 100 LY), stellar members are so closely packed that spacings between nearest neighbors may average one tenth of a light year.

Using Spica, in Virgo, 36 degrees to the north, as a guide, observers in the southern United States may view Omega during the spring and summer, when it lies just above the horizon. Small instruments are helpful in locating this beautiful cluster, but larger telescopes are necessary to resolve its stars.

A Peculiar Galaxy

If we shift our telescope to an area just 4½ degrees north of Omega, we encounter an amazing object, the peculiar galaxy *NGC 5128 (CENTAURUS A)* (RA 13 h 22 m δ −42° 45'), which appears to be a large, bright elliptical galaxy crossed by a wide obscuration band. As only spiral galaxies have dark (equatorial) lanes, and the bands of spirals appear narrower and less chaotic, the strange dark formation jutting across NGC 5128 must be regarded as extraordinary and unique (fig. 149). (We therefore suggest that NGC 5128 be named the "Karate galaxy," because it wears a black belt!) The exact distance of this galaxy is not known, although estimates have ranged from 7 million light years to 25 million; the lack of resolution into stars on photographic plates made with the largest telescopes suggests a figure of at least 15 million light years. This accords with the 5 million parsecs tentatively adopted by Geoffrey and Margaret Burbidge in 1958, and implies a luminosity of about 20 billion $L \odot$ and an absolute magnitude of nearly −21. Thus, with a mass of about 200 billion Suns, NGC 5128 is one of the most luminous and massive galaxies known.

It is also an intense source of radio emission, called "Centaurus A," with a pattern of radiation similar to Cygnus A's, suggesting that highly explosive events are occurring within the galaxy core. Most of this radiation comes from the dark band, believed to be the remnant of these outbursts. Contour maps of the radio emission of Centaurus A indicate two radio sources near the central source, and in photographs, the outer parts of the bright disk seem to indicate a spiral form; it has therefore been suggested that this strange object might really be two colliding galaxies: an elliptical and a spiral. Walter Baade and Rudolf Minkowski suggested that the dark band is actually a spiral galaxy seen edge on and that it presents a silhouette against the bright elliptical because of the spiral's much higher content of interstellar material and the foreshortening effect of the spiral's edge-on position. The collision theory has been discarded, however, because radio energy emitted by two large lobes on either side of the core

148. NGC 5139, Omega Centauri, the most spectacular globular cluster in the skies, visible to the naked eye for viewers in the southern hemisphere. Shows resolution of the stars in outer regions; photographed with the 60-inch reflector. (Harvard Boyden Station, South Africa)

149. *NGC 5128, Centaurus A (the "Karate" Galaxy!);* an unusual type of elliptical galaxy and an intense source of radio emission, most of it coming from the black encircling band, probably the remnant of explosive activity in the galactic core. Photographed with the 200-inch telescope (*Hale Observatories*).

implies a strong magnetic field, which is confirmed by the extensive polarization in some areas. Exceedingly powerful explosions in the nucleus have apparently ejected material at very high speeds along magnetic lines of force to regions beyond the outer rim of the galaxy. These violent events may also justify our title the "Karate galaxy"!

The primary goal of research on galaxies with active nuclei is to learn how their enormous luminosity is generated. A wealth of spectral information is available on Centaurus A, but it was obtained from observations made at different epochs, with various types of instruments; thus it has been difficult to obtain a single spectrum over a wide range of X-ray and gamma-ray energies (frequencies), which is important in constructing models of the X-ray and gamma-ray emission from such galactic nuclei. In January and July 1978, the High Energy Astrophysics Orbiter 1 Spacecraft (HEAO) obtained new spectra of Centaurus A, indicating that its X rays are not heat-related but are traceable to synchrotron photons undergoing inverse Compton scattering; high-speed electrons in a magnetic field emit photons, which are then scattered by other such electrons and thus acquire higher frequencies (p. 141). This scattering process (the Synchrotron Self-Compton process) can reproduce the remarkable constancy of the X-ray spectrum of Centaurus A even though its intensity changes.

CRUX

When the earliest voyagers sailed around the Cape of Good Hope, the first constellation to appear in the evening sky was a glowing cross formed by four bright stars, three of these of the first magnitude. In ancient times, the stars of Crux, the Southern Cross, belonged to Centaurus, which surrounds it on three sides: to the north, east, and west. Amerigo Vespucci, on his voyage to South America in 1501, called the four stars *Mandorla,* or "Almond" (a term in Italian art referring to the aureole of saints ascending to heaven); in 1517, the Italian navigator Andrea Corsali referred to this figure as a "marvelous cross," as did Antonio Pigafetta (Magellan's companion), who said it was used to determine altitudes, and who originated the widely used name "Crucero." English writers and explorers of the sixteenth and seventeenth centuries gave it similar titles; as the "Southern Celestial Clock" of the Portuguese naturalist Cristoval D'Acosta, it became a timepiece for many nations, because it stands erect the moment it passes the meridian, and thereafter the exact hour may be determined visually from its inclination.

The precession of the Earth's axis has moved the Cross ever farther to the south; visible as far north as Persia in ancient times, it was an object of feasting and veneration, but was replaced by Delphinus in later centuries. About the time of the Crucifixion, the Southern Cross last appeared on the horizon of Jerusalem, above 30 degrees north latitude. This constellation cannot be seen in its entirety today at latitudes farther north than 25 degrees. In the words of Dante,

O! thou septentrional and widowed site
Because thou art deprived of seeing these!
Purgatorio

Principal Stars of the Southern Cross

The foot of the Cross is defined by *ACRUX (α) ALPHA Cruxis* (RA 12 h 23.8 m δ −62° 49') m 0.87, approximately 360 to 370 light years distant and the fourteenth-brightest star. In 1685, Jesuit missionaries in Siam discovered its duplicity; the two blue-white components, 4.4 seconds apart, constitute a magnificent double star. Both components are rich in helium; α1, the more luminous, with 3,000 $L \odot$, is a class-B1 subgiant, and α2, about 1,900 $L \odot$, has a peculiar B3 spectrum of broad, hazy lines, indicating that this star is rapidly rotating. Both stars are spectroscopic binaries with periods of approximately two months. Acrux lies about 40 degrees due south from Corvus the Crow, and observers no farther north than latitude 25 degrees may look for it on May 13, at its culmination, as it peeps above the horizon.

The end of the eastern arm of the Cross is marked by *(β) BETA* (RA 12 h 44.8 m δ −59° 25') m 1.28, a B-type giant, its distance estimated variously between 425 and 490 light years; calculated from the larger figure, its luminosity is approximately 5,800 times the Sun's. Beta is also a pulsating variable of very small range, similar in type to β Canis Majoris. At the top of the Cross is *GACRUX (γ) GAMMA Crucis* (RA 12 h 28 m δ −56° 50') m 1.63, a red giant, of spectral type M3, about 220 light years distant, with a luminosity of about 900 $L \odot$. The nineteenth-century American astronomer B. A. Gould believed that γ varied, by about one-half magnitude, but his claim is unsupported by modern observations. The optical companion γ₂ (SAO240022), magnitude 6.4, class A2, lies 111 seconds to the northeast of Gacrux.

The Cross is completed by *(δ) DELTA* (RA 12 h 12.5 m   δ −58° 28′)   m 2.82, at the end of the western arm. A B2 subgiant, it has been estimated at widely varying distances, from 257 to 570 light years, the latter figure yielding a luminosity of about 1,900 $L \odot$. About 1½ degrees south-southeast of δ lies a fifth star, *(ε) EPSILON,* magnitude 3.59, spectral class K2, about 58.7 light years distant, somewhat spoiling the perfection of the Cross and giving it a kite-like appearance.

A Box of Jewels

About 1½ degrees southeast of β and surrounding the sixth-magnitude star (κ) Kappa, is the celebrated open star cluster *NGC 4755*, the *JEWEL BOX CLUSTER* (RA 12 h 50.6 m   δ −60° 05′), discovered by Sir John Herschel, who likened it to "a gorgeous piece of fancy jewellery." Lying in a brilliant part of the southern Milky Way, it contains, within its 10-minute field, about fifty extremely luminous supergiants of class B, each of them thousands of times the luminosity of the Sun; centered in this glittering nest of stars, like a ruby in a setting of diamonds and sapphires, is a red supergiant comparable to Betelgeuse, with an absolute magnitude of about −5.7 and a luminosity of about 16,000 $L \odot$. The Jewel Box, at an estimated distance of some 7,700 light years, is probably no more than a few million years old, with its massive and short-lived supergiants about to leave the main sequence as they rapidly evolve and expand into the subgiant phase. About 10 minutes from the cluster, and probably also a member, is the star *HD 111613* (SAO 252054); an A-type bright supergiant of absolute magnitude −7.3 and visible to the naked eye at magnitude 5.74, it has the amazing luminosity of 83,000 $L \odot$.

A Sack of Coal

Immediately to the south of the Jewel Box and east of Acrux is the dark, pear-shaped obscuration nebula called the *COAL SACK* (fig. 150), or Soot Bag, 7 degrees long by 5 degrees wide. Observed in 1499 by the explorer Vicenti Pinzón, it was Vespucci's *Il Canopo Fosco*, "The Dark Canopus," in the sense of a "dark star," alluding to bright α *Carinae*, whose name "Canopus" was applied in many ways. Later on, the Coal Sack was called *Macula Magellani*, "Magellan's Spot," referred to in the nineteenth century by the English astronomer William Henry Smyth as the "Black Magellanic Cloud." Australians saw it as an evil spirit that has treed an opposum taking refuge among the stars of the Cross, whereas Peruvians identified it as a heavenly doe suckling its fawn. Astronomers at the turn of the century, mystified by the Coal Sack and other dark nebulae, believed that they were empty holes in space; however, since Trumpler's discovery, in 1930, of interstellar extinction (p. 101), the true nature of these intervening clouds of gas and dust is finally understood. Thus, as descriptive titles, "Coal Sack" and the earlier "Black Cloud" are more apt than their originators realized!

150. *The Coal Sack Nebula*, a strange dark formation near the Southern Cross, seen against the bright stellar masses of the Milky Way in Crux and the Eta Carinae nebulosity. Photographed by Miss Harwood at Arequipa, Peru. (*Harvard Observatory*)

LUPUS

When Zeus in high Olympus learned about the impiety and wickedness of King Lycaon of Arcadia and his fifty sons, the great god decided to investigate these rumors of evil, and, disguised as a day laborer, he descended to the Earth. Received with mock hospitality at the tyrant's palace, Zeus was offered food, at a sacrificial banquet, into which Lycaon had secretly mixed human flesh; in disgust, Zeus upset the table and blasted the monarch's household with thunderbolts. Terrified, Lycaon fled to the fields, where he tried to cry out but could only produce howling noises; on all fours, his royal robes transformed into shaggy hair, the king bared savage fangs and turned upon his own sheep with wolfish bestiality.

Scholars of Greek mythology believe that Lycaon is represented by Lupus the Wolf, a constellation lying to the immediate east of Centaurus' upper body. Its history may be as ancient as the Akkadians, to whom scholars attribute the title *Urbat*, the "Beast of Death." This stellar figure was known to the Arabians as *Al Asadah*, "The Lioness," and it was repeated on a Turkish planisphere as *Al Sabu'*, "The Wild Beast." Similarly, the Greeks and Romans thought of it as a "Wild Animal," and the poet Aratos referred to it as a sacrificial beast clutched in the Centaur's hand, whereas the astronomer Eratosthenes described Lupus' long, narrow form as a "Wine Skin." Arabian astronomers had earlier combined some stars of Lupus and Centaurus to form *Al Shamārīh*, "The Palm Branches."

A Few of Its Stars

About 6½ degrees east of Zeta Centauri is *(α) ALPHA Lupi* (RA 14 h 38.6 m　δ −47° 10′)　m 2.28, which marks the Wolf's tail. Alpha is a B-type giant with the slight variability and ultrashort cycle typical of β Canis Majoris stars; its parallax, spectroscopically determined, places it at a distance of 430 light years, indicating a luminosity of about 1,700 $L \odot$. A faint companion lies 27.6 seconds distant from α, possibly sharing its motion.

An interesting quadruple system is *(ε) EPSILON* (RA 15 h 19 m　δ −44° 31′)　m 3.36, marking the Wolf's underbelly. A subgiant of class B3, between 300 and 450 light years distant, ε is a close binary whose fourth- and sixth-magnitude components, separated by 60 astronomical units, take at least seven hundred years to circle their common center of mass; the primary component is also a spectroscopic binary, with a period of less than a day. A ninth-magnitude common-proper-motion companion is visible at 26.6 seconds from the major pair.

The southernmost major star of the Wolf, marking its hind foot, is *(ζ) ZETA* (RA 15 h 8.7 m　δ −51° 55′)　m 3.4, a yellow giant (class G8), variously estimated at between 90 and 137 light years distant and forming a wide double with a seventh-magnitude companion at 71.9 seconds. About halfway between ε and ζ lies *(μ) MU,* magnitude 4.27, a rapidly rotating class-B8 main-sequence star and triple system, the components consisting of two close fifth-magnitude components and a seventh-magnitude companion at 23.7 seconds. Lupus' forepaw is marked by *(η) ETA* (RA 15 h 56.8 m　δ −38° 15′)　m 3.45, a color-contrast double suitable for small telescopes, the primary a luminous B-type main-sequence star.

Of great interest is a radio source 1½ degrees northeast of β and quite close to Kappa Centauri, near the Lupus-Centauri border; detected in 1964, this source is a possible remnant of *SN A.D. 1006,* the earliest supernova in our Galaxy to be recorded and witnessed by observers the world over. This highly luminous "golden star" appeared forty-eight years before the famous supernova of A.D. 1054, the "Guest Star" that produced the Crab Nebula, in Taurus. The Lupus radio source appears as an irregularly shaped circle, 40 minutes in diameter, with a layered structure of concentric shells and two inner lobes. In 1976, very faint filaments of nebulosity were detected visually near the radio source. In the continued search for the residual star of the supernova, a helium-poor subdwarf was located near the center of the supernova remnant, suggesting a possible evolutionary connection. Subdwarfs, hot stars that may represent a phase between the red-giant and white-dwarf stages, have been under intensive scrutiny, owing to recent improvements in computers and observations in new wavelength bands by orbiting satellites. In contrast to the hotter O-type subdwarfs, B-type subdwarfs show an underabundance of helium; this may result from the strong surface gravity of these stars, which causes the heavier helium atoms to drop past the hydrogen atoms and leave a helium-poor atmosphere. That subdwarfs may be close binaries exchanging mass may also be a factor in their evolution, but it is not clear how such a mass transfer would affect the components, nor are astronomers certain what process has stripped the precursors of the helium-rich O-type subdwarfs of their outer, hydrogen-rich layers.

NORMA

To the southeast of Lupus, on a clear night, the southern observer may notice a faint figure about 13 degrees long, shaped somewhat like a shovel or dipper. This is *Norma et Regula,* the "Level and Square," which earlier had stars of the Southern Triangle until Lacaille redefined Norma as a carpenter's square, appropriately associated with the Compasses of Circinus, to the south. Norma was also *Quadra Euclidis,* i.e., "Euclid's Square," but frequently translated incorrectly as "Quadrant." However, Norma's irregular form might as easily represent the quadrant, an important astronomical instrument formerly used to measure the altitudes of stars, since it lost its namesake in the north, the now obsolete "Mural Quadrant" (a showpiece of Tycho's observatory).

Norma's brightest star is fourth-magnitude *(γ) GAMMA 2,* a deep yellow giant marking the southeast corner of the Square and forming a wide visual double with *(γ) GAMMA 1,* a cream-colored supergiant just to its west. Two degrees to the northeast lies the binary *(ε) EPSILON,* its components a magnitude 4.5 main-sequence star of spectral class B3, and an A-type star, magnitude 7.5. About 7 degrees west of the two Gammas and just east of the faint globular cluster NGC 5927, in the southeastern part of Lupus, a seventh-magnitude nova occurred, photographically detected, on July 1, 1893, by Margaret Fleming, at Harvard Observatory. Its spectrum was similar to that of the famous Nova Aurigae, which had appeared in the northern skies only a year previously.

About 4 degrees south-southwest of Gamma Normae, and just north of the fifth-magnitude yellow giant *(κ) KAPPA,* look for *NGC 6067 (Δ 360),* a large, rich open cluster, 20 minutes in diameter, containing over a hundred stars of the tenth magnitude.

ARA

'neath the glowing sting of that huge sign
The Scorpion, near the south, the Altar hangs. . . .

Aratos

To the east of Norma lies another trapezoidal figure, which, when on its side, resembles a short-handled scoop. In contrast to the modern groups, Ara the Altar was known to the Greeks and Romans as a constellation and may have originated as *Tul-Ku,* the "Holy

Altar," an archaic Euphratean figure formerly in the zodiac but replaced by Libra when the Greeks adopted Ara and moved it to its present position. The Romans identified it as *Thymele,* the altar of Dionysus (Bacchus), the god who rescued Ariadne (p. 210), and as *Ara Centauri,* the "Centaur's Altar," because in classical times, centuries before the invention of Norma, Ara was joined to both Centaurus and Lupus. Borrowing from the Greeks, Arabian astronomers called Ara *Al Mijmarah,* "The Censer" (incense vessel); several Roman titles signify an altar, also a hearth, and a brazier (for burning incense), and, in fact, Ara was identified as *Turibulum,* a "Censer," until the eighteenth century. The Latin poet Manilius associated Ara with the altar on which Jove and the Olympian gods took their vows before defeating the Titans, but authors of the biblical school regarded Ara as the altar of Moses or that of Noah, erected after the Flood.

Objects of Interest

To view the chief stars of this constellation, observers must be no farther north than 30 degrees latitude; for those as far north as Washington, D.C., Ara's faint outline remains above the southern horizon no more than four hours. Its brightest star is *(β) BETA* (RA 17 h 21 m $\delta -55°29'$) m 2.87, an orange supergiant approximately 1,030 light years distant, and 5,700 times as luminous as the Sun. About 2½ degrees north-east of β (about 10½° south of θ Scorpii), is *NGC 6397,* a bright globular cluster discovered by Lacaille in 1755; with a diameter of about 19 minutes, its loose structure allows stellar resolution in small telescopes. Revised estimates of its distance indicate that it may be the closest globular cluster to the Sun. Although it contains bright red giants, this cluster seems to lack the short-period pulsating variables usually found in globulars. NGC 6397 is not visible from northern latitudes greater than 30 degrees.

In the northwest part of Ara lies *NGC 6188,* a spectacular mixture of bright and dark nebulosity, similar to the famous Horsehead Nebula, in Orion, and the Cone nebula, in Monoceros. Situated about 2 degrees southeast of ε Normae, and roughly 7 degrees south-southwest of Zeta Scorpii, the nebula in Ara is the site of the open star cluster *NGC 6193,* which contains about thirty luminous sixth-magnitude stars; these young, hot blue giants illuminate the bright portion of the nebula. Various theories have been advanced to account for the turbulent events occurring in these nebulae, such as advancing shock fronts of expanding gas/dust clouds, the effects of light pressure from newly born stars, or of even more violent phenomena, but these fascinating mysteries, closely related to star formation, still remain unexplained.

TELESCOPIUM

The instrument that has taken us on our journey through the heavens found its own place among the stars in the mid-eighteenth century, thanks to Lacaille. Its presently recognized boundaries, south of the Southern Crown and Sagittarius, and east of Ara, were established by the astronomers Baily and Gould. The brightest star, *(α) ALPHA,* is a blue-white giant, magnitude 3.5. Just to its east lies the optical double *(δ) DELTA 1* and *2* (RA 17 h 5 m   δ −22° 91'), consisting of two very similar stars, subgiants of spectral classes B5 and B6.

Of historic interest is **RR TELESCOPII** (RA 20 h 00 m   δ −55° 52'), in the southeastern part of the constellation, a periodic variable with a magnitude range of 12.5 to 15; in 1944 it became one of the slowest novae known, increasing in luminosity by several magnitudes and requiring five additional years to reach a maximum of magnitude 6.5. In 1951, its spectrum first revealed the characteristic features of a nova, with large approach velocities of up to 536 miles per second, indicating the eruption of material from the star's surface.

INDUS

Although deprived of his earthly territories, the American Indian now occupies a long corridor in the heavens, thanks to Bayer. Included among the twelve new southern constellations in his *Uranometria,* Indus the Indian is depicted as a nude figure, carrying arrows in both hands, but no bow. Its major stars lie more or less between Telescopium and Grus, but the official boundaries of Indus reach deep into the south circumpolar zone between Pavo and Tucana; Julius Schiller, in his attempt to re-form the constellations according to the Bible, combined Indus and Pavo to create the patriarch Job!

The brightest star of Indus is third-magnitude *(α) ALPHA,* an orange giant occupying its northwestern corner. In the eastern part of this figure, near the borders of Grus and Tucana, is *(ε) EPSILON* (RA 21 h 59 m   δ −57° 00')   m 4.7, of special interest because it is only 11.5 light years distant from us, the sixth-nearest naked-eye star. A class-K5 main-sequence object, ε has four-fifths the Sun's radius and 13 percent its luminosity. Epsilon's distance from the Solar System is gradually decreasing, at the rate of 25 miles per second. About 6 degrees to the northwest we find an attractive double star, *(θ) THETA,* 91 light years distant; its 4.5- and seventh-magnitude components are increasing their separation of 6 seconds and are therefore suitable for moderate instruments. If you wish to catch a glimpse of θ on the southern horizon, your latitude cannot be more than 35 degrees north. In fact, only the northernmost stars of the American Indian are visible in the United States!

MICROSCOPIUM

Immediately north of Indus and southwest of Piscis Austrinus lies the faint figure of Microscopium, formed by Lacaille to honor Telescopium's counterpart in the science of the microcosm. An earlier figure which occupied Microscopium's approximate position was referred to by some astronomers as *Neper* or *Bohrer,* the "Auger" or "Drill." To view this Microscope, a telescope is helpful, because it has no star brighter than magnitude 4.7, an honor held by the yellow giant *(γ) GAMMA* (class G4), 228 light years distant, more or less indicating the eyepiece at the top of the instrument. The objective and specimen slide, at the bottom, are marked by *(θ) THETA 1* and *2,* magnitudes 4.8 and 5.7, respectively, a very wide naked-eye pair; θ1 is about 68.5 light years distant, and each star is an A-type magnetic-spectrum variable with abnormally strong metallic lines, like α Canum (Cor Caroli, in Canes Venatici).

GRUS

Formerly a part of Piscis Austrinus, to which Arabian astronomy assigned it centuries ago, Grus the Crane acquired its modern identity in Bayer's charts and now occupies a large region south of the Fish and east of Microscopium and Indus. An obsolete title is *Phoenicopterus,* the "Flamingo" or "Bittern." Because of its high flight, the crane was an ancient Egyptian symbol for a star observer, which, if not the inspiration for Bayer's title, nevertheless adds to its appropriateness.

Marking the Crane's body is blue-white *AL NĀ'IR (α) ALPHA* (RA 22 h 5 m   δ −47° 12′)   m 1.76, meaning "The Bright One" (of the Fish's tail), a title retained from the Arabians. When Alpha reaches its highest point in the heavens, on October 11, spring in the southern hemisphere, it can be seen about 17 degrees southwest of famous Fomalhaut (α Piscis Austrini) from latitudes up to 40 degrees north. Five degrees east of Al Nā'ir is *(β) BETA,* magnitude 2.8, a red giant 280 light years distant and eight hundred times more luminous than the Sun.

The Crane thrusts its long neck far to the northwest, and there we find *(γ) GAMMA* (RA 21 h 51 m   δ −37° 36′)   m 3.03, formerly the Fish's tail but now the Bird's eye, a white class-B8 giant 540 light years distant, with a luminosity of about $1,450 L \odot$. At the base of the Crane's neck, forming a triangle with α and β is the wide optical pair **(δ)** *DELTA 1* and *2,* about fourth magnitude; δ1 is a G5 (solar) type giant of uncertain distance, and δ2, which lies 45 seconds to its southeast, is a red giant (class M4) at least 88 light years distant.

About 5 degrees east of the Deltas is the binary *(θ) THETA,* an F6 subgiant; a couple of degrees northeast of θ is the group of spiral galaxies including *NGC 7552 (Δ 475),* 3 by 3 minutes, magnitude 11.6, a barred spiral with a very small, bright center and tightly wound arms; *NGC 7582 (Δ 476),* an elongated barred spiral of intermediate (b) type with its arms more unwound; the faint barred galaxy *NGC 7599,* an example of a c system with the arms well unwound, and *NGC 7590 (Δ477),* an ordinary spiral galaxy of the intermediate (b) type, fairly elongated. Grus contains, in fact, numerous spirals of every classification, but none is much brighter than the twelfth magnitude, and thus they are suitable only for very large telescopes.

FOURTEEN

Phoenix to Antlia

PHOENIX

All these things
Have their beginning in some other creature,
But there is one bird which renews itself
Out of itself. . . .

Ovid

One of Bayer's modern creations, Phoenix is a somewhat batlike figure lying to the east of Grus and west of Eridanus' southernmost stars. Recognized in earlier times, its birdlike pattern was identified as a "Griffin" or "Eagle," although the curving line formed by several of its stars reminded the Arabs of a primitive boat, which they named *Al Zaurak;* another Arabic title was *Al Ri'āl,* "The Young Ostriches." After the Chinese adopted the western constellations, this figure became their "Fire Bird," which seems to allude to a universal emblem of renewal and immortality, the legendary Egyptian Phoenix, which could return to life after dying by fire; in fact, the Phoenix is an astronomical symbol associated with cyclic periods in ancient China, Egypt, India, and Persia. The name "Phoenix" also occurs in Greek mythology, as that of several princely characters, one of them a founder of Phoenicia, and another a participant in the siege of Troy.

For observers in the southern hemisphere, Phoenix is a constellation of the late spring. Northernmost and brightest of its stars, *ANKAA (α) ALPHA Phoenicis* (RA 0 h 26 m δ −42° 18′) m 2.4, is easily located about 18 degrees southeast of Fomalhaut, reaching its highest point in the heavens on November 17. Ankaa was the Arabs' *Nā'ir al Zaurak,* the "Bright One in the Boat"; we know it, however, as an orange giant about 78 light years distant and a spectroscopic binary. Other multiple systems in Phoenix are as follows: *(β) BETA,* in the central part of the figure, a third-magnitude yellow giant of late class G, 130 light years distant, the components at a separation of 1.4 seconds (with a very faint distant neighbor, not physically bound to it, at 57″.5), and *(ζ) ZETA,* a triple system that marks the Bird's tail, some 8 degrees directly to the south of β. About 218 light years distant, ζ consists of two B6 main-sequence stars, 0.8 second apart, circled by a more distant companion at 6.4 seconds. The fourth-magnitude primary is also an eclipsing variable.

A "Dwarf Cepheid"

About 6½ degrees west of α lies **SX PHOENICIS** (RA 23 h 44 m   δ −41° 51')   m 7.1 to m 7.5, a well known pulsating A-type subdwarf, unusual because of its 79-minute light cycle. Although SX is two or three times as luminous as the Sun, its mass is only about 0.24 $\mathcal{M}$ ☉, an unusually small mass for an A-type star; rapid variations in its spectral lines indicate the presence of two or more components. At least 140 light years distant, SX Phoenicis has the large proper motion of 0.89 second per annum in a more or less southerly direction, with a radial velocity of about 9.5 miles per second in approach.

FORNAX

Lacaille borrowed a few stars from the long and winding River Eridanus to create *Fornax Chemica,* the "Chemical Furnace," a faint zigzag figure tucked in among Eridanus, Cetus, and Sculptor. In honor of the chemist Lavoisier, Bode renamed this new constellation *Apparatus Chemicus,* but today we know it simply as Fornax. Its brightest star, *(α) ALPHA* (RA 3 h 10 m   δ −29° 12')   m 3.86, culminates on December 19, ushering in the summer for observers in the southern hemisphere. A long-period binary, about 40 light years distant, α's duplicity was first recorded by Sir John Herschel at Cape Town in 1835; the primary, a cream-colored subgiant of spectral class F8, is now about 2 seconds from its seventh-magnitude variable companion. These components are often difficult to resolve as separate stars, especially when the secondary fades in magnitude. Alpha's proper motion toward the northeast is also carrying it in our direction at a velocity of about 13.5 miles per second.

An Outstanding Galaxy Cluster

Viewers blessed with clear skies and wide-angle eyepieces should focus upon the **FORNAX GALAXY CLUSTER** (approximate location RA 3 h 35 m   δ −35° 40'), eighteen bright galaxies crowded together within a 3-degree field in the extreme southeast part of Fornax, spilling across the border of Eridanus. The cluster includes a large number of elliptical galaxies of various types and many barred spirals. The brightest member is tenth-magnitude **NGC 1316,** classed either as an elliptical or an S0 spiral galaxy (which combines the features of elliptical and spiral systems). Dense clusters like the Fornax cluster appear to contain primarily elliptical or S0 systems, owing perhaps to the more frequent collisions among the closely packed galaxies in such clusters, events that sweep dust and gaseous matter out from the colliding galaxies into intergalactic space.

The galaxy *NGC 1316,* now identified with a powerful radio source called *Fornax A,* was, in fact, sometimes regarded as an example of two colliding galaxies, rather than a single system; however, the data obtained with refined photographic techniques raised serious doubts concerning this explanation, revealing instead a bright round object spotted with a symmetrically aligned chain of dark globules. The latest observations with radio telescopes indicate that the centers of such galaxies are in violent activity, which often produces cosmic jets, i.e., narrow streams of ionized gas more than 1 million light years in length (see Cygnus A, p. 288, Centaurus A, p. 360, and Virgo A, p. 195).

The Fornax galaxy cluster also includes the excellent barred spiral *NGC 1365,* an S-shaped SBc structure (the spiral arms clearly defined), that is the finest example of its type in the southern heavens, and also one of the most luminous known.

A Galactic Freak

Approximately 10 degrees to the west of the Fornax galaxy cluster lies a very peculiar spheroidal dwarf galaxy called the *FORNAX SYSTEM* (RA 2 h 37 m δ −34° 40′). About 630,000 light years distant and a member of the Local Group of galaxies, it contains five faint globular clusters and is about fifty times the size of the largest globular known. (Refer also to the Sculptor System, p. 298.)

HOROLOGIUM

Journeying southward again from Fornax, we cross a portion of the River Eridanus to arrive within the boundaries of *Horologium Oscillatorium,* the "Pendulum Clock," another modern constellation, whose lengthy form is marked by stars so faint that only a keen-eyed observer on a clear night can hope to identify it. The brightest star is *(α) ALPHA* (RA 4 h 12 m δ −42° 17′) m 3.86, a K-type giant at the northeastern end of the figure, 59 parsecs (192 LY) distant. Just half a degree to its northwest is *(δ) DELTA,* a fifth-magnitude F-type "normal" dwarf, about 91 light years distant. At the southern end of Horologium is *(β) BETA,* also of the fifth magnitude, a white "Sirian" class-A5 giant some 284 light years distant. Several variables are present, only two or three reaching naked-eye visibility, notably the long-period *R HOROLOGII* (RA 2 h 52 m δ −50° 06′), about 13 degrees southwest of α and δ, a red class-M7 emission star ranging from magnitude 5 to 14.

CAELUM

Immediately to the northeast of Horologium is another of Lacaille's modern creations, Caelum the "Engraving Tool" ("Scalpel"), sometimes called Scalptorium. Although visible to observers up to 40 degrees north latitude, Caelum has no star brighter than magnitude 4.5, that of *(α) ALPHA* (RA 4 h 39 m δ −41° 57′), an F2 main-sequence

star and binary, about 65 light years distant, its faint secondary at a separation of 6.6 seconds. About 7 degrees to the northeast, close to the border of Columba, is the binary *(γ) GAMMA,* 169.5 light years distant, its components an orange giant of class K3, magnitude 4.5, and an eighth-magnitude companion at a separation of 2.9 seconds.

COLUMBA

East of Caelum and south of Lepus (the Hare, at Orion's feet), we find the modern constellation Columba Noae, which is "Noah's Dove." Having appeared as a part of Canis Major on one of Bayer's charts in 1603, it was first formally published by the French astronomer Augustin Royer in 1679 and generally recognized from that time onward. Owing to its proximity to Argo Navis, "Noah's Ark" of the biblicists, it was named the attendant dove. An amusing tradition arose concerning α and β Columbae when they still belonged to Canis Major: ancient desert dwellers confused one of these stars with the brilliant Canopus (the Arabs' *Suhail),* and their arguing—and swearing— resulted in the title *Al Suhail al Muḥlif,* "The Suhail of the Oath," thus *Al Muḥlīfain,* which was applied to the present α and β Columbae and to various other stars.

PHACT, also Phad, *(α) ALPHA* (RA 5 h 38 m δ −34° 06′) m 2.64, is in the upper part of the Dove's figure (some chartists place it on the back), its name probably derived from Ḥaḍ'ar, or "Ground"; *Ḥaḍ'ar* and *Wezn,* i.e., "Ground" and "Weight," respectively, were titles applied to several pairs of stars, apparently because of their closeness to the horizon. About 140 light years distant, α is an emission-line main-sequence star of class B8; an eleventh-magnitude optical companion lies at 13.5 seconds, with the separation slowly increasing. The English astronomer Lockyer suggested that Phact was involved in Egyptian worship of the god Amon in archaic times, long before the precession of the Earth's axis brought this star nearer the southern horizon, causing it to lose its prominence by about 3000 B.C., when the incomparably brighter Sirius took Phact's place as a sacred object.

About 2½ degrees to α's southeast is *WAZN (β) BETA,* its title explained above, a third-magnitude K-type giant just slightly farther away than α; two degrees to α's northeast, on the Dove's wing, is *(μ) MU,* one of the three "runaway" stars already described (pp. 310, 350), its space velocity about 74 miles per second.

PICTOR

In paying tribute to the fine arts, Lacaille created the "Painter's Easel" from a group of stars to the south of Columba and just west of the bright star Canopus (α Carinae). Shortened today to Pictor, the full name that Lacaille gave to this small figure was *Equuleus Pictoris,* but it was also called *Pluteum Pictoris,* both of which mean "Painter's Easel."

Although none of its stars is brighter than *(α) ALPHA Pictoris,* a white Sirius-type star of magnitude 3.27 situated near the southern border, within this faint constellation we find some very interesting and historic objects. One of these, the variable *RR,* known as *NOVA PICTORIS 1925* (RA 0 h 35 m　δ −62° 36′), about 1½ degrees west-southwest of α and about 1,000 light years distant, was discovered by R. Watson on May 25, 1925, when its magnitude was 2.3. Photographed and thoroughly studied at the Royal Observatory at the Cape of Good Hope, it reached magnitude 1.2 on June 9, fading to magnitude 4 on July 4, but rising again on August 9 to magnitude 1.9; after a third and last "comeback" from magnitude 3.7 to magnitude 2.3, it gradually faded, reaching magnitude 12.5 in 1975, almost its original magnitude, of 12.7. At peak luminosity, RR had an absolute magnitude of at least −6.3, or 30,000 $L \odot$, with some estimates even higher. Very large radial velocities of approach, which increased as RR faded, indicated that the star was shedding its outer layers in the outburst. A present light cycle of about .145 day indicates the possibility that RR Pictoris has a very close companion, a factor in its nova-like eruption.

About 8½ degrees northwest of the A5 giant *(β) BETA* is famous *KAPTEYN'S STAR* (RA 5 h 9 m　δ −45° 00′)　m 8.8, a red dwarf with the largest proper motion next to Barnard's star, in Ophiuchus. With about 1/250 the Sun's luminosity, it lies 12.7 light years distant, as it speeds through space relative to the earth at the unusual velocity of some 175 miles per second.

ARGO NAVIS

. . . And out of the clouds Zeus answering called back a mantic
peal of thunder; and the bright branches of sheer lightning
　　broke into flame.
The heroes, trusting the signs apparent of God
drew breath; and the prophet cried aloud,
bespeaking glad expectations, to bend to their sweeps;
and slakeless the oars went dipping from the speed in their
　　hands. . . .

Pindar, *Pythia,* IV

Pelias, king of Iolcus, fearing that his life was threatened by the young hero Jason, sent him on a dangerous mission to fetch the Golden Fleece, at Colchis, where Phrixus had nailed it to a sacred oak after sacrificing the ram that had borne him safely across the sea (p. 308). Jason summoned the help of Argus, son of Phrixus; under Athena's guidance, Argus built a ship of fifty oars named *Argo* and fitted it at the prow with a speaking timber from the oracle of Dodona. With the oracle's permission, Jason assembled a crew called Argonauts, consisting of fifty of the most distinguished nobles of Greece, among them Orpheus, Castor and Pollux, Hercules, Theseus, Cepheus, and numerous sons of kings and divinities. When, after many difficulties, the Argonauts arrived at Colchis, Aeëtes, the king, promised to give Jason the Fleece if he single-handedly yoked two enormous fire-exhaling, brazen-footed bulls. Aided by the magic powers of Aeëtes' love-smitten daughter Medea (a descendant of the Ocean), Jason did so, but King Aeëtes, secretly planning to burn the *Argo* and kill the crew, broke his word and withheld the Fleece. During the night, however, Medea (accompanied by Jason) secured the Fleece by drugging the dragon that guarded it, and she then fled with Jason on board the *Argo*. Jason's ultimate betrayal of Medea reminds us of the story of his crewman Theseus, wherein he deserts Ariadne; but the vengeful spirit of powerful Medea contrasts strongly with Ariadne's gentler nature.

The goddess Athena memorialized the voyage of the Argonauts by transferring their ship to the heavens, where it became the constellation Argo Navis, occupying a region nearly 75 degrees long, east of Canis Major and south of Monoceros and Hydra. Lacaille divided the great Ship into three constellations: Carina the Keel, Puppis the Stern, and Vela the Sails, and he also formed two more subdivisions, Malus the Mast and Pyxis Nautica, the Mariner's Compass, of which only "Pyxis" survives. In the three divisions of Argo, only one sequence of Greek letters is used.

An alternate myth describes Argo as the first oceangoing ship, built by the Egyptian king Danaus, carrying the king and his fifty daughters from Libya to Rhodes following a dispute over the kingdom with his brother Egyptus. The *Argo* is also associated with the Deluge, identified as the Ark by many ancient cultures; the name *Argo* may have been derived from the Phoenician *arek*, or "long." According to Ptolemy, however, inhabitants of northeastern Africa saw it as a "Horse," whereas the Hindus called it *Sata Vaēsa*, the "One Hundred Creators," one of four "Nourishers of the World" (four quarters of the heavens). Argo's other Roman titles include *Fatis Heroum*, the "Heroes' Raft"; *Currus Maris*, the "Sea Chariot"; and the intriguing *Navigium Praedatorium*, or "Pirate Ship"!

CARINA

. . . the star
Which pours his light in a glance of fire,
When he disperses the morning dew. . . .
　　　　　　　　ancient Egyptian priest

East of Pictor and south of Canis Major, Vela, and Puppis, lies Carina, the Ship's Keel, an oddly shaped and sprawling constellation, distinguished by the brightest star of the far southern heavens, *CANOPUS (α) ALPHA* (RA 6 h 23 m　δ −52° 40′)　m −0.72, some 36 degrees south of Sirius and second only to it in magnitude. Culminating at midnight on December 27, Canopus signals the beginning of the summer season for stargazers in the southern hemisphere. At 9 P.M. on February 6, it peeps above the horizon for those at the 37th parallel north and is easily viewed by observers in the southern United States, particularly the Gulf Coast and Florida. According to the second-century Asian astronomer Scylax, it was named after the Canopus of mythology, Menelaus' chief pilot, who died in Egypt after the destruction of Troy and in whose honor both the brilliant star and the ancient city of Canopus (northeast of Alexandria) were named. However, according to the fifth-century Athenian general Aristides, α's name was derived from the Coptic or Egyptian *Kahi Nub*, "Golden Earth," apparently a reference to the star's supposedly "golden" color, an illusion caused by atmospheric scattering of the blue and green rays near the Earth's horizon (see p. 101).

Canopus was widely worshiped in ancient Egypt, and many temples were oriented to its rising with the Sun at the autumnal equinox in 6400 B.C. as a symbol of Khons, the first southern star god; thousands of years later, a temple built by Rameses III at Thebes was dedicated to its setting, and through its association on the Nile with the leading god of Egypt, it became the "Star of Osiris." Canopus was also sacred in early India, where it was "Agastya," an inspired sage and helmsman to the son of Varuna, the water goddess; as in Egypt, its heliacal rising was greeted with religious ceremonies. Of tremendous importance to all desert dwellers, it was the Arabians' "Suhail," from *Al Sahl,* "The Plain," also a personal title signifying "brilliant," "glorious," and "beautiful." An old fable describes Suhail's unsuccessful courtship of the feminine giant "Al Jauzah" (the early Arabian Orion), and the suitor's subsequent expulsion far to the south! Down through the ages until at least 100 B.C., Canopus was worshiped in China as *Laou Jon,* the "Old Man"; centuries later, it became the "Star of Saint Catherine" for Greek and Russian pilgrims journeying to her convent and shrine at Sinai.

The Alexandrian astronomer Posidonius (260 B.C., not to be confused with the first-century stoic philosopher Posidonius) made astronomical and geographical history when he utilized Canopus to measure a degree on the Earth's surface between Alexandria and Rhodes from the watchtower of Eudoxus, the earliest known observatory in classical days; Eudoxus (409–356 B.C.), a mathematician and student of Plato, had devised a system describing the apparent movements of the Sun and the planets. In his writings, the poet Manilius (1st century A.D.) used Canopus and Ursa Major to prove that the Earth is round. Canopus, an F-type supergiant with a diameter at least thirty times the Sun's, is much larger than Sirius (1.8 dia. $\odot$). Estimates of the distance of Canopus range from slightly more than 600 light years to as much as 1,152 light years; thus its luminosity may be as great as 10,500 $L \odot$. Difficulties in obtaining a reliable trigonometric parallax for Canopus indicate a distance of more than 400 light years, the limit for an accurate parallax, thus resulting in many conflicting estimates, but, surprisingly, the Cape Observatory has reported a distance of only 120 light years, from its spectroscopic parallax, so that its luminosity is about 1,400 $L \odot$, and its absolute magnitude is −3.1.

Some Other Stars in the Keel

Far to the southeast of Canopus, west of Crux and Centaurus, among a group of stars charted by Halley as a separate figure, is the bright star ***MIAPLACIDUS (β) BETA*** (RA 9 h 13 m δ −69° 31') m 1.67, about 65 light years distant and classed as an A-type subgiant or giant. Its name may have been derived from the Arabic *Mi'ah,* or "Waters" (thus translatable as "Placid Waters"), appropriate because of β's position in the Ship's hold! Roughly 10 degrees to the north lies second-magnitude ***ASPIDISKE (ι) IOTA,*** (RA 9 h 16 m δ −59° 04'), the Roman *Scutulum,* or "Little Shield," and the Arabian *Turnais,* all of which are names for the ornamental aplustre at the ship's stern. An F-type supergiant, Iota has a luminosity about fifty-two hundred times the Sun's. Look for the 5.5-magnitude yellow class-G5 (solar) star ***HD 79698 (SAO 236723),*** 5 minutes to its west. A bit farther west and slightly north, about 1 degree from Iota, forming a wide naked-eye double with it, is the spectroscopic binary *a* (not Alpha!), **HD 79351,** magnitude 3.4, formerly given the Greek letter ξ (now belonging to a star in Puppis) and the name "Asmidiske," which is erroneous but rhymes pleasantly with that of its neighbor Aspidiske! A class-B2 subgiant about 600 light years distant, Asmidiske consists of twin components, nearly equal in mass and luminosity, which revolve around their common center of mass in less than a week. Some 6 degrees to Iota's west and prominently situated in the middle of the Ship's Keel is **AVIOR (ε) EPSILON** (RA 8 h 22 m   δ −59° 21′)   m 1.86, 340 light years distant, with a composite spectrum of classes K0 II (bright giant) and B. Needless to say, the foregoing stars in Carina are advantageously viewed only by our colleagues in the southern hemisphere.

Objects of the Milky Way

The southern Milky Way passes through the northeastern portion of Carina, rich in variable stars, nebulae, and open clusters. A noted variable with an extremely ancient tradition is **ETA CARINAE** (RA 10 h 43 m   δ −59° 25′), lying within the large diffuse nebulosity **NGC 3372,** the famed **KEYHOLE NEBULA** (fig. 151). Eta Carinae's variability may have been observed by the Babylonians of near-prehistoric times, possibly referred to in ancient inscriptions (according to the orientalist Jensen) and associated in temple worship with Ea, their greatest god. Halley is credited with first recording it, in 1677, when it was a fourth-magnitude star; until 1814, η oscillated between the fourth and second magnitudes, but then it rose to the first magnitude, in 1827, after which it fell to the second magnitude, where it remained for five years. Sir John Herschel observed η from the Cape in December 1837, when it was about as bright as Canopus. After many dramatic fluctuations, including the attainment of a new maximum luminosity at an apparent magnitude of about −0.8, in 1843, η declined in brightness until, in March of 1886, it was recorded below naked-eye visibility, at magnitude 7.6, dropping to the eighth magnitude by 1900. In 1941 Eta Carinae began to brighten again, attaining the seventh magnitude in 1953 and promising future excitement for stargazers south of the 30th parallel. Eta became newsworthy in the spring of 1982, when scientists announced that its spectrum reveals an overabundance of nitrogen, thus making it a likely candidate for membership in the "Supernova Club," although the promised event may not occur for another ten thousand years!

Often classed as a galactic nova, η has a peculiar spectrum with bright emission lines, including the hydrogen alpha line, which gives the star a reddish color, although its spectral type, initially classified as F5, remains uncertain. Bright lines of ionized metals have been present for many years, and a nebulous shell surrounding η shows the surprising expansion rate of 270 miles per second. Eta Carinae's distance is also uncertain, but if a proposed figure of 3,912 light years is accepted, this star must have the incredible luminosity at maximum of over 1 million Suns and an absolute magnitude of about −11.

The Keyhole Nebula, surrounding Eta, was described in detail by Sir John Herschel, who drew its most brilliant portion (which, according to R. H. Allen, apparently faded to invisibility between 1837 and 1871) and counted 1,203 stars within its intricate branching structure, which covers more than a square degree in the sky. The popular

151. NGC 3372, the Keyhole Nebula, in the vicinity of Eta Carinae. The dark "keyhole" is seen in the bright mass to the upper right. Photographed at Harvard's Boyden Station, South Africa. (Harvard Observatory)

name "Keyhole" refers to the two rounded parts of an elongated dark mass that cuts across one of the bright nebular clouds, as though inviting us to open this glowing celestial door with the key of knowledge!

Less than 3 degrees to the east-northeast of Eta Carinae is the spectacular open star cluster *NGC 3532* (RA 11 h 03 m δ −58° 24′), which both Sir John Herschel and William Pickering described as the finest of its type in the heavens. A-type stars predominate among its two hundred brightest members in a 60-by-30-minute field.

Carina extends toward the south pole, its southern border marking the 75th parallel; objects at 60 degrees south latitude or greater cannot be viewed from the United States or Europe. Lying close to the 64th parallel is the large scattered open cluster *IC 2602*, containing thirty bright stars within a 12-minute field, suitable for wide-angle low-power instruments. In its center lies *(θ) THETA* (RA 10 h 41 m δ −64° 8′) m 2.7, a B-type main-sequence star at a distance of some 700 light years, indicating that IC 2602 may be one of the close galactic clusters.

Approximately 10 degrees west of θ is *(υ) UPSILON,* a third-magnitude A7 "bright giant," 340 light years distant, with 630 L ⊙, forming a fine double with an F-type sixth-magnitude companion at 5 seconds, easily resolved in small telescopes. About 2.5 degrees north of υ is the star lettered "l," notable as one of the largest and most luminous of all known Cepheid variables, with a magnitude range of 4.3 to 5.1, a period of 35.556 days, and a spectrum that varies from F8 to K0. With a diameter about two hundred times the Sun's, this supergiant, 3,000 light years distant, radiates with the intensity of more than twelve thousand Suns!

Some 15 degrees southeast of Canopus lies the naked-eye open cluster *NGC 2516*

(RA 8 h 0 m   δ −60° 44′), 1,200 light years distant; with a 60-minute diameter (the equivalent of about 20 light years), this attractive galactic cluster contains more than a hundred stars, among them a centrally positioned red giant and three prominent double stars.

PUPPIS

The Ship's Stern occupies a long north-to-south area between Canis Major and Vela. The southern part of the Milky Way covers the entire length of Puppis, filling its borders with a rich assortment of deep-sky objects, including many interesting variables and rich open clusters. The brightest star in this constellation is the blue supergiant *(ζ) ZETA* (RA 8 h 2 m   δ −39° 52′)   m 2.25, another of the "Muḥlīfaïn" mistaken for Canopus by early desert dwellers, and designated *Suhail Ḥaḍar* by Al Sufi; *Ḥaḍar,* or "Ground," was applied to various stars of Argo, Centaurus, Columba, and Canis Major because of their closeness to the horizon for viewers in the northern latitudes. Zeta is about 2,400 light years distant; its luminosity, nearly 60,000 $L \odot$, is exceeded by those of few other known stars in our Galaxy. Just 2½ degrees to ζ's northwest is the rich open cluster *NGC 2477,* containing three hundred stars in a 20-minute field. Another degree farther to the northwest is the large scattered cluster of bright stars *NGC 2451.*

Continuing toward the west and slightly north for another 4 degrees, we find the orange giant *(π) PI* (RA 7 h 15 m   δ −37° 00′)   m 2.8, about 140 light years distant and forming a triangle with η and ε Canis Majoris. The blue-white pair *(υ) UPSILON 1* and *2 Puppis,* magnitudes 4.7 and 5.1, respectively, only 26 minutes north-northeast of π, contrast attractively with π's orange color.

A Historic Nova

On November 8, 1942, one of the brightest novae of this century suddenly appeared about 5 degrees north-northeast of Zeta; it was widely observed in both North and South America as a star of magnitude 0.3. *CP PUPPIS,* or *NOVA PUPPIS 1942,* as it is called, faded very quickly, and was no longer visible to the naked eye by the end of November. Its apparent prenova magnitude may have been as large as +18; its luminosity at maximum probably equaled that of 1.3 million Suns, and its absolute magnitude was −10.5, an increase in luminosity by a factor of about 15 million. "CP" has also been tentatively classed as a supernova, and its postnova spectral class, Oe, indicates a very hot blue star with emission lines. The large and rapid variations in its radial velocity led R. P. Kraft, in 1964, to suspect that CP Puppis is a very close binary, with its components in rapid revolution.

Also of Interest

About 7 degrees southwest of Zeta is *(σ) SIGMA* (RA 7 h 27.6 m   δ −43° 12′)   m 3.3, a beautiful contrast double consisting of an orange giant and a yellow (class-G5) main-sequence star of magnitude 8.5 that seems white in comparison with the primary.

The apparent separation is 22.4 seconds; the solar-type companion has a luminosity about equal to the Sun's, while the primary, with 120 times the Sun's luminosity, is also a spectroscopic binary, thus forming a triple system with the companion.

About 2.7 degrees southwest of σ is the red giant *L₂(GC 9604)*, a third-magnitude semiregular variable, discovered by Gould in 1872 and of especial interest to stargazers, because it remains a naked-eye object throughout its 141-day period, ranging from the sixth magnitude to the third magnitude and even brighter. About 8 degrees southeast of L₂, in the far southeast corner of Puppis, lies *V PUPPIS* (RA 7 h 57 m   δ −49° 7'), a bright eclipsing variable of the Beta Lyrae type. The components, two egg-shaped B-type giants nearly in contact, complete one revolution approximately every 1 1/2 days, alternately eclipsing each other during their cycle. A stream of gas flowing from the fainter component (580 L ☉) to the brighter one (1,100 L ☉), is slowly altering their masses and influencing their evolution.

Clusters in the Far North of Puppis

When Sirius rises, the northern portions of Puppis come into view and can be easily seen from the northern hemisphere. Some 14 1/2 degrees east of the great "Dog Star" lies the rich open cluster *M46, NGC 2437* (RA 7 h 39.6 m   δ −14° 42'), containing about 150 stars from the ninth to the thirteenth magnitudes in a 25-minute field (although the total stellar population may exceed 500). In small telescopes, M46 appears as a circular cloud; estimates of its distance vary between 3,200 light years and 5,400 light years, implying a diameter of about thirty LY. On the cluster's northern edge lies the small annular planetary nebula *NGC 2438*, a faint object of the twelfth magnitude with an apparent diameter of only about 1 minute. Its radial velocity differs considerably from that of the cluster, and astronomers have therefore concluded that the planetary is not a cluster member. Of note is the sixteenth-magnitude central star of the nebula, which has an estimated surface temperature of 75,000 degrees K., thus placing it among the hottest known stars.

Approximately 1 1/2 degrees west of M46 lies *M47, NGC 2422*, a bright but more scattered cluster of stars, at a distance variously estimated between 1,540 and 1,789 light years and containing about forty-five stars in a 20-minute field. The easily resolved double star *Σ 1121* lies near the center, its magnitude 7 and 7.5 components separated by about 4,100 astronomical units.

VELA

We now add the Sails to our Ship, and Argo Navis is ready to embark upon its legendary voyage. A fairly large constellation, Vela lies to the southeast of Puppis and west of Centaurus. Its brightest star, close to its western border, is the notable *(γ) GAMMA Velorum* (RA 8 h 8 m   δ −47° 12')   m 1.88; sometimes called *REGOR*, it is another of the *Muḥlīfain*, and thus *Al Suhail al Muḥlīf*, "The Suhail of the Oath," already referred to under α and β Columbae (p. 374). If we are below the 40th parallel, we can easily resolve this magnificent double, which consists of a bluish primary and its fainter blue-white B3 companion at 41 seconds, in small telescopes (a more distant, m-8.5 component may also be noted). Dubbed the "Spectral Gem" of the southern skies, γ Velorum is admired by astronomers because of its vivid and colorful spectral pattern; it is also of great scientific interest as the classic example of a *Wolf-Rayet star.* The first known stars of this type were discovered spectroscopically in 1867, by C. Wolf and G. Rayet, at the Paris Observatory; Wolf-Rayet stars are unusually hot, luminous blue giants with continuous spectra crossed by many extremely broad, bright lines (about 60–100 Å wide in some cases). These emission lines, which are many times brighter than the continuous background spectra, stem from doubly ionized helium and highly ionized atoms of carbon, nitrogen, oxygen, and silicon; they are at times accompanied at their violet end by absorption lines. The wide lines in the spectrum of such a star indicate that it is surrounded by a shell of radially expanding gas, which originates in material ejected from the star's surface at tremendous speeds (up to 1,800 miles per second). The various Doppler shifts arising from all the radial motions of these various parts of the shell are superimposed, so that the spectral lines are broadened. The absorption lines at the violet end of the spectrum indicate an expanding shell, because it arises from radiation coming from the core of the star, which is then absorbed in the approaching shell.

Wolf-Rayet stars were formerly classed as O stars, because they seemed similar to O-type helium giants of the same effective temperature, but now they have been given a special class, W, in the categories WC and WN, indicating that either carbon or nitrogen, respectively, dominates the spectrum; γ Velorum is classed as WC7. No more than a few hundred Wolf-Rayet stars have been observed (many in the Magellanic Clouds and other nearby galaxies), and all at very great distances, with absolute magnitudes ranging from −4 to −8. Estimates of the distance of γ Velorum itself are uncertain, but it is probably the nearest Wolf-Rayet star; based upon a computed distance of 520 light years, the luminosity of the primary is about 3,900 $L \odot$, and the absolute magnitude is about −4.1. Periodic changes in Gamma's radial velocity indicate duplicity of the primary, and recent studies, in fact, reveal the presence of an O7 giant as the very close invisible component.

Several interesting objects lie more or less in Gamma's field, including the open cluster **NGC 2547,** 2 degrees to its south, visible with binoculars or a small telescope; with a diameter of 15 minutes, it contains some fifty stars of magnitudes 7 to 15. Two Cepheids also lie within the vicinity of γ Velorum: less than a degree to its northeast is **AH,** a pale yellow variable (class F8p) with a magnitude range of 5.5 to 5.9 in a period of 4.23 days; and about 2.8 degrees north and slightly to the east is the peculiar short-period variable **AI,** a rather nontypical dwarf Cepheid discovered by Hertzsprung in 1931 from photographs made at Johannesburg. Within a period of some 2 hours 41 minutes, it may range from magnitude 6.4 to 7.1, and its highly erratic light curve indicates that it is pulsating in a complicated manner, producing a pattern of multiple oscillations.

Other Stars

About 8 degrees to the southeast of γ, at the lower right-hand corner of the Sails and almost on the Vela-Carina border, is *(δ) DELTA,* an A-type main-sequence star of about second magnitude, with a sixth-magnitude companion at 2.6 seconds. A distant tenth-magnitude companion at 1 minute 9 seconds is itself double, the magnitude 10.5 and 13 components 6 seconds apart. Some 5 degrees east of δ is the blue-white subgiant and spectroscopic binary *(κ) KAPPA,* magnitude 2.45, 470 light years distant, with the luminosity of nineteen hundred Suns; a rich field surrounds κ; look for the eighth-magnitude open clusters NGC 2910 and 2925, about 2.2 degrees to its northeast, each containing thirty stars in fields of 6 minutes and 11 minutes, respectively.

About 10 degrees to the northeast of Gamma is *(λ) LAMBDA,* magnitude 2.24, sometimes given the name of *SUHAIL,* from Al Sufi's *Al Suhail al Wazn,* "The Suhail of the Weight," although the title "Suhail" was properly a name for Canopus. Lambda is also one of the *Muḥlīfain,* along with γ and ζ. The confusion over nomenclature in the vicinity of Argo Navis seems profound, with many stars vying for the same titles! Lambda, 750 light years distant, is a class-K5 orange supergiant with the luminosity of fifty-eight hundred Suns; a companion of about fifteenth magnitude lies 18 seconds to its southeast.

A few degrees northeast of λ is the magnitude 3.6 cream-colored subgiant and binary *(ψ) PSI,* which the Chinese called *Tseen Ke,* or "Heaven's Record"; another 6 degrees to the west, and some 10 degrees east-northeast of λ, is the bright, large planetary nebula *NGC 3132* (RA 10 h 5 m   δ −40° 11')   m 8.2, lying directly on the Vela-Antlia border. Comparable in size to the great Ring nebula, in Lyra, it appears even brighter because of its tenth-magnitude central star, and is sometimes called the "Eight-Burst nebula" because its structure seems to give the impression of many rings. C. R. O'Dell reported a distance of about 2,800 light years.

PYXIS

Lacaille's Pyxis Nautica, the "Nautical Box" or "Mariner's Compass" is a small, faint figure just east of Puppis and south of Hydra. Its brightest stars are α, a B2 giant only of magnitude 3.69, the yellow giant β, and the orange giant γ, of about fourth magnitude. In May of 1902, as if to call attention to this faint group of stars, a fourteenth-magnitude star about 4 degrees east-northeast of α Pyxidis suddenly flared up in a nova-like outburst, reaching a peak brightness of magnitude 7.3, and then slowly fading back to the fourteenth magnitude by the following January. The Harvard photographic plates reveal a previous flareup in the spring of 1890. In 1920 it again flared up, to reach a magnitude of 6.4; examination of its spectrum at that time led to its classification as a nova, the earliest to be recognized as a "recurrent nova." It is now designated as *T PYXIDIS,* or *NOVA PYXIDIS 1920* (RA 9 h 2.6 m   δ −32° 11'). Since 1920, two more outbursts have been recorded, in 1944 and again in 1967; photographs made during the 1944 outburst revealed expansion velocities up to 1,250 miles per second (comparable to those of classical novae). A spectrum obtained when Nova Pyxidis was at minimum luminosity in 1938 indicated a type-O star with bright lines superimposed over a continuous background. T Pyxidis differs from other recurrent novae owing to its gradual fading after an outburst, a characteristic it shares with slow novae. The interval between maxima is also shorter than that of other recurrent novae, although this interval appears to be increasing, also a mysterious feature. It has changed through all colors during observed eruptions from bluest near maximum luminosity, then yellowish, and, finally, reddish as it faded. Rapid fluctuations of 0.1 magnitude indicate that Nova Pyxidis may be a close and rapid binary, like many others.

ANTLIA

Immediately east of Pyxis is the faint constellation Antlia Pneumatica, the "Air Pump," Lacaille's *Machine Pneumatique.* Some colorful stars form its outlines; the brightest, both red giants of class M0, are *(α) ALPHA,* in the northeastern corner of the Pump, magnitude 4.25, about 277 light years distant, and *(ε) EPSILON,* at the southwest corner, magnitude 4.5, at a distance of 313 light years. Near the southeastern border of Antlia, marking the end of the Air Pump's handle, is *(ι) IOTA,* magnitude 4.6, a yellow giant of spectral class G5, about 182 light years distant.

Cream-colored *(θ) THETA* (RA 9 h 42 m δ −27° 46') m 4.79, at the northern corner of the Pump, is a class-F7 main-sequence star only about 45.5 light years distant. In its neighborhood are some interesting objects; about 4.5 degrees to the southwest (and 3 degrees southeast of Lambda Pyxidis), look for the naked-eye pair *(ζ) ZETA 1* and *2,* A-type stars close to the sixth magnitude. Zeta 1 is double, its components, both of type A, at a separation of 8 seconds; the spectrum of Zeta 2 is characterized by abnormally strong metallic lines (indicating a lower temperature than would a hydrogen spectrum).

About 2 degrees to Theta's west-southwest lies the variable *S ANTLIAE* (RA 9 h 30 m δ −28° 24'), magnitude range 6.3 to 6.8. Discovered in 1888 by the American astronomer H. M. Paul, it created great interest because of its short light cycle, about 7 hours 46 minutes; S Antliae is now classed as an F0 dwarf eclipsing binary of the W Ursae Majoris type (p. 44).

FIFTEEN

South Circumpolar Constellations

The constellations from approximately the 60th declination parallel south to the south celestial pole never set for observers in latitudes greater than 50 degrees south, and scientists in experimental stations on Antarctica can observe the constellations in their heavens describe circular paths about the pole. The official boundaries of many of these modern figures enclose neat quadrilaterals, but some of them are quite irregular, their long, narrow corridors or zigzag patterns extending far into the polar regions in a crazy quilt of interlocking borders. Most of their stars are always below the southern horizon for those of us at northern latitudes above 30 degrees; thus we depend upon the studies made by observatories in Australia, South Africa, and South America for pertinent information.

MUSCA

Having completed our circular tour of the near southern heavens, we now go directly south of the Cross to observe the major stars of *Musca Australis vel Indica,* the "Southern (or Indian) Fly," known simply as Musca. It has also been called *Apis,* the "Bee" (don't confuse this title with *Apus,* the "Bird of Paradise," to its southeast). Apus was combined with Musca and nearby Chameleon by the biblicist Schiller to create "Eve." (The "Northern Fly," *Musca Borealis,* now obsolete, was situated above the back of Aries.)

In mid-May (mid-autumn in the southern hemisphere), when the Southern Cross culminates, so does Musca; its brightest star, *(α) ALPHA* (RA 12 h 34 m   δ −68° 52′)   m 2.7, lies directly south and slightly to the east of α Crucis. Alpha Muscae, about 430 light years distant, is a B2 subgiant with a luminosity about equal to that of twelve hundred Suns, and it is tentatively classed as a multiple system. Close by, to its southeast, is the Cepheid variable *R,* a yellowish star that ranges in magnitude from 6.4 to 7.3 in a period of 7.5 days.

About 2 degrees to α's northeast is the third-magnitude blue-white double star *(β) BETA.* Estimates of its distance vary from 290 light years to 470 light years, determined spectroscopically; thus its luminosity is about 580 $L\odot$. According to current figures, β

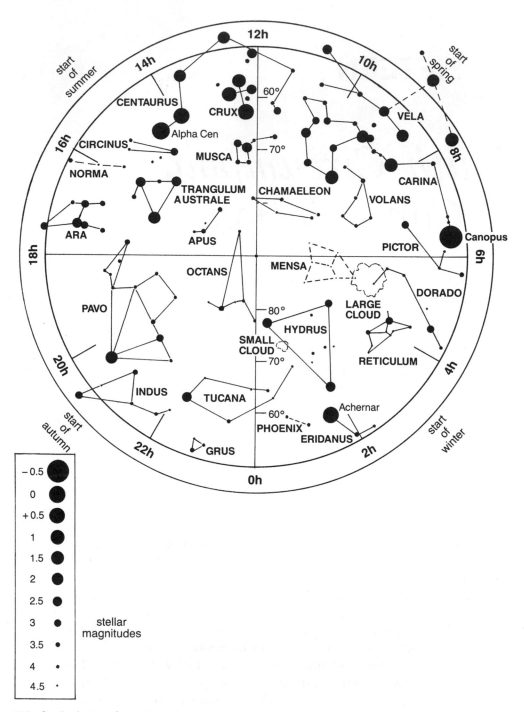

152. South circumpolar zone.

is receding at about 26 miles per second. The fourth-magnitude red giant *(ε) EPSILON* lies a few degrees to Beta's west, marking the "right wing"; Musca's left wing is occupied by *(δ) DELTA,* magnitude 3.6, an orange giant lying a few degrees to the southeast of Alpha. Some 3 degrees to the south and slightly to the west of α, *(γ) GAMMA,* a magnitude-3.8 class-B5 main-sequence star, marks Musca's tail; about 1 degree to Gamma's southwest is the large, eighth-magnitude globular cluster *NGC 4372,* diameter 18 minutes.

CIRCINUS

East of Musca, we find the faint figure of Circinus, Lacaille's "Pair of Compasses," associated with Norma the Level, to its northeast, as we mentioned. Its brightest star, *(α) ALPHA* (RA 14 h 38 m   δ −64° 46′)   m 3.2, lying about 4 degrees to the south of brilliant Alpha Centauri, is an F-type main-sequence star with a ninth-magnitude companion of class K5 at a separation of 15.7 seconds, or 320 astronomical units. In the field of α Circini, immediately to its south, is *X,* the *NOVA CIRCINI 1926,* which reached a magnitude of 6.5 at maximum luminosity, fading to the prenova state of magnitude 17.

TRIANGULUM AUSTRALE

East of Circinus, south of Norma, and southeast of α and β Centauri, is Triangulum Australe, the Southern Triangle, first published in Bayer's *Uranometria.* Its three bright stars were Caesius' "Three Patriarchs" (Abraham, Isaac, and Jacob). The southeastern corner of the Triangle is occupied by the golden star *ATRIA (α) ALPHA* (RA 16 h 43 m   δ −68° 56′)   m 1.9, "Abraham" of the foregoing trio, a K2 giant 55 light years distant. At the northern corner of the Triangle, about 8 degrees northwest of α, is cream-colored *(β) BETA,* magnitude 2.85, class F2 V, which current trigonometric and spectroscopic measurements place at a distance of about 10 parsecs, the standard for determining absolute magnitude (p. 18). Thus Beta's absolute magnitude of +3.0 is close to its apparent magnitude. The coppery-colored Cepheid *S Trianguli,* of sixth magnitude, lies 35 minutes to β's east-southeast, varying from magnitude 6.1 to 6.7 in about 6⅓ days.

About 6½ degrees southwest of β is white *(γ) GAMMA,* an A-type main-sequence star of about third magnitude and approximately 91 light years distant, with strong spectral lines of the rare metal europium.

APUS

South of the Southern Triangle lies Apus, the "Bird of Paradise," another of Bayer's twelve new constellations, its name derived from *apous,* meaning "without feet," because the swallow of ancient Greek legend was legless. Bayer, who credited its formation to the sixteenth-century navigators who explored the southern seas, called it *Apus Indica,* the "Bird of India," but it was also "Avis," or "Avis Indica," to other astronomers. In his planisphere, the English astronomer J. Ellard Gore called it the "House Swallow"; similarly, in distant China Apus was *E Cho,* the "Curious Sparrow," and it was also the "Little Wonder Bird." This coincidence of names aroused much controversy concerning the origins of Phoenix, Indus, and Apus among astronomers from the sixteenth century onward.

The Bird of Paradise's two brightest stars, both of nearly the fourth magnitude, are *(α) ALPHA,* on the southwestern tail feathers, a K5 giant, and *(γ) GAMMA,* on its Head, several degrees to Alpha's east, a subgiant and also a K star. Just to the west of γ is the colorful double *DELTA 1* and *2* (RA 16 h 13 m   δ −78° 34′), a wide pair consisting of a red giant and an orange giant, magnitudes 4.7 and 5.3, respectively.

PAVO

When Jove, on one of his many sexual adventures, pursued and overcame the maiden Io, his furious wife, Juno, turned Io into a heifer and placed her under the guard of Argus, a giant with one hundred eyes. At Jove's command, however, Mercury slew Argus, whereupon Juno took the giant's eyes and fastened them to the feathers of her bird, the peacock, so that its tail is spread with jewels. According to R. H. Allen, the peacock is an appropriate symbol for a constellation, because it symbolizes immortality, owing to its starry tail, which rendered it sacred to Juno, Queen of the Heavens.

One of the twelve figures adopted by Bayer, Pavo the Peacock occupies a region northeast of the Bird of Paradise and extends from the 56th parallel at the borders of Telescopium and Indus all the way to Octans at the 75th parallel. In the far northeastern corner, just at the border of Telescopium and about 4 degrees west-northwest from β Indi, is Pavo's brightest star, *(α) ALPHA PAVONIS* (RA 20 h 22 m   δ −56° 54′)   m 1.9, between 231 and 310 light years distant, according to its "spectroscopic parallax." With a luminosity of about 1,200 $L \odot$, α is also a spectroscopic binary, its period of revolution about 11 days and 18 hours.

About 10 degrees to Alpha Pavonis' south-southeast is the A5 subgiant *(β) BETA,* magnitude 3.4, 91 to 160 light years distant, according to varying estimates. A few degrees west of β is the yellow G5 subgiant *(δ) DELTA,* magnitude 3.6, only 18.6 light years distant, with the large proper motion usual for nearby stars: μ + .199 (seconds of time per year to the east) and δ −1.14″ (seconds of arc per year to the south); and a radial velocity of 22 kilometers per second (about 13.66 mi./sec.) in approach.

Farther west, marking the northwest corner of the quadrangle β, δ, ε, ζ and κ is the variable star *(κ) KAPPA* (RA 18 h 52 m   δ −67° 18′), a W Virginis Cepheid, its magnitude ranging from 3.9 to 4.8 in a period of 9.1 days and its spectrum changing from F5 to G5 in the process.

Observers in the southern hemisphere should look for the very large and bright globular cluster *NGC 6752,* ∆ *295* (RA 19 h 6 m   δ −60° 4′)   m 7.2, about 10 degrees west-southwest of α; at approximately 20,000 light years, this is one of the closer globulars, its luminosity about 100,000 *L* ⊙. Some 3 degrees to its south is the large barred-spiral galaxy *NGC 6744.*

In Pavo's far northwest is the red giant *(ξ) XI* (RA 18 h 19 m   δ −62° 37′)   m 4.4, an interesting color-contrast binary 306 light years distant. Stargazers visiting observatories in countries far to the south have a chance to view two small planetary nebulae in the same field, 2 degrees southeast of Xi: *NGC 6630* and *IC 4723,* both of fifteenth magnitude, with diameters of 19 seconds.

TUCANA

Continuing on toward the east and passing across the narrow channel that separates the northern and southern sections of Indus, we arrive at the constellation of Tucana the Toucan, a Brazilian bird with a very large beak, keeping company with Apus, Pavo, and Phoenix. One of Bayer's figures, it was adopted by the Chinese as their *Neaou Chuy,* the "Beak Bird," and in England as the "Brasilian Pye," whereas Kepler, Riccioli, and later authorities called it the *Anser Americanus,* or "American Goose." Tucana is a long east-west figure lying south of Grus and Phoenix and north of Hydrus. Its brightest star is *(α) ALPHA* (RA 22 h 15 m   δ −60° 31′)   m 2.85, marking the bird's head; its spectroscopic "parallax" gives a distance of about 114 light years, but the true, trigonometric parallax yields the smaller figure of 62 light years and a corresponding luminosity of about 20 *L* ⊙. Alpha Tucanae is tentatively classed as a spectroscopic binary because of variations in its radial velocity, with a possible period of revolution of 11½ years, although the companion has never been seen. The fifth-magnitude M5 red giant *(ν) NU,* which R. H. Allen calls "strongly red," is some 2 degrees to the southeast.

Near the tail of the Bird we find an excellent common-proper-motion pair, the main-sequence stars *(β) BETA 1* and *2* (RA 0 h 29 m   δ −63° 14′), about magnitude 4.5, of the late-B and early-A classes, respectively; some 108 light years distant, the Betas are receding at the moderate velocity of about 10 kilometers (6.21 mi.) per second. About 9 degrees south and slightly to their west lies the outstanding globular cluster *NGC 104, 47 TUCANAE* (RA 0 h 22 m   δ −72° 21′)   m 4.5, which appears like a hazy star to the naked eye but in large telescopes is revealed as a rich, highly compressed stellar center surrounded by thousands of twelfth- to fourteenth-magnitude and fainter stars, its diameter about 25 minutes. The cluster lies in a relatively "starless," or dark, part of the heavens; this adds to its striking appearance. Current estimates place NGC 104 at a distance of 15,000 to 20,000 light years, the probable distance of Omega Centauri. The absolute magnitude of the cluster is about −8.8; it contains a number of red giants, including five Mira-type variables. Spectral lines of heavier metals are unusually pronounced for a globular cluster, because these are population-II stars, normally deficient in such metals.

Magellan's Lesser Galaxy

About 2½ degrees east of NGC 104 lies *NUBECULA MINOR, THE SMALL MAGEL-*

LANIC CLOUD, NGC 292 (RA 0 h 50 m　δ −73° 30′)　m 1.5 (fig. 153), Magellan's "Lesser Cloud," an irregular galaxy in our own galactic neighborhood, and the companion of Nubecula Major, the Large Cloud, in Dorado. Visible to the naked eye as a bright patch of mist, the Small Cloud lies about 17 degrees from the south celestial pole, exactly on the opposite side of it from the Southern Cross. About 190,000 light years distant from us, and 80,000 light years from the Great Cloud, Nubecula Minor has an apparent diameter of 3½ degrees, but its true diameter is about 30,000 light years; to cross the bright portion alone, a ray of light must travel for ten thousand years! Radio telescopes have detected a large hydrogen envelope surrounding both Magellanic Clouds, and a tidal streamer originating in the Small Cloud also appears to link them.

Shapley was the first to classify the Small Cloud as a typical irregular dwarf galaxy; its particular claim to fame is Henrietta Leavitt's investigation of its many Cepheid variables, from which she discovered the period-luminosity relation (p. 73). In contrast to the Large Cloud, the Small Magellanic Cloud appears to be dust-free and contains only a few nebulous regions, which implies that fewer population-I stars are present. Typical of irregular galaxies, it shows no well-defined nucleus and no definite spiral structure, as seen in its highly inclined position (30° from edge on); its flattened form implies possible rotation, and both clouds may also be slowly revolving about their common center of mass, as well as orbiting our own Galaxy, in periods of many millions of years.

Splendid diffuse nebulosities, numerous open star clusters of every type, and also many fine globulars have been observed in the Small Cloud; among these objects are the following: *NGC 371,* a large open cluster heavily populated with giant stars, lying about 1 degree northeast of the main body of the Cloud; *NGC 346,* a nebulous star cluster about 7 minutes to the south of NGC 371; and farther south, *NGC 456–65,* a chain of clusters 20 minutes long, about 2 degrees east of the Cloud's main mass.

153. *Region of the Magellanic Clouds.* NGC 292, the Small Magellanic Cloud (bottom center), an irregular galaxy of the Local Group, is seen directly below its companion, the Large Magellanic Cloud. The bright star lying to the north of the Small Cloud is Achernar, α Tucani. (Refer to the star chart at the beginning of Chapter Fifteen.) Photographed at Harvard's Boyden Station, South Africa. (*Harvard University*)

HYDRUS

East of Tucana lies the irregular form of Hydrus, the little Water Snake, one of Bayer's figures, not to be confused with the ancient Hydra, its great counterpart to the north. Schiller combined it with Tucana and the Small Magellanic Cloud to create the archangel Raphael. The Chinese created four asterisms out of Hydrus and neighboring areas to show various parts of the Serpent's anatomy.

About 7½ degrees south-southeast of brilliant Achernar (α Eridani), we find *(α) ALPHA Hydri* (RA 1 h 57 m δ −61° 49′) m 2.8, supposedly the head of the Serpent.* About 30 light years distant, Alpha is an F-type main-sequence star, some six times the Sun's luminosity. Marking the Serpent's tail, in the far southwest part of Hydrus, is *(β) BETA* (RA 0 h 23 m δ −77° 32′) m 2.78, a yellow G-type subgiant slightly brighter than Alpha and the nearest bright star to the south celestial pole, close to 3 degrees away. At 21 light years, β is one of the fairly close stars; in cosmic terms, practically a neighbor. Similar in spectral type to our Sun, it has a luminosity of about 2.7 $L \odot$.

In the southeastern corner of Hydrus' twisting form is *(γ) GAMMA* (RA 3 h 48 m δ −74° 24′) m 3.27, its distance variously estimated from 160 to 300 light years, the latter figure indicating a luminosity of about 330 $L \odot$. Lying between the Magellanic Clouds, Gamma Hydri is referred to by the navigator Corsali as a star that accompanies "two clouds of reasonable bigness moving about the place of the pole . . ." in a description of the Clouds' circumpolar course.

Two pairs of interesting wide optical color-contrast doubles are *(η) ETA 1* and *2* and *(π) PI 1* and *2,* approximately 6 degrees to 7 degrees south of α. The Eta pair consists of a magnitude-4.7 orange giant and a faint white magnitude-6.7 A-type star; to their east we find the Pi pair, about magnitude 5.5, a red giant and an orange one.

RETICULUM

Immediately northeast of Hydrus and about 7 degrees northwest from the Large Magellanic Cloud is Lacaille's *Reticulum Rhomboidalis,* the "Rhomboidal Net" (the reticle he used in his southern observations); according to R. H. Allen, however, it was first drawn as the "Rhombus" by Isaak Habrecht of Strassburg and adopted by Lacaille. Its stars also seem to suggest the form of a butterfly. Although it lies next to the southern end of Horologium (which extends far to the north), we include Reticulum in our circumpolar chapter because its principal stars lie just south of the 60th parallel.

The brightest star is *(α) ALPHA Reticuli* (RA 4 h 14 m δ −62° 28′) m 3.35, a luminous yellow giant, of class G6, about 391 light years distant, receding with a velocity of 36 kilometers (22.36 mi.) per second. *(β) BETA,* to its southwest (RA 3 h 44 m δ −64° 48′) m 3.85, is a deep golden K-type subgiant with a radial velocity of 51 kilometers (31.67 mi.) per second in recession.

About 3 degrees northwest of β is the solar-type wide common-proper-motion pair

---

* *In a rare instance of confusion, R. H. Allen refers to the tail as "almost reaching Achernar," and then later identifies β, near the southern border of Hydrus-Octans, as "in the tail."*

(ζ) ZETA 1 and *2* (RA 3 h 17 m　δ −62° 34′), magnitudes 5.5 and 5.2, respectively; both stars are class-G2 main-sequence objects, with the moderate rate of recession of 12 kilometers (7.45 mi.) per second.

DORADO

Like Horologium, Dorado the Goldfish is a long, twisting figure, its official boundaries zigzagging from the 49th to the 70th parallels south. First published in Bayer's *Uranometria,* its name refers to a large tropical fish and not the small pet that graces our home aquarium. Caesius, of the biblical school, combined Dorado with the Greater Magellanic Cloud and Volans the Flying Fish to form "Abel the Just." Some of the older star charts refer to it as *Xiphias,* the "Swordfish," and the translation "Swordfish" appears along with "Dorado" in modern listings.

Near the northern end of Dorado lies its brightest star, *(α) ALPHA* (RA 4 h 34 m　δ −55° 03′)　m 3.8, a white A-type giant some 192 light years distant. Two or three degrees to its west is a rich field of spiral galaxies, barred spirals, and ellipticals.

Nine degrees to the southeast of α is the Cepheid variable *(β) BETA* (RA 5 h 33 m　δ −62° 31′), magnitude range 3.3 to 4.7, one of the brightest Cepheids known; Canopus lies 12 degrees to α's northeast, and the Large Cloud is only about 8 degrees to the south. At maximum luminosity, Beta's spectrum is that of a supergiant, class F6, but it changes to G5 at minimum as the creamy color of the star deepens to gold in a cycle of 9.84 days. Beta Doradūs is far too distant for an accurate determination of its parallax, estimates varying from 1,700 light years to as much as 7,948 light years.

Magellan's Greater Galaxy

About 8 degrees south of β we reach the outer regions of *NUBECULA MAJOR, THE LARGE MAGELLANIC CLOUD* (RA 5 h 20 m　δ −69°)　m 1.0 (fig. 154), one of the nearest external galaxies of the Local Group and the larger companion of Nubecula Minor, in Tucana. The Large Cloud lies about 20 degrees from the south celestial pole, with its southern regions spilling over into the neighboring constellation of Mensa. The earliest navigators to sail around the Cape of Good Hope named these two conspicuous patches of light the "Cape Clouds," but, long before, the Clouds were already known to Polynesian islanders as Upper and Lower *Mahu,* or "Mist." The Large Cloud was also *Al Bakr,* "The White Ox," of southern Arabia, translated as the "Oxen of Tehama" (a Red Sea province) by the German astronomer Ideler and probably applied to both clouds.

With an angular diameter of 6 degrees, the Large Magellanic Cloud is at least 50,000 light years in diameter, including its outer regions; and it is about 190,000 light years distant, somewhat nearer than the Small Cloud; the apparent separation of the Clouds is about 22 degrees, the true distance approximately 80,000 light years. Although both Clouds are generally classified as irregular galaxies, they differ considerably in structure. The Large Cloud has an elongated core suggesting the nucleus of a barred spiral

154. The Large Magellanic Cloud, an irregular galaxy of the Local Group visible to the naked eye as a patch of light. Contains supergiant S Doradūs, the most luminous star known to us in the universe. Attached to the eastern (right-hand) end of the Large Cloud is the Tarantula Nebula; many other bright nebulae, star clouds, and clusters lie farther out. (North is to the bottom.) Photographed at Harvard's Boyden Station, South Africa. (*Harvard University*)

galaxy at the ends of which are traces of a spiral pattern that appears to be in the early stages of formation.

The Large Magellanic Cloud contains about 25 billion solar masses and has a luminosity of about 2 billion Suns. It is predominantly composed of population-I stars and contains an abundance of cosmic dust; important features are extensive HII regions, where, as in the Orion nebula, the interstellar material is illuminated by young, hot supergiants and protostars. A total of 750 supergiants with absolute magnitudes smaller than −6.5 have been counted, and among them is the most luminous star known to us in the entire universe, *S DORADŪS* (RA 5 h 18.6 m δ −69° 18′) m 8.4 to m 9.5, lying in the irregular star cluster *NGC 1910* on the northern edge of the central "bar" of the Large Cloud. The peculiar A-type spectrum of S Doradūs contains bright hydrogen lines, indicating the star's high surface temperature and great luminosity, which averages about 500,000 $L \odot$, but has been known to rise above 1 million $L \odot$. The cause of its variability is uncertain.

The stellar population of the Large Cloud also includes four red and four blue supergiants brighter than the tenth apparent magnitude, hundreds of population-I Cepheids, and a number of RR Lyrae stars in its globular clusters. A tremendous number of open star clusters are present, including the huge association of blue giants in *NGC 1936,* a bright gaseous nebula centered in a vast region of neutral hydrogen, yet another "baby-star cradle" (pp. 112–13). Among the globular clusters are a few unusual ones that contain luminous blue stars and seem to be as young as typical open clusters. The Large Magellanic Cloud is also a happy hunting ground for searchers for planetary nebulae; at least four hundred planetaries have been recorded.

A Great Spider

In the eastern part of the Large Cloud is the most outstanding bright diffuse nebula known, the vast *TARANTULA NEBULA,* or *GREAT LOOPED NEBULA, NGC 2070* (fig. 155), enclosing *30 DORADŪS,* a cluster that contains more than a hundred supergiants in an area 100 light years wide. W. H. Smyth called NGC 2070 the "True Lover's Knot," and Sir John Herschel described its intricate structure as one of the most extraordinary objects in the heavens. Its diameter is about 800 light years, and if filaments extending from the nebular rim are included, the total dimensions are some 1,800 by 1,700 light years. The Tarantula is also a radio source, and such observations imply a mass of nearly 500,000 $\mathcal{M}\odot$.

155. *NGC 2070, the Tarantula Nebula,* a gigantic bright gaseous nebula enclosing the extremely compact star cluster 30 Doradūs, in the eastern tip of the Large Magellanic Cloud. Photographed with the 60-inch reflector at Harvard's Boyden Station, South Africa. (*Harvard Observatory*)

In the field of NGC 2070 is the region *N 159;* near it the organic molecule carbon monoxide was detected by astronomers using the 13-foot Anglo-Australian radio telescope; interstellar organic molecules support the theory that extraterrestrial life exists.

The Brightest Supernova Since Kepler's Star

Of the vast number of exploding stars in our universe, a supernova of naked-eye magnitude rarely occurs more than once in centuries. Therefore, ever since Johannes Kepler's student discovered Supernova 1604, in the southern part of Ophiuchus, astronomers have been hoping to witness the repetition of such an exciting and important event. On the night of February 24, 1987, their hopes were finally fulfilled when Ian Shelton, of the University of Toronto, photographing the Large Magellanic Cloud with the 10-inch Bruce telescope at Las Campanas Observatory, in Chile, discovered a fifth-magnitude star in the eastern part of the Cloud where no star of that brightness should have been found. Walking outside, he visually confirmed the existence of this "new star," and the news of his discovery made headlines around the world.

Supernova (SN) 1987A (RA 5 h 36 m δ −69° 18′) magnitude 4.5, just west of 30 Doradūs and 170,000 light years distant, was initially expected to increase its luminosity considerably, because of the rapid expansion of its gas shell (some 17,000 mi./sec.), the strong hydrogen lines in its spectrum, and the detection of neutrinos (five pulses over a 7-second interval) emitted by the star prior to the detection of photons, features suggesting that SN 1987A is a type II supernova, a massive star which collapsed and whose core was thus compressed to form a neutron star (pulsar). Until the end of March 1987, the supernova remained at a magnitude no smaller than 4.25, but by late May of that year its magnitude had decreased (its luminosity increased) to about 2.9. The supernova then slowly began to fade, and two months later its magnitude was about 4.6.

The identity of the antecedent star of the supernova has been established as a type B3 blue supergiant called "Sanduleak −69° 202", which was no brighter than the twelfth magnitude and had some 15 solar masses before its explosion. Puzzling spectral features of SN 1987A obtained by the International Ultraviolet Explorer (IUE) resemble those of a type I supernova, a white dwarf which explodes when it reaches a critical limit (the Chandresekhar limit). Added to its unusual spectrum, certain strange aspects of the supernova's behavior may lead astronomers to new theories about the birth of supernovae. In the meanwhile, further events concerning the remarkable "guest star" of our century are eagerly awaited by the scientific world.

MENSA

Beneath the Cloud stands the Mountain. Immediately south of Dorado and the Large Magellanic Cloud are the faint stars of *Mons Mensae,* the "Table Mountain"; known today simply as Mensa, it was formed by Lacaille in honor of the cloud-capped Table Mountain at Cape Town, which, as he explained, "has witnessed my nightly vigils and daily toils."

The eastern corner of the mountaintop is marked by *(α) ALPHA,* a class-G5 main-sequence star, fairly close at a distance of about 28 light years, but of only the fifth magnitude. *(β) BETA,* lying approximately 12 degrees to its northwest, is a foreground object in the southern rim of the Large Cloud. At least 55 light years distant, β is a deep yellow class-G8 star, apparent magnitude 5.3. Neatly centered in the upper part of the Mountain is the optical double *(γ) GAMMA Mensae* (RA 5 h 34 m   δ − 76° 23′)   m 5.2, a class-K4 orange giant about 424 light years distant, with an eleventh-magnitude companion at 38.2 seconds, a separation that is presently increasing, for the convenience of stargazers far to the south.

VOLANS

East of the Large Magellanic Cloud and south and west of Carina is Bayer's "Flying Fish," *Piscis Volans,* known today simply as Volans. In his *Rudolphine Tables,* Kepler referred to it as *Passer,* the Sparrow, which became the Chinese *Fe Yu.* The spectroscopic binary *(α) ALPHA Volantis,* on the Carina-Volans border, a white A5 main-sequence star about 78 light years distant, is of only the fourth magnitude. About 10 degrees directly to its west is the brightest star in the constellation, *(β) BETA,* an orange K2 giant of magnitude 3.8, some 192 light years distant.

Approximately 20 degrees east of the borders of the Large Magellanic Cloud lies the westernmost bright star marking the figure of the Flying Fish, the fine common-proper-motion pair *(γ) GAMMA 1* and *2* (RA 7 h 9 m   δ − 70° 25′)   m 5.7 and 3.8, respectively, about 75 light years distant; the components are a creamy F4 dwarf and a deep yellow G8 giant, at a separation of 13.6 seconds. Centered in the figure of the Fish is the common-proper-motion double and spectroscopic binary *(ε) EPSILON,* magnitude 4.4, a blue-white main-sequence star 391 light years distant, its components 6.1 seconds apart. Several degrees to its west lies the barred spiral galaxy *NGC 2442,* its angular dimensions about 6 by 5 minutes.

CHAMAELEON

Southeast of the Flying Fish we enter the territory of Bayer's Chamaeleon, whose official boundaries circle the polar area of Octans, from about 7 hours 32 minutes to 13 hours 48 minutes in right ascension. Carina and Musca lie to its north. The south pole, too, deserves a "Dipper," a figure the Chinese perceived in their *Seaou Tow,* clearly delineated by several of the Chamaeleon's stars.

In the far western part of faint Chamaeleon, at the end of the small dipper's handle, is *(α) ALPHA,* 78 light years distant, a cream-colored F6 subgiant and the brightest star in the constellation, although of only fourth magnitude. Some 30 degrees to α's east, at the northern corner of the Chinese dipper, is *(γ) GAMMA,* magnitude 4.1, a red M-type giant 251 light years distant. At the eastern corner of the dipper is the triple star *(δ)*

DELTA 1 and *2* (RA 10 h 45 m   δ −89° 12′). Delta 1 is an orange giant, magnitude 5.4, 359 light years distant, its close components only 0.6 second apart; δ 2, its blue-white optical companion, magnitude 4.5, presents an attractive color contrast for small telescopes. Centered within the dipper is *NGC 3195* (RA 10 h 10 m   δ −80° 37′), a fairly bright but small planetary nebula, the apparent diameter 40 by 30 seconds.

OCTANS

The area of the south celestial pole is occupied by Octans Hadleianus, the octant invented by John Hadley in 1730 and transferred to the skies by Lacaille in 1752. There is little tradition associated with the skies of frozen Antarctica; however, the Arabs believed that both polestars, when attentively observed by the afflicted, exert a healing power.

Stars for Penguins

Most of the stars of Octans are quite faint, giving an impression to the naked eye that this region of the heavens is rather empty. A group of its brighter stars extends northward toward Pavo and Indus, thus the official borders of Octans coincide almost with the 75th parallel on that side. The brightest star, approximately 40 degrees west of β Hydri, is *(ν) NU Octantis* (RA 21 h 41 m   δ −77° 23′)   m 3.76, an orange K-type giant about 104 light years distant. Some 10 degrees to its west lies *(α) ALPHA*, (RA 21 h 05 m   δ −77° 01′)   magnitude 5.15, spectral-class F4 III, thus also a giant, its distance about 228 light years; about 20 degrees southeast of α is another F-type star but a "normal dwarf," *(β) BETA* (RA 22 h 46 m   δ −81° 23′)   m 4.15, 65 light years distant.

An interesting formation of three yellow stars, each similar to our Sun in spectral type, is *(γ) GAMMA 1, 2,* and *3* (RA 23 h 52 m   δ −82° 01′), about 18 degrees east-southeast of Beta. Gamma 1 and Gamma 2 are about 1 degree 21 minutes apart, classes G7 and G8, respectively, both of them giants; they are not gravitationally bound. Gamma 1 is about 300 light years distant, with a radial velocity in recession of 15 kilometers per second (9.31 mi./sec.), and Gamma 2, somewhat farther away at 326 light years, is receding at a velocity of 27 kilometers (16.77 mi.) per second. All three stars have rather similar proper motions toward the southwest, and this, plus their almost identical spectral types, suggest that they may have begun their lives at the same time, in a common gas/dust envelope, having been accelerated from it toward the southwest by some explosive force.

A small wide-angle eyepiece includes the three Gammas in one field.

The Southern Polaris

Having begun our journey through the heavens with the northern Pole Star, we end it with the naked-eye star most nearly corresponding to it in the south, faint *(σ) SIGMA*

156. *The Cone Nebula,* lying at the southern tip of open cluster NGC 2264, in Monoceros, is a striking example of dark nebulosity that conceals the formation of new stars. (*Hale Observatories*)

Octanis (RA 21 h 8 m δ −88° 57′) m 5.5, which lies about 7 degrees south and slightly to the west from γ. Sigma is catalogued as a class-A7 star with broad spectral lines, indicating that it is in rapid rotation, but it is also listed as a type-F main-sequence star; its parallax has not been recorded. Sigma is departing from the south celestial pole with a proper motion of 0.095 and receding from us at 7½ miles per second.

Who shall understand the mysteries of thy creations?
For Thou hast exalted above the ninth sphere the sphere
   of intelligence.

It is the Temple confronting us
"The tenth that shall be sacred to the Lord."
It is the Sphere transcending height,
To which conception cannot reach,
And there stands the veiled palanquin of Thy glory. . . .

Bibliography

TEXTBOOKS

Abell, George W. *Exploration of the Universe*. New York: Holt, Rinehart and Winston, 1973.

Hodge, Paul W. *Concepts of Contemporary Astronomy*. New York: McGraw-Hill, 1974.

Jastrow, Robert; and Thompson, Malcolm H. *Astronomy: Fundamentals and Frontiers*. New York: John Wiley, 1974.

Motz, Lloyd; and Duveen, Anneta. *Essentials of Astronomy*. New York: Columbia University Press, 1977.

GENERAL AND HISTORICAL

Ferris, Timothy. *Galaxies*. New York: Stewart, Tabori and Chang/Workman, 1980.

Martin, Martha Evans; and Menzel, Donald H. *The Friendly Stars*. New York: Harper, 1907; reprinted by Dover. Revised by Don Rice and Dr. Craig Foltz. New York: Van Nostrand Reinhold, 1982.

Moore, Patrick. *The Unfolding Universe*. New York: Crown, 1982.

Motz, Lloyd. *Astronomy A to Z*. New York: Grosset and Dunlap, 1964.

———. *On the Path of Venus*. New York: Pantheon/Random House, 1976.

———. *This Is Astronomy*. New York: Columbia University Press, 1963.

———. *This Is Outer Space*. New York: New American Library, 1963.

Sagan, Carl. *The Cosmic Connection: An Extraterrestial Perspective*. New York: Doubleday, 1973.

Sagan, Carl. *Cosmos*. New York, Toronto: Random House, 1980.

Smart, W. M. *Some Famous Stars*. New York: Longmans, Green, 1950.

Tauber, Gerald E. *Man's View of the Universe*. New York: Crown, 1979.

SPECIAL TOPICS

Abetti, Giorgio, and Hack, Margherita. *Nebulas and Galaxies* (translation). London: Faber & Faber, 1964.

Allen, David A. *Infrared—The New Astronomy*. New York and Toronto: Halsted Press, Div. of John Wiley, 1975.

Lynds, Beverly T. (ed.). *Dark Nebulae, Globules and Protostars*. Tucson: The University of Arizona Press, 1971.

Motz, Lloyd. *The Universe—Its Beginning and End*. New York: Charles Scribner's Sons, 1975.

Osterbrock, D. E., and O'Dell, C. R. *IAU Symposium No. 34 (Planetary Nebulae)*. (Holland: D. Reidel); New York: Springer-Verlag, 1968.

Osterbrock, Donald E. *Astrophysics of Gaseous Nebulae*. San Francisco: W. H. Freeman, 1974.

Shklovskii, I. S. *Stars—Their Birth, Life and Death* (Revised translation by Richard B. Rodman). San Francisco: W. H. Freeman, 1978.

Thackeray, A. D. *Astronomical Spectroscopy*. New York: Macmillan, 1961.

Verschuur, Gerrit L. *The Invisible Universe: The Story of Radio Astronomy*. New York: Springer Verlag, 1974.

STAR LORE AND MYTHOLOGY

Allen, Richard Hinckley. *Star Names: Their Lore and Meaning* (formerly *Star-Names and Their Meanings*, New York: Stechert, 1899). Revised version New York: Dover Publications, 1963.

Gayley, Charles Mills. *The Classic Myths in English Literature and in Art* (Waltham, Mass.: Blaiskell). Reprinted Lexington, Mass.: Xerox College Publication, 1939.

Graves, Robert. *The Greek Myths*. Garden City, N.Y.: George Braziller, 1981.

Jagendorf, M. A. *Stories and Lore of the Zodiac*. New York: The Vanguard Press, 1977.

Ovid (Naso, Publius Ovidius). *Metamorphoses*. Translation of Rolfe Humphries. Bloomington, Inc.: Indiana University Press, 1967.

Pindar. Translation of Richard Lattimore. Chicago: Phoenix Books, University of Chicago Press, 7th ed., 1968.

Seznec, Jean. *Survival of the Pagan Gods.* Translation of Barbara F. Sessions. New York: Pantheon Books, 1953.

GUIDEBOOKS

Brown, Peter Lancaster. *What Star Is That?* New York: Viking Press, 1971.

Burnham, Robert Jr. *Burnham's Celestial Handbook: An Observer's Guide to the Universe Beyond the Solar System.* Three volumes. New York: Dover Publications, 1978.

Howard, Neale E. *The Telescope Handbook and Star Atlas.* New York: Thomas Y. Crowell, 1975.

Kyselka, Will; and Lanterman, Ray. *North Star to Southern Cross.* Honolulu: University Press of Hawaii, 1976.

Moore, Patrick. *Naked-Eye Astronomy.* New York: W. W. Norton, 1974.

————. *The New Guide to the Stars.* New York: W. W. Norton, 1974.

Neely, James. *A Primer for Star-Gazers.* New York: Harper, 1970.

Olcott, W. T.; and Putnam, E. W. *Field Book of the Skies.* Revised by R. Newton and Margaret W. Mayall, New York: G. P. Putnam's Sons, 1954.

TABLES AND CHARTS

Allen, C. W. *Astrophysical Quantities.* (3rd ed.). London: Athlone Press, University of London, 1973.

Becvar, A. *Skalnate Pleso Atlas of the Heavens, 1950.0.* Desk ed. Cambridge, Mass.: Sky Publishing Corp., 1969.

Edmund Scientific Co. *Edmund Mag 5 Star Atlas.* Barrington, N.J.: Edmund Scientific Co., 1974.

Hirshfeld, Alan; and Sinnott, Roger W. (eds.). *Sky Catalogue 2000.0.* Two volumes. Cambridge, Mass.: Sky Publishing Corp. and Cambridge University Press, 1982.

Norton, Arthur P. *A Star Atlas and Reference Handbook* (17th ed.). Edinburgh and Cambridge, Mass.: Gall and Inglis, and Sky Publishing Corp., 1978.

JOURNALS

Astronomical Journal. New York: The American Institute of Physics.

Astrophysical Journal. Chicago: University of Chicago Press.

Monthly Notices. London, Burlington House: Royal Astronomical Society.

The Observatory. England, Hailsham, Sussex: Royal Greenwich Observatory, Herstmonceux Castle.

Sky and Telescope. Cambridge, Mass.: Sky Publishing Corp., Bay State Rd.

Appendix

CONSTELLATIONS

Constellation	genitive ending	Meaning	Contractions	α	δ	Area
				h	°	(°)²
Andromeda	-dae	Chained maiden	And	1	40 N	722
Antlia	-liae	Air pump	Ant	10	35 S	239
Apus	-podis	Bird of paradise	Aps	16	75 S	206
Aquarius	-rii	Water bearer	Aqr	23	15 S	980
Aquila	-lae	Eagle	Aql	20	5 N	652
Ara	-rae	Altar	Ara	17	55 S	237
Aries	-ietis	Ram	Ari	3	20 N	441
Auriga	-gae	Charioteer	Aur	6	40 N	657
Boötes	-tis	Herdsman	Boo	15	30 N	907
Caelum	-aeli	Chisel	Cae	5	40 S	125
Camelopardus	-di	Giraffe	Cam	6	70 N	757
Cancer	-cri	Crab	Cnc	9	20 N	506
Canes Venatici	-num -corum	Hunting dogs	CVn	13	40 N	465
Canis Major	-is -ris	Great dog	CMa	7	20 S	380
Canis Minor	-is -ris	Small dog	CMi	8	5 N	183
Capricornus	-ni	Sea goat	Cap	21	20 S	414
Carina	-nae	Keel	Car	9	60 S	494
Cassiopeia	-peiae	Lady in chair	Cas	1	60 N	598
Centaurus	-ri	Centaur	Cen	13	50 S	1 060
Cepheus	-phei	King	Cep	22	70 N	588
Cetus	-ti	Whale	Cet	2	10 S	1 231
Chamaeleon	-ntis	Chamaeleon	Cha	11	80 S	132
Circinus	-ni	Compasses	Cir	15	60 S	93
Columba	-bae	Dove	Col	6	35 S	270
Coma Berenices	-mae -cis	Berenice's hair	Com	13	20 N	386
Corona Australis	-nae -lis	S crown	CrA	19	40 S	128
Corona Borealis	-nae -lis	N crown	CrB	16	30 N	179
Corvus	-vi	Crow	Crv	12	20 S	184
Crater	-eris	Cup	Crt	11	15 S	282
Crux	-ucis	S cross	Cru	12	60 S	68
Cygnus	-gni	Swan	Cyg	21	40 N	804
Delphinus	-ni	Dolphin	Del	21	10 N	189
Dorado	-dus	Swordfish	Dor	5	65 S	179
Draco	-onis	Dragon	Dra	17	65 N	1 083
Equuleus	-lei	Small horse	Equ	21	10 N	72
Eridanus	-ni	River Eridanus	Eri	3	20 S	1 138
Fornax	-acis	Furnace	For	3	30 S	398
Gemini	-norum	Heavenly twins	Gem	7	20 N	514
Grus	-ruis	Crane	Gru	22	45 S	366
Hercules	-lis	Kneeling giant	Her	17	30 N	1 225
Horologium	-gii	Clock	Hor	3	60 S	249
Hydra	-drae	Water monster	Hya	10	20 S	1 303
Hydrus	-dri	Sea-serpent	Hyi	2	75 S	243
Indus	-di	Indian	Ind	21	55 S	294
Lacerta	-tae	Lizard	Lac	22	45 N	201

Constellation	genitive ending	Meaning	Con-tractions	α	δ	Area
				h	°	(°)²
Leo	-onis	Lion	Leo	11	15 N	947
Leo Minor	-onis -ris	Small lion	LMi	10	35 N	232
Lepus	-poris	Hare	Lep	6	20 S	290
Libra	-rae	Scales	Lib	15	15 S	538
Lupus	-pi	Wolf	Lup	15	45 S	334
Lynx	-ncis	Lyria	Lyn	8	45 N	545
Lyra	-rae	Lyre	Lyr	19	40 N	286
Mensa	-sae	Table (mountain)	Men	5	80 S	153
Microscopium	-pii	Microscope	Mic	21	35 S	210
Moroceros	-rotis	Unicorn	Mon	7	5 S	482
Musca	-cae	Fly	Mus	12	70 S	138
Norma	-mae	Square	Nor	16	50 S	165
Octans	-ntis	Octant	Oct	22	85 S	291
Ophiuchus	-chi	Serpent bearer	Oph	17	0	948
Orion	-nis	Hunter	Ori	5	5 N	594
Pavo	-vonis	Peacock	Pav	20	65 S	378
Pegasus	-si	Winged horse	Peg	22	20 N	1 121
Perseus	-sei	Champion	Per	3	45 N	615
Phoenix	-nicis	Phoenix	Phe	1	50 S	469
Pictor	-ris	Painter's easel	Pic	6	55 S	247
Pisces	-cium	Fishes	Psc	1	15 N	889
Piscis Austrinus	-is -ni	S fish	PsA	22	30 S	245
Puppis	-ppis	Poop (stern)	Pup	8	40 S	673
Pyxis (= Malus)	-xidis	Compass	Pyx	9	30 S	221
Reticulum	-li	Net	Ret	4	60 S	114
Sagitta	-tae	Arrow	Sge	20	10 N	80
Sagittarius	-rii	Archer	Sgr	19	25 S	867
Scorpius	-pii	Scorpion	Sco	17	40 S	497
Sculptor	-ris	Sculptor	Scl	0	30 S	475
Scutum	-ti	Shield	Sct	19	10 S	109
Serpens (Caput and Cauda)	-ntis	Serpent. Head Tail	Ser	16 18	10 N 5 S	429 +208
Sextans	-ntis	Sextant	Sex	10	0	314
Taurus	-ri	Bull	Tau	4	15 N	797
Telescopium	-pii	Telescope	Tel	19	50 S	252
Triangulum	-li	Triangle	Tri	2	30 N	132
Triangulum Australe	-li -lis	S Triangle	TrA	16	65 S	110
Tucana	-nae	Toucan	Tuc	0	65 S	295
Ursa Major	-sea -ris	Great bear	UMa	11	50 N	1 280
Ursa Minor	-sea -ris	Small bear	UMi	15	70 N	256
Vela	-lorum	Sails	Vel	9	50 S	500
Virgo	-ginis	Virgin	Vir	13	0	1 294
Volans	-ntis	Flying fish	Vol	8	70 S	141
Vupecula	-lae	Small fox	Vul	20	25 N	268

THE GREEK ALPHABET

Alpha	A	α	Iota	I	ι	Rho	P	ρ
Beta	B	β	Kappa	K	$\kappa.$	Sigma	Σ	$\sigma.$
Gamma	Γ	γ	Lambda	Λ	λ	Tau	T	τ
Delta	Δ	δ	Mu	M	μ	Upsilon	Υ	υ
Epsilon	E	$\epsilon,$	Nu	N	ν	Phi	Φ	$\phi.\ \varphi$
Zeta	Z	ζ	Xi	Ξ	ξ	Chi	X	χ
Eta	H	η	Omicron	O	o	Psi	Ψ	ψ
Theta	Θ	$\theta,\ \vartheta$	Pi	Π	$\pi.$	Omega	Ω	ω

Index

★
★   ★